THE HARVARD COLLEGE
OBSERVATORY

Harvard College Observatory in 1851, as viewed from the corner of Concord Avenue and Bond Street in Cambridge.

THE HARVARD COLLEGE OBSERVATORY

The First Four Directorships, 1839-1919

BESSIE ZABAN JONES
and
LYLE GIFFORD BOYD

Foreword by Donald H. Menzel

THE BELKNAP PRESS OF
HARVARD UNIVERSITY PRESS
CAMBRIDGE, MASSACHUSETTS

1971

PREFACE

1700075

This history of the Harvard College Observatory reviews the story from the mid-seventeenth century, when the first astronomical observations were made at Harvard; describes in detail the developments from the founding of the Observatory in 1839 through the first four administrations; and ends with the death of the fourth director, Edward Pickering, in 1919.

Of the several other accounts that have appeared in print, three deserve special mention. The earliest, which dealt with the period from about 1815, when the Harvard Corporation took its first official step toward the establishment of an Observatory, to the year 1855, was written by the first director, William Cranch Bond, and appeared in volume 1 of the *Annals of the Harvard College Observatory.* The next account, written by Professor Arthur Searle, covered the years 1855–1876 and appeared in volume 8 of the *Annals.* The third, written by Professor Solon I. Bailey and published in 1931 as Harvard College Observatory Monograph No. 4, recapitulated the events of the earlier periods and brought the narrative up to the year 1927.

Unlike the previous histories, which were largely impersonal, the present volume proposes to give more than the bare statistics of physical growth, endowment, instrumentation, and research projects. By liberal quotation from correspondence, both personal and scientific in nature, we attempt also to portray something of the human qualities of the men and women who were closely involved with the Observatory. In addition to Observatory publications such as the *Annals,* the Annual Reports,

Preface

and the Circulars, we have drawn on hitherto unexplored primary material, the letters and documents relating to the Observatory that are preserved in the Archives of Harvard University. Although this book does not purport to be a history of American astronomy, preparing the work has entailed much supplementary reading to provide the necessary background. The sources used, other than the Archives, are cited in the notes to the individual chapters.

The idea of preparing a new history was conceived by Professor Donald H. Menzel, the sixth director of the Observatory. To make the book possible, he secured a grant from the National Science Foundation and, when that had expired, ensured the completion of the work by use of a gift from the late Richard Perkin of New Canaan, Connecticut. We acknowledge a special debt to Dr. Menzel, who has generously and patiently advised, criticized, encouraged, and given invaluable help at every stage of the project, especially with reference to the scientific background.

Our thanks go also to Mr. Douglas W. Bryant, University Librarian, for providing working facilities in the early months of this project; to Mr. Kimball C. Elkins of the Harvard University Archives, Mrs. Thomasine C. Brooks of the Phillips Library of the Observatory, and Miss Carolyn E. Jakeman of the Houghton Library for their aid in locating elusive material; to Miss Margaret Harwood and Mr. Leon Campbell for furnishing unpublished documents, rare photographs, and personal reminiscences; to Professor Owen J. Gingerich and Dr. Jesse L. Greenstein for reading parts of the manuscript and making helpful suggestions; to Mr. Roger Lowell Putnam, Mrs. Wayland S. Bailey, Dr. I. John Danziger, Miss Eleanor S. Upton, Mrs. Martha H. Liller, and Sr. Fernando L. de Romaña for supplying certain data and photographs; to Professor I. Bernard Cohen for obtaining the photograph of the Hancock telescope; to Mrs. Catherine Coté and Mr. James K. Ufford of the Fogg Art Museum for help in assembling pictures from the Harvard University Portrait Collection, especially Figures 2, 3, 4, 8, 11, 12, 14, 15, 19, 21, 23, and 45; to Mr. Eugen E. Franck of the Observatory's photographic laboratory for securing usable prints of early astronomical photographs; and to Mrs. Abbie McLaughlin, Mrs. Linda McDermott, Miss Olga Choruzi, and Miss Joan Drees, who cheerfully typed a large part of the manuscript.

vi

Preface

In writing the book, each of us has assumed chief responsibility for individual sections. Chapters 1, 2, 3, 7, 11, and 12 were written by B. Z. J.; Chapters 4, 5, 6, 8, 9, and 10 by L. G. B.

BESSIE ZABAN JONES

Cambridge, Massachusetts　　　　　　LYLE GIFFORD BOYD
November 1969

CONTENTS

ix

ILLUSTRATIONS

Illustrations

FOREWORD

Although Harvard College Observatory was not the first astronomical observatory in the United States, it has been the center of a large number of pioneer projects in astronomy. Many of the great figures in that science—European as well as American—have had more than casual contact with the Harvard Observatory. In the absence of a history of American astronomy, the story of scientific developments within one leading institution helps fill that gap. Harvard astronomers were among the first to recognize the potential of photography for astronomical research. The Harvard collection of more than half a million photographic plates provides a unique record of what has been happening in the sky over the past 90 years. Astronomers come to Harvard from all over the world just to consult this record, to see what this particular star or celestial object was doing in the distant past, changing brightness or drifting across the sky. Such studies have contributed materially to the understanding of such unusual objects as the recently discovered quasars. Space exploration, a field in which both the Harvard College Observatory and the Smithsonian Astrophysical Observatory have played major roles, has a link to the past as well as to the future. For, as orbiting astronomical telescopes reveal the existence of certain celestial objects with unusual or abnormal characteristics, records of the past will often provide important data for understanding or interpreting these mysteries. The history of past science is the paved road along which present-day science must travel into the future. I can only hope that this volume will inspire other historians to write a more detailed, full-scale history of American astronomy.

I wish to thank personally the authors and the many other people and organizations that made this volume possible.

Donald H. Menzel

November 21, 1970

THE HARVARD COLLEGE
OBSERVATORY

I

BEFORE THE OBSERVATORY

1642–1839

Until the third decade of the last century it was still true that an American observatory was little more than a "tube with an eye at one end and a star at the other."[1] Astronomy at Harvard, however, did not wait upon the possession of tubes, much less upon an observatory building. As early as 1642 the "Laws, Liberties and Orders of Harvard College" listed astronomy as one of the subjects in the curriculum. Students heard lectures in their third (then senior) year; after Harvard added the fourth year in 1652, "Junior" and "Senior sophisters" studied from special texts, supplemented by disputations once a week on "phylosophical and astronomical questions."

At what point Copernicus displaced Ptolemy in the curriculum is not exactly clear, but Samuel Eliot Morison concludes from his great study of the question that the "Copernican system was well established at Harvard before the 'New England Almanac for 1659' appeared."[2] These almanacs, prepared by tutors who were usually candidates for the ministry and chosen for the task by the President, were printed annually on a small press in the College Yard. That for 1659, produced by Zechariah Brigden, included an exposition of the Copernican system in an essay Morison regards as "the earliest extant . . . by a Harvard graduate," perhaps even the "first in popular science published in the English colonies" (Fig. 1). In another almanac following shortly on this example one author wrote gaily that the planets "dance 'illipticall Sallyes, Ebbs and flowes,' by reason of 'Magneticall Charmes' emanating from the Sun." (This description contains more truth than the writer could have known at the time, for although there is no connection with plane-

Aſtronomia Inſtaurata:

OR,

A new and compendious Reſtauration

OF

ASTRONOMIE.

In Four Parts.

Wherein is contained,

1. Logiſtica Aſtronomica: Aſtronomicall Arithmetick, ſhewing how to reſolve the four Species thereof in Aſtronomical Fractions, and to finde the Part Proportionall in any queſtion thereunto belonging.

2. Doctrina Sphærica: The Doctrine of the Sphere, practically propounded, and applyed in all queſtions tending to the Diurnall Motion.

3. Doctrina Theorica: The Theorie of the Cœleſtiall Motions, repreſenting the true Face of the Viſible World; the proper Motions of the Luminaries, and other Planets; and how to inveſtigate their true places ſpeedily, by the admirable Doctrine of *Trigonometrie*, after a new Order.

4. Tabulæ Aſtronomicæ: Aſtronomicall Tables, Briefly exhibiting the true Motions of the Sun, Moon, and other Planets, and their Eclipſes for any time, either Paſt, Preſent, or to come.

Whereunto is added, a ſhort Catalogue of all the moſt Accurate aud Remarkable Cœleſtiall Obſervations, that have been made by *Tycho*, *Longomontanus*, *Gaſſendus*, the Landgrave of *Haſſia*, and others.

BY

VINCENT WING.

S L O N D O N: C
Printed by R. and W. Leybourn, for the Company of Stationers.

M DC LVI.

Fig. 1. Title page of Vincent Wing's *Astronomia Instaurata* (1656), which first brought to English readers the discoveries of Copernicus, Galileo, and Kepler; used by Zechariah Brigden for his New England Almanac of 1659.

tary motions, George Ellery Hale in 1908 definitely established the existence of strong magnetic fields in sunspots!)[3] The chief contribution of these young astronomers, according to Morison, was in helping persuade "the New England farmer that he lived on a revolving planet . . . and the knowledge that the pious and learned presidents of Harvard College sponsored this new interpretation of the cosmos as consistent with Holy Writ, must have given the New Astronomy impeccable authority, in the minds of New England church members."

Latin *theses* and *quaestiones* by Harvard students provide further evidence of astronomical interest, though they clearly pretended to more learning than their authors could have possessed. For example, one finds such titles as "Gravitas est Vis Attractrix Terrae" (Gravity is the Attracting Force of the Earth), or "Terra non est Centrum Mundi" (The Earth is not the Center of the World), or that of the budding cosmogonist who, in defending the provocative theory that the world is not infinite but indefinitely extended ("Mundus non est Infinitus sed Indefinité Extensus") seems to have foreshadowed Einstein's proposition that the "world is finite but unbounded." No one claimed either great originality or scientific profundity in such exercises, yet in this period, Morison concludes, astronomy "was the only subject in which the College made some contributions toward the advancement of learning." Indeed, to a more extensive treatment of that advance he gave the title, "The Harvard School of Astronomy in the Seventeenth Century."[4]

The tribute to Harvard astronomy by no means suggests a lack of other sciences taught at the College. One has only to examine Professor Morison's account in Volume I of his *Harvard College in the Seventeenth Century* to see what place the sciences held in the curriculum, and Professor I. Bernard Cohen's survey of the apparatus provided for them convincingly reveals the importance placed on the best possible mechanical means for studying them. "For at least half a century," Cohen writes, "the prevailing opinion among many American scholars has been that the sciences were but little cultivated in American institutions of higher learning during the first half of the 19th century, that what there was was badly taught, and that during the 18th century the sciences barely even existed in our colleges." His review of the beginning of physics, astronomy, chemistry, biology, and geology at Harvard alone and his illuminating comments on the general nature of scientific activity in the earlier years refute that judgment. "The most cursory examination

3

of the development of science teaching at Harvard," he adds, "shows
that natural philosophy, including physics and astronomy, was the earliest of the sciences to appear on the Harvard scene as a full-fledged
and independent department of study." He concludes with the statement
that, examined in the context of the intellectual interests of the time,
"the scientific culture present in our Colonial colleges is most impressive,
even when compared to that of many European institutions. And the
scientific instruments that were used in that early formative period are
thus the tangible memorial to the largeness of the intellectual horizon—
which included science—that characterized our academic forefathers."[5]

For astronomy one tangible memorial came as early as 1671/2 in
the form of Harvard's first "tube." Even without a special building
to house it, perhaps its acquisition may be viewed as the beginning
of a Harvard observatory. The telescope was the gift of John Winthrop,
Governor of Connecticut, son of the first Governor of Massachusetts,
first colonial Fellow of the Royal Society, and great-uncle of Harvard's
famous Hollis Professor, John Winthrop IV. With this instrument the
donor believed (erroneously) that he might have observed the fifth satellite of Jupiter[6] (actually not seen until 1892 by the noted astronomer
E. E. Barnard). When the telescope, "3 foote and a halfe with a concave
ey-glasse," arrived, three grateful young tutors, Alexander Nowell,
Joseph Browne, and John Richardson, acknowledged it:

Cambridge, Febr: 2, 1671/2

Right Worshipful,—
Wee cannot but thankfully acknowledge that great & undeserved love
& respect manifested towards us in that large & learned letter sent unto
us by Mr. Martin, wherein your Worship has been pleased to prescribe
many usefull directions to instruct us in our fitting the Telescope for use,
according to the rules of art. As alsoe, in sending therewith severall instruments, whereby wee might be enabled to reduce the former precepts into
practice. The eye-glasse sent by Mr. Greene we have received in safety.
Wee have not as yet had an opportunity of doeing anything considerable
with it (the two last nights being cloudy), but wee hope (God willing)
to employ it shortly in the service of Urania . . . Honoured sr, wee have
received all the aforenamed particulars as a sure witnesse of your unfeigned
love to Learning, and a clear demonstration of your hearty desire eminently
to promote the same in this schoole of the Prophets.[7]

Any considerable service to Urania performed with the new telescope
by these earnest young men could not have continued very long, for

two of the signers of the letter, Nowell and Browne, died early. Increase and Cotton Mather observed Halley's comet with the Winthrop instrument in 1682, but perhaps the most rewarding observations were made by Thomas Brattle, graduate of 1676, and for twenty years treasurer of Harvard College.

After preparing the Almanac for 1678, "or Coelestial Motions of the Sun & Planets with their Principal Aspects," Brattle observed the comet of 1680. His report, sent to the Royal Society, received favorable notice by Newton in his *Principia*. The observations were useful to him, Professor Morison explains, because "being based on fixed stars rather than on azimuths and altitudes, they helped him to determine orbital elements of a comet moving in an ellipse. Hence our little college telescope, used by a careful and intelligent observer, contributed its mite toward helping Newton to test Kepler's three laws, to work out the law of gravitation, and to write his great work."[8] Brattle made other observations on lunar and solar eclipses which, duly communicated to the Royal Society, appeared in the *Philosophical Transactions* in 1704 and 1707.

Now less well known than others in the annals of early American astronomy, though at one time called "the most famous New Englander in science in his day," Thomas Robie[9] during his years as tutor (1712–1723) prepared ten or eleven almanacs. Many of them reflected his intense interest in direct observation with the Harvard apparatus, which increased in the period sufficiently to require a special "Philosophy Chamber" in Massachusetts Hall. Among the instruments were an 8-foot telescope received in 1712 and a 24-foot refractor in 1722 with which Robie observed a solar eclipse described in 1723 in a delightfully modest letter. He had made the observations, he wrote, "at Cambridge, N. E. when I lived at ye College there" and, while asking pardon for "my present writing," wished to explain that "I could not refuse doing it, from a desire I have of advancing myself in Astronomy." In an almanac of 1720 he appended "an account of the Solar System, according to Copernicus and the Modern Astronomers," with tables of the "middle Distances of the Planets from the Sun" and of "The Diameter of the Sun and Planets with the Moon." These were all calculated "from the latest observations by Sir Isaac Newton's Rules." Robie wrote also on Jupiter's satellites, lunar eclipses, the zodiacal light, the aurora borealis, and an earthquake in 1715. Of a supposed meteor he declared in 1719, "As to Prognostications from it, I utterly abhor and detest

5

them all, and look upon these to be but the Effect of Ignorance and Fancy; for I have not so learned Philosophy or Divinity, as to be dismayed at the Signs of Heaven; this would be to act the Part of an Heathen not of a Christian Philosopher." After leaving Harvard in 1723 he settled in Salem to concentrate on the practice of medicine, though he continued his astronomical observations and in 1725 was elected a Fellow of the Royal Society.

The greatest advance in astronomy at Harvard, however, came in 1727 with the establishment of the Hollis Professorship of Mathematics and Natural Philosophy, accompanied by a generous collection of books and instruments. Though not the first benefaction of Thomas Hollis[10] (Fig. 2), the professorship seems to have been inspired by his encounter in London with young Isaac Greenwood, who became the first incumbent of this post. Greenwood's story is an unhappy one, and but for his undeniable competence the difficulties he created at Harvard might have cast doubt on the whole enterprise. After graduation from Harvard in 1721, where he had been taught by Thomas Robie, he went to London for further study. There he attended the lectures of J. T. Desaguliers, called "the greatest demonstrator of scientific experiments of his age," and had the good fortune also to meet and impress Hollis. "Accordingly," writes Josiah Quincy in his circumstantial account of Greenwood's appointment and subsequent career, "early in 1727 Hollis transmitted twelve hundred pounds sterling to the treasurer of the College, and directed that the funds should be applied to the instituting and settling a Professor of the Mathematics and Experimental Philosophy." At the same time he specifically mentioned Greenwood, "one well qualified as an instructor of youth in these sciences." Before the nomination could be confirmed, however, "a cloud soon spread over his prospects." Hollis had to send the unhappy report that Greenwood "left us on a sudden without paying a debt of 300£." It was a reprehensible act itself, but, even more shocking, among the extravagances listed was the item "three pairs of pearl-colored silk stockings," not exactly suited to "the character of philosopher much less to that of clergyman." There were also dark hints of "rakish company," and, though Hollis did not specify it, Greenwood's chief weakness, intemperance, ultimately proved his undoing.

Nevertheless, unwilling to prejudice the chances of the otherwise promising young man, Hollis held out hopes for his repentance and

FIG. 2. Thomas Hollis, from a painting (1766) by John Singleton Copley. (Harvard University Portrait Collection.)

declared him still acceptable if elected *unanimously*. These warnings delayed decision about six months, but when it appeared that in his impatience Hollis might choose another candidate, a *Baptist*, the Corporation acted (Hollis was himself a Baptist, but the denomination was apt to arouse old feuds at Harvard). Ignoring the donor's plaintive query, "What has the dispute of baptism to do to enter into one Professorship or another?" the alarmed Corporation hastily forestalled the threatened election of a believer in the mystical power of water only to find itself saddled with a votary of stronger stuff. The vote on May 12, 1727, which twice repeated the necessary word *unanimously,* "declared the choice of Mr. Greenwood." The Corporation's act, Quincy remarks severely, indicates that "in this perversity or blindness of sectarian zeal, want of character or want of morals had little weight in the scale against what it called heresy or schism." Yet even Quincy, while careful to exonerate Hollis of any blame for this unfortunate appointment, had to admit that in his scientific attainments Greenwood was well qualified for his professorship.

The requirements of the post included a knowledge of "several parts of the Mathematics & of natural and Experimental Phylosophy," and instruction "in a System of natural Phylosophy & a course of Experimental," that was to embrace, besides "Hydrostaticks, Mechanicks, opticks," geometry, algebra, trigonometry, and so on, the general "Principles of Astronomy & Geography, viz., the doctrine of the Sphere, the use of the Globes, the Motions of the Heavenly Bodies according to the different Hypotheses of Ptolemy, Tycho Brahe & Copernicus, with the general Principles of Dialling, the Division of the world into its various Kingdoms, with the use of the Maps, etc." These provisions Greenwood fulfilled well for ten years, in the course of which he not only contributed to the *Philosophical Transactions* but wrote an *Arithmetick* (1729), the first such work by a native American. He also served as keeper and cataloguer of the Hollis and other instruments. Besides his course in mathematics he taught astronomy as provided in the rules laid down by Hollis, and in 1734 outlined an ambitious course of lectures in which he offered to show to his audience representations of sunspots, mountains of the moon, eclipses of the satellites of Jupiter, and some of the nebulae.

Even before being made Hollis Professor he had prepared an elaborate outline of lectures that were evidently intended to do for students here what he had learned in London: "An Experimental Course of Mechanical Philosophy, Whereby Such a Competent Skill in Natural Knowledge May be Attained to, (by Means of Various instruments, and Machines, with Which There Are above Three Hundred Curious, and Useful Experiments Performed) That Such Persons as Are Desirous Thereof, May, in a Few Weeks Time, Make Themselves Better Acquainted with the

Principles of Nature, and the Wonderful Discoveries of the Incomparable Sir Isaac Newton, than by a Years Application to Books and Schemes." None of his genuinely excellent teaching, however, prevailed against his repeated lapses from grace, and, after six months of postponing the dreadful decision, a reluctant Board of Overseers finally consented to his dismissal. In Professor Theodore Hornberger's view, Greenwood closes the transition stage from the formative period of college curricula in the sciences and begins the final triumph of Newtonianism:

His tragic life should not be allowed to obscure his revolutionary influence upon science at Harvard. He found the instruction almost wholly book learning; he left it marked by awareness of the prime importance of observation and experiment . . . He gave American students their first sight of the demonstration of physical principles by laboratory apparatus, of the planetarium as a means of visual apprehension of the facts of astronomy. There is no better example of the swift transmission of teaching method from London lecture halls to an American College . . . When Greenwood was discharged in 1738, he left behind him at Harvard a new way of doing things, and students who were ready to take up, with more sobriety, where he left off. For the rest of the century philosophical apparatus and a course of demonstrations were among the chief ambitions of every college in the country.[11]

John Winthrop (Fig. 3) became the second Hollis Professor in 1738/9, when he was just twenty-four years old.[12] During his brilliant forty-year tenure he taught with great effectiveness, furthered the advancement of science within the college—especially astronomy, mathematics, and physics—and enhanced his personal prestige abroad through his observations and contributions to the Royal Society, of which he became a Fellow in 1766. He wrote on sunspots, transits of Mercury, lunar eclipses, meteors, determinations of longitude, comets, and meteorology. His course of lectures embraced subjects directly related to mathematics and physics (Newton's laws of motion, gravity, magnetism, fluids, optics) and astronomy ("the Motions & Phaenomina of the Planets & Planetary System"). These he illustrated, in his own words, with the use of the "Orary, a Machine Whereon was admirably shown the motion of the Moon around the Earth, & of both round the Sun, as their Centre." (The orrery in this period served the purpose of the more elaborate planetarium of the 20th century.) His elaborate notes on "The Method of Astronomical Calculations"[13] he outlined for his

Fig. 3. John Winthrop, from a painting (about 1766) by John Singleton Copley. Winthrop's telescope, made by James Short (1740), is now in the Harvard Collection of Historical Scientific Instruments. (Harvard University Portrait Collection.)

own guidance in teaching, but a set of problems put before his students evidently influenced his immediate successors and others. A bibliography[14] of the *Mathematical Theses of Junior and Senior Classes, 1782–1839,* for example, contains 406 titles, an appreciable number of them with the kind of lunar and solar projections and related astronomical calculations that Winthrop had expected his students to master. Some of the authors, heirs of the Winthrop method though writing for later Hollis professors, achieved distinction in astronomy, mathematics, academic life, and public service. One finds among them the names of Benjamin Peirce, John Davis, James Walker, William Prescott, Joseph Lovering, George Bancroft. The list includes two Pickerings, a name of highest honor in the annals of Harvard astronomy, and three future presidents of the University—Edward Everett, Jared Sparks, and James Walker—who were actively interested in the development of astronomy at Harvard and especially in the Observatory. It is worth noting also that the young Benjamin Thompson,[15] who heard Winthrop lecture, as Count Rumford became a benefactor of the American Academy of Arts and Sciences and, with the endowment of a professorship, of Harvard.

That not all students loved their tedious mathematical assignments, however, is obvious from this amusing "Lament" by a student who regretted time lost from the humanities:

Harvard Student's Lament, 1744

Now algebra, geometry,
Arithmetick, astronomy,
Opticks, chronology, and staticks,
All tiresome parts of mathematicks,
With twenty harder names than these
Disturb my brains, and break my peace.
All seeming inconsistencies
Are solv'd by A's or solved by B's;
Our senses are depriv'd by prisms,
Our arguments by syllogisms.
If I should confidently write,
This ink is black, this paper white,
They'd contradict it, and perplex one
With motion, light, and it's reflection,
And solve th' apparent falsehood by
The curious structure of the eye.

Shou'd you the poker want, and take it,
Glowing as red as fire can make it,
And burn your finger, or your coat,
They'd falsly tell you, 'tis not hot.
The fire they say has in't, 'tis true,
The power of causing pain in you,
But no more heat's in fire, that heats you,
Than there is pain i' th' stick that beats you.

11

The Harvard College Observatory

. . . . .

We're told how planets roll on high,
How large their orbits, and how nigh;
I hope in little time to know,
Whether the moon's a cheese, or no;
Whether the man in't (as some tell ye),
With beef and pudding fills his belly;
Why, like a lunatick confin'd,
He lives at distance from mankind;
Who at one resolute attack
Might whirl the prison off his back;
Or like a maggot in a nut
Full bravely eat his passage out.[16]

Nevertheless, none who had sat under Winthrop or, for that matter, his predecessor could escape some knowledge of Newtonian theory, awareness of advancing scientific developments, or actual observation and demonstration by means of the excellent scientific apparatus Harvard owned up to that fatal night in January 1764, when everything but one instrument burned in the fire of Harvard Hall.[17] Free of students during the winter recess, but occupied by the General Court taking refuge from an outbreak of smallpox in Boston, the building that housed the College Library, the museum, students' personal property, and the philosophical apparatus burned to the ground and threatened to take with it three other structures—the new Hollis Hall, Stoughton, and Massachusetts, as well as the president's house. The histories of Harvard have made the story familiar, but of all accounts perhaps the most vivid is the first-hand report of President Holyoke's daughter to her husband, John Mascarene (then in London), written six days after the disaster:

On Tuesday night about 12 o'clock, in the severest snow storm I ever remember I heard the cry of Fire, one moment brought me to the window, when [I] saw the old Harvard College on fire, and it was with the utmost difficulty they savd the other Buildings. Stoughton was on fire an hour, Massachusetts catchd in three places, and Hollis Hall is burnt much . . . there was nothing saved in old College, except a bed or two, the whole Library, except some Books lent out and Mr. Hollis's last donation, were demolished, the whole apparatus . . . this is a most terrible accident, this Library in which were so many valuable Books, ancient manuscripts, the Labour of the Learned, and the work of ages, in a few hours turnd to ashes. Our College is now poorer than any on the Continent— we are all real mourners on this occasion, and I doubt not your attachment to alma mater, will make you feel sorrowful upon this conflagration.

12

Fig. 4. President Edward Holyoke, from a painting by John Singleton Copley. (Harvard University Portrait Collection.)

As to Father he had very near lost his life on the occasion, the snow was in drifts in many places four and five feet high, papa went thro it all with nothing more upon him than he sits in the house, the President's house was in great danger the wind was strong at N west the latter part of the time, and in short if Stoughton had gone all the houses in town to the Eastward of the College would have gone. I think I never saw so great a strife of elements before, it is supposed the Fire began in the Beam under the hearth of the Library, the Gov'r & a great number of the court assisted in extinguishing the Fire. . . .

13

The Harvard College Observatory

Despite his exertions of the night before, the seventy-five-year-old President Holyoke (Fig. 4) on January 25 sent a letter to the press describing the catastrophe. Printed in the *Massachusetts Gazette* on February 2 and reissued as a most lugubrious broadside (Fig. 5), it gives virtually the only public account to appear, for the Stamp Act agitation dominated the news at the moment. After naming the most serious losses in the library collections, the document continues with a description of the precious philosophical apparatus, much of it given by Thomas Hollis and other benefactors for experiments in mechanics, hydrostatics, pneumatics, optics, and astronomy:

> For *Astronomy,* we had before been supplied with Telescopes of different lengths; one of 24 feet; and a brass Quadrant of 2 feet radius, carrying a Telescope of a greater length; which formerly belonged to the celebrated Dr. Halley. We had also the most useful instruments for *Dialling;*—and for *Surveying,* a brass semicircle, with plain sights and magnetic needle. Also, a curious Telescope, with a complete apparatus for taking the difference of Level . . . Many very valuable additions have of late years been made to this apparatus by several generous benefactors . . . From these Gentlemen we received fine reflecting Telescopes of different magnifying powers; and adapted to different observations; Microscopes of the several sorts now in use; Hadley's Quadrant fitted in a new manner; a nice Variation Compass, and Dipping needle; with instruments for the several magnetical and electrical experiments—all new, and of excellent workmanship———ALL DESTROYED!

Even unhappier, if possible, was one whose

> . . . sad remembrance of that fatal night
> When Science fell a victim to the Flames

led him to this poetic outburst:

> But now, how chang'd the Scene! Behold the walls,
> Not long ago the fam'd Repository
> Of Solid Learning, levell'd to the Dust
> Ye flames, more merciless than the fell Hand
> Of all-devouring Time, more savage far
> Than Earthquake's horrid Shocks, why did ye not
> Recoil with Shame, when near the Sacred Volumes
> Arrang'd with Care, your pointed spires approach'd?
> Why could ye not, the famed Museum spare,
> Unrival'd in Columbia, where my Sons
> Beheld, unveil'd by Winthrop's artful Hand,
> The face of Nature, beautiful and Fair?[18]

14

An Account of the Fire at *Harvard-College,*
in *Cambridge* ; with the Loſs ſuſtained thereby.

CAMBRIDGE, Jan. 25. 1764.

LAST night HARVARD COLLEGE, ſuffered the moſt ruinous loſs it ever met with ſince its foundation. In the middle of a very tempeſtuous night, a ſevere cold ſtorm of ſnow attended with high wind, we were awaked by the alarm of fire. *Harvard*-Hall, the only one of our antient buildings which ſtill remained,[*] and the repoſitory of our moſt valuable treaſures, the public LIBRARY and Philoſophical APPARATUS, was ſeen in flames. As it was a time of vacation, in which the ſtudents were all diſperſed, not a ſingle perſon was left in any of the Colleges, except two or three in that part of *Maſſachuſetts* moſt diſtant from *Harvard*, where the fire could not be perceived till the whole ſurrounding air began to be illuminated by it : When it was diſcovered from the town, it had riſen to a degree of violence that defied all oppoſition. It is conjectured to have begun in a beam under the hearth in the library, where a fire had been kept for the uſe of the General Court, now reſiding and fitting there, by reaſon of the Small-Pox at Boſton : from thence it burſt out into the Library. The books eaſily ſubmitted to the fury of the flame, which with a rapid and irreſiſtable progreſs made its way into the Apparatus-Chamber, and ſpread thro' the whole building. In a very ſhort time, this venerable Monument of the Piety of our Anceſtors was turn'd into an heap of ruins. The other Colleges, *Stoughton*-Hall and *Maſſachuſetts*-Hall, were in the utmoſt hazard of ſharing the ſame fate. The wind driving the flaming cinders directly upon their roofs, they blazed out ſeveral times in different places ; nor could they have been ſaved by all the help the Town could afford, had it not been for the aſſiſtance of the Gentlemen of the General Court, among whom his Excellency the Governor was very active ; who, notwithſtanding the extreme rigor of the ſeaſon, exerted themſelves in ſupplying the town Engine with water, which they were obliged to fetch at laſt from a diſtance, two of the College pumps being then rendered uſeleſs. Even the new and beautiful *Hollis*-Hall, though it was on the windward ſide, hardly eſcaped. It ſtood ſo near to *Harvard*, that the flames actually ſeized it, and, if they had not been immediately ſuppreſſed, muſt have carried it.

But by the Bleſſing of God on the vigorous efforts of the aſſiſtants, the ruin was confined to *Harvard*-Hall ; and there, beſides the deſtruction of the private property of thoſe who had chambers in it, the public loſs is very great ; perhaps, irreparable. The Library and the Apparatus, which for many years had been growing, and were now judged to be the beſt furniſhed in America, are annihilated. But to give the public a more diſtinct idea of the loſs, we ſhall exhibit a ſummary view of the general contents of each, as far as we can, on a ſudden, recollect them.

Of the LIBRARY.

IT contained—The Holy Scriptures in almoſt all languages, with the moſt valuable Expoſitors and Commentators, antient and modern :—The whole Library of the late learned Dr. Lightfoot, which at his death he bequeathed to this College, and contained the Targums, Talmuds, Rabbins, Polygot, and other valuable tracts relative to oriental literature, which is taught here : The library of the late eminent Dr. Theophilus Gale : —

[*] *Harvard*-Hall, 42 feet broad, 97 long, and four ſtories high, was founded A. D. 1672.

—All the Fathers, Greek and Latin, in their beſt editions. — A great number of tracts in defence of revealed religion, wrote by the moſt maſterly hands, in the laſt and preſent century.— Sermons of the moſt celebrated Engliſh divines, both of the eſtabliſhed national church and proteſtant diſſenters :—Tracts upon all the branches of polemic divinity :—The donation of the venerable Society for propagating the Goſpel in foreign parts, conſiſting of a great many volumes of tracts againſt Popery, publiſhed in the Reigns of Charles II. and James II. the Boylean lectures, and other the moſt eſteemed Engliſh ſermons :—A valuab'e collection of modern theological treatiſes, preſented by the Right Rev. Dr. Sherlock, late Lord Biſhop of London, the Rev. Dr. Hales, F. R. S. and Dr. Wilſon of London :—A vaſt number of philological tracts, containing the rudiments of almoſt all languages, antient and modern :—The Hebrew, Greek and Roman antiquities:—The Greek and Roman Claſſics, preſented by the late excellent and catholic-ſpirited Biſhop Berkeley ; moſt of them the beſt editions :—A large Collection of Hiſtory and biographical tracts, antient and modern.—Diſſertations on various Political ſubjects —The Tranſactions of the Royal Society, Academy of Sciences in France, Acta Eruditorum, Miſcellanea curioſa, the works of Boyle and Newton, with a great variety of other mathematical and philoſophical treatiſes.—A collection of the moſt approved Medical Authors, chiefly preſented by Mr. James, of the iſland of Jamaica ; to which Dr. Mead and other Gentlemen made very conſiderable additions : Alſo Anatomical cuts and two compleat Skeletons of different ſexes. This collection would have been very ſerviceable to a Profeſſor of Phyſic and Anatomy, when the revenues of the College ſhould have been ſufficient to ſubſiſt a gentleman in this character.—A few antient and valuable Manuſcripts in different languages.—A pair of excellent new Globes of the largeſt ſize, preſented by Andrew Oliver, jun. Eſq;—A variety of Curioſities natural and artificial, both of American and foreign produce.—A font of Greek types (which, as we had not yet a printing-office, was repoſited in the library) preſented by our great benefactor the late worthy Thomas Hollis, Eſq; of London ; whoſe picture, as large as the life, and inſtitutions for two Profeſſorſhips and ten Scholarſhips, periſhed in the flames.——Some of the moſt conſiderable additions that had been made of late years to the library, came from other branches of this generous Family.

The library contained above five thouſand volumes, all which were conſumed, except a few books in the hands of the members of the houſe ; and two donations, one made by our late honorable Lieutenant Governor Dummer, to the value of 50 l. ſterling, the other of 56 volumes, by the preſent worthy Thomas Hollis, Eſq; F. R. S. of London, to whom we have been annually obliged for valuable additions to our late library : Which donations, being but lately received, have not the proper boxes prepared for them ; and ſo eſcaped the general ruin.

As the library records are burnt, no doubt ſome valuable benefactions have been omitted in this account, which was drawn up only by memory.

Of the APPARATUS.

WHEN the late worthy THOMAS HOLLIS, Eſq; of London founded a Profeſſorſhip of Mathematics and Philoſophy in Harvard-College, he ſent a fine Apparatus for Experimental Philoſophy in its ſeveral Branches.

Under the head of *Mechanics*, there were machines for experiments of falling bodies, of the centre of gravity, and of centrifugal forces ;—the ſeveral mechanical powers, balances of different ſorts, levers, pullies, axes in peritrochio, wedges, compound engines ; with curious models of each in braſs.

In *Hydroſtatics*, very nice balances, jars and bottles of various ſizes fitted with braſs caps, veſſels for proving the grand hydroſtatic Paradox, ſiphons, glaſs models of pumps, hydroſtatic balance, &c.

In *Pneumatics*, there was a number of different tubes for the Torricellian experiment, a large double-barrell'd Air-pump, with a great variety of receivers of different ſizes and ſhapes ; ſyringes, exhauſting and condenſing ; Barometer, Thermometer ;—with many other articles.

In *Optics*, there were ſeveral ſorts of mirrors, concave, convex, cylindric ; Lenſes of different foci ; inſtruments for proving the fundamental law of refraction ; Priſms, with the whole apparatus for the Newtonian theory of light and colors; the camera obſcura, &c.

And a variety of inſtruments for miſcellaneous purpoſes.

THE following articles were afterwards ſent us by Mr. Thomas Hollis, Nephew to that generous Gentleman, viz. an Orrery, an armillary Sphere, and a box of Microſcopes ; all of exquiſite workmanſhip.

For *Aſtronomy*, we had before been ſupplied with Teleſcopes of different lengths ; one of 24 feet ; and a braſs Quadrant of 2 feet radius, carrying a Teleſcope of a greater length ; which formerly belonged to the celebrated Dr. Halley. We had alſo the moſt uſeful inſtruments for Dialling ;— and for Surveying, a braſs ſemicircle, with plain ſights and magnetic needle. Alſo, a curious teleſcope, with a complete apparatus for taking the difference of Level ; lately preſented by Chriſtopher Kilby, Eſq;

Many very valuable additions have of late years been made to this apparatus by ſeveral generous benefactors, whom it would be ingratitude not to commemorate here, as no veſtiges of their donations remain. We are under obligation to mention particularly, the late Sir Peter Warren, Knt. Sir Henry Frankland, Bart. Hon. Jonathan Belcher, Eſq; Lt. Governor of Nova-Scotia ; Thomas Hancock, Eſq; James Bowdoin, Eſq; Ezekiel Goldthwait, Eſq; John Hancock, A. M. of Boſton, and Mr. Gilbert Harriſon of London, Merchant. From theſe Gentlemen we received fine reflecting Teleſcopes of different magnifying powers ; and adapted to different obſervations ; Microſcopes of the ſeveral ſorts now in uſe ; Hadley's Quadrant fitted in a new manner ; a nice Variation Compaſs, and Dipping needle ; with inſtruments for the ſeveral magnetical and electrical experiments—all new, and of excellent workmanſhip.——ALL DESTROYED !

Cambridge, Jan. 26. 1764. As the General Aſſembly have this day chearfully and unanimouſly voted to rebuild *Harvard*-Hall, it encourages us to hope, that the LIBRARY and APPARATUS will alſo be repaired by the private munificence of thoſe who with well to America, have a regard for New-England, and know the importance of literature to the Church and State.

BOSTON: PRINTED BY R. AND S. DRAPER.

1764.

FIG. 5. Broadside describing the fire of 1764, published in the *Massachusetts Gazette,* February 2, 1764. (From a copy in the Houghton Library.)

The Harvard College Observatory

Happily, all of science did not fall victim to the flames. One instrument, "a fine reflecting telescope," given by Thomas Hancock, escaped because it had been borrowed by Winthrop. It has had a strange history. The Corporation voted thanks[19] to the donor on April 6, 1761, a most opportune time for its acquisition, as Winthrop was then in the midst of preparations to observe the transit of Venus[20] of June 6. The last such event had occurred in 1639; it had been observed then by William Crabtree and by the young curate Jeremiah Horrocks, though he had been seriously hampered by its falling on Sunday, when church duties interfered. Now, Venus would cross the sun's disk for the first time since Halley in 1716 had pointed out the usefulness of the phenomenon in determining the sun's distance, and astronomers all over the world were stirred to test his recommendations of sites and methods. The opportunities for doing so were limited by the fact that these transits occur only at long intervals. Two such events take place eight years apart, and then do not recur for more than a century. (The last pair of transits of Venus were observed in 1874 and 1882; the next will not come until 2004 and 2012.) Halley had suggested that a comparison of the times required for the planet to complete the crossing, as determined from widely separated observation posts, would provide a good approximation of the sun's distance. In joining the British program for observation on this side of the Atlantic, Winthrop led the first such enterprise in the colonies and anticipated the kind of collaboration with governments and other astronomers that has characterized the spirit of the Harvard Observatory from its beginning.

Since the transit would not be visible in the latitude of New England, Winthrop decided to observe it from Newfoundland. Accordingly, as Quincy wrote, "Winthrop addressed a memorial to Governor Bernard, who, entering cordially into his view, by a special message on the subject obtained from the Massachusetts Legislature leave to place the Province sloop at his service for the purpose."[21] On May 5, 1761, the Harvard Corporation, assured that the General Court would "bear the charges" of the expedition, voted that Winthrop "may take with him . . . such astronomical instruments as he thinks he shall need . . . provided that he shall see to it that all such instruments shall be insur'd so that the College may not suffer either by the loss of any of them or any damage that shall happen to them on the voyage."[22]

Winthrop's first choice was the new Hancock telescope (Fig. 6). The handiwork of the celebrated English instrument maker Benjamin Martin, it served the expedition remarkably well. In his own account[23] of the

FIG. 6. The "Hancock" telescope, made by B. Martin, presented to Harvard in 1761 by Thomas Hancock, and taken by John Winthrop that year to Newfoundland to observe the transit of Venus. The only Harvard-owned telescope to survive the fire of 1764, it is now in the Science Museum, South Kensington, London. (Reproduced by permission of the Museum.)

journey and its results (Fig. 7) Winthrop described the instrument as a "curious reflecting telescope, adjusted with spirit-levels at right angles to each other, and having horizontal and vertical wires for taking corresponding altitudes." He carried on operations "every fair day and many times a day . . . with an assiduity which the swarms of insects, that were in possession of the hill, were not able to abate, tho' they persecuted us severely, and without intermission, both by day and by night, with their venemous stings." His results, sent to the Royal Society (1764), were considered "the only American figures made available to the world of science."[24] When James Short computed the total results of the British contingent by comparing observations made at selected stations, he found that those made at Newfoundland and at the Cape of Good Hope gave the value 8.25 seconds of arc for the solar parallax, though the over-all figure was 8.56. Two main obstacles prevented a decisive determination, however—the atmosphere of Venus which, in diffusing the light from the sun, made it difficult to discover the exact moment of internal contact, and the uncertainty about the longitude of the various stations.[25]

Under Winthrop's scrupulous care the Hancock telescope met with no damage. Ironically, by means not yet or ever likely to be known, the College later suffered the total loss of the one instrument known to have survived the disaster of 1764. Whether carried off in the transfer of apparatus from Cambridge during the Revolution or afterward to London by a Loyalist, the telescope finally landed in the Science Museum at South Kensington. An exchange of letters recently uncovered in the Observatory archives does not explain the mystery of its disappearance, but reveals that in the course of its wanderings it came into the possession of Thomas H. Court, a collector of early instruments in England and coauthor of a *History of the Microscope*. His plan to bequeath the Hancock gift to Harvard, expressly promised in a letter to Edward C. Pickering, dated October 12, 1918, was obviously never carried out.[26]

By 1769, when the next transit of Venus occurred, Winthrop used other instruments made by Short, part of the new equipment replacing that destroyed in the fire. Eloquent appeals to restore the losses had met with a munificent shower of gifts from many widely scattered sources. The list is amazing: books, money, instruments (including an orrery given by James Bowdoin in 1764 but not delivered until several

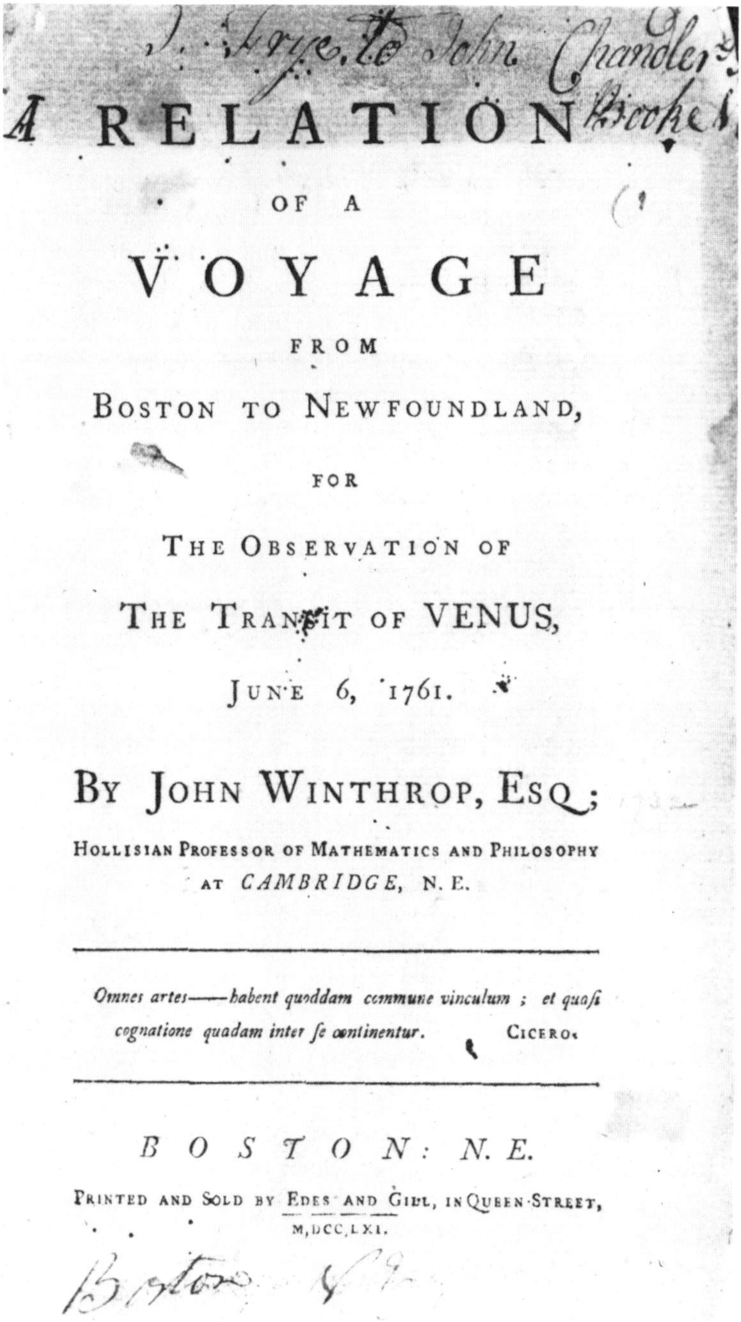

J. Frye to John Chandler Book (?)

A RELATION

OF A

VOYAGE

FROM

BOSTON TO NEWFOUNDLAND,

FOR

THE OBSERVATION OF

THE TRANSIT OF VENUS,

JUNE 6, 1761.

BY JOHN WINTHROP, ESQ;

HOLLISIAN PROFESSOR OF MATHEMATICS AND PHILOSOPHY
AT *CAMBRIDGE*, N. E.

---

*Omnes artes——habent quæddam commune vinculum ; et quaſi
cognatione quadam inter ſe continentur.* CICERO.

B O S T O N: N. E.

PRINTED AND SOLD BY EDES AND GILL, IN QUEEN-STREET,
M,DCC,LXI.

Boston

FIG. 7. Title page of a report of Winthrop's scientific expedition to New-foundland, the first sponsored by a college in the New World. (From the copy in the Houghton Library.)

years later), Hadley's quadrant, globes, telescopes, lenses, solar microscope, and so on. Well-wishers sent donations not only from Boston, Cambridge, Gloucester, Newbury, Salem, Barnstable, Charlestown, Concord, and Beverly, but from the "Province of New Hampshire," Nova Scotia, Rhode Island, and even "residences unknown." Benjamin Franklin supplied "valuable instruments for the apparatus," along with a bust of Lord Chatham, and from Great Britain came many excellent volumes and money, as well as "two curious Egyptian mummies for the Museum." In a remarkably short time the losses, thought to have been irreparable at the time, seemed fairly insignificant when compared with the new scientific facilities, which were housed in a building especially provided by the General Court.[27]

Winthrop's poor health and the unwillingness of the Governor and Council of Massachusetts to support the enterprise prevented a second Harvard expedition to view the transit of 1769, even though the excitement and preparations elsewhere were more intense than in 1761. As a result, Winthrop had to watch Venus move across the Cambridge sun. He had a number of instruments, but depended chiefly on a 2-foot telescope by Short, the size of the telescopes used by other observers. Besides making simple arrangements for his own observations, he gave two public lectures to explain their astronomical importance:

On account of their rarity alone [the transits] must afford an exquisite entertainment to an astronomical taste. But this is not all. There is another circumstance which strongly recommends them. They furnish the only adequate means of solving a most difficult Problem,—that of determining the true distance of the Sun from the Earth. This has always been a principal object of astronomical inquiry. Without this, we can never ascertain the true dimensions of the solar system and the several orbs of which it is composed, nor assign the magnitudes and densities of the Sun, the planets and comets; nor, of consequence attain a just idea of the grandeur of the works of GOD.[28]

Though his observations were necessarily more limited than before, Winthrop still made a careful report which he sent to England for publication in the *Philosophical Transactions*,[29] but this time he was not the only observer in the colonies. At least nineteen independently initiated groups took part, out of a total world list of 150—an indication to one enthusiastic historian of the event that American astronomy had "come of age . . . Rarely," he asserts, "has a single event so affected the course of a science."[30]

Before the Observatory

One of the more elaborately prepared parties was sponsored by the American Philosophical Society; William Smith, a member of it, loftily commiserated with astronomers abroad who, for reasons of weather or other mishaps, fared less well than those on this side of the Atlantic:

And here, as Dr. Halley expresses it, "Since Venus, like her sex, is exceeding *coy,* and deigns but in certain ages to come to the eyes of men, divested of her borrowed dress;" an American cannot help lamenting for his Brother Astronomers in Europe—men of fame and great abilities—that they were condemnd, amid horizontal vapors only to a transient glimpse of this rare phenomenon, and that they could not have shared with us some part, at least, of that *luxury of gazing,* which we enjoyed here.[31]

Despite such confidence, Smith's calculations were not conclusive (he arrived at the mean value of 8.6045 seconds of arc), but the failure was not his alone. Again the differences among the observers were too wide, partly owing to the kind of difficulties encountered in 1761, partly also from some distortion of image in the instruments used. After 1769 astronomers looked to the next pair of transits in 1874 and 1882, for which vast preparations were made, but hopes for a more decisive result again proved elusive, and eventually other methods superseded that of reliance on transits.

On his first expedition in 1761 Winthrop had taken with him as assistants three of his pupils of "good proficiency in mathematical studies," one of whom, Samuel Williams,[32] succeeded him as Hollis Professor in November 1779. After graduation from Harvard in 1761, Williams, like many students who had shown an interest in science, went into the ministry and at the time of his appointment was pastor of a church in Bradford, Massachusetts. Negotiations with him took several months.

The choice for the Hollis professorship, because of its donor, its purposes, and the prestige acquired during the preceding forty years, demanded no mere routine treatment, and a young man, however well qualified, might reasonably hesitate to step into Winthrop's shoes. The Corporation named Williams on November 23, 1779, then appointed a committee to wait on him "to urge his acceptance of the Professorship of the Mathematics." It also voted that "he be licensed to borrow such books as he shall desire from the College during the time he shall have the aforesaid election under consideration." But not until February could the President report to the Fellows that the Reverend Samuel Williams had signified his acceptance. They then voted

That he be desired to enter upon the duties of his office at the end of the present vacation, which will be the first day of March next; and that the President be desired to acquaint Mr. Williams that the Corporation will appoint as early a day for the solemnities of his public inauguration as the season of the year and other circumstances will permit.

Williams was also to receive an inventory of the apparatus and a key to it as soon as he arrived in Cambridge.

Nor was this all. Something had to be done about that still unsolved Harvard faculty problem, a place to live in. It is worth noting that the petition of April 3, 1780, addressed on behalf of Williams by the President and Fellows to the Honorable Council of the Massachusetts House of Representatives, supported the appeal not for a dwelling merely, but for a house that could also serve as an *observatory:*

It has been represented to your Memorialists by the Revd Samuel Williams lately appointed Hollisian Professor of Mathematics & Natural Philosophy in this University & who is now acting in the duties of that important office, that he has been endeavoring ever since his coming to Cambridge to procure a house into which he might remove his family; but that the only house he can procure is one in which he shall not be able to make astronomical observations of any sort; that he is extremely unwilling to be in a situation incompatible with any part of the duty incumbent upon him, & in possession of the best astronomical apparatus in America, of which in such a situation he can make no use—a situation in which he must be obliged to disappoint the expectations of the public & dishonor the University by inactivity & a neglect of those observations which will be expected of us in Europe.

The petition went on to request that one of several estates, euphemistically described as "lately the property of absentees" (that is, Tories) be allowed Mr. Williams. With this matter settled, the date for the "solemnities" of the inauguration was set for the first Tuesday in May. (The General Court met the sizable charges of moving Williams to Cambridge!)

Although Williams was hardly another Winthrop, his strong interest in teaching led him to devise an elaborate course in astronomy for seniors. Instituted in 1785, it included lectures on solar, sidereal, and lunar astronomy, eclipses, the planets, satellites, methods of observing apparent diameters, places, distances, parallax, refraction, comets, and so forth—all designed to "establish the truth and certainty of the Copernican system," and, in so far as he could arrange it, to be accompanied by direct observation with the "best astronomical apparatus in America." He lectured also on heat, air, electricity, and magnetism. Nor did he lag behind Winthrop in promoting an expedition, on this occasion to observe the eclipse of October 27, 1780. In September the Corporation on his behalf appointed a committee "to confer with a committee of the American Academy of Arts and Sciences upon measures to procure an accurate observation of the Solar Eclipse, in the eastern parts of the state . . . and if expedient to join in an application to the General Court for assistance." On September 15, 1780, Williams requested leave "to repair to Penobscot" (Maine was then a part of Massachusetts) and for permission "to take such instruments & books belonging to the College, as may be necessary in making the observation." The Academy was new, but in joining Harvard on this occasion it established an early tradition of collaboration that was to prove of incalculable benefit when the Observatory was established more than a half century later.

As the American Revolution was still in progress, Williams had also to secure permission from the commander of the British garrison to land at the site that promised the best view of totality. In his own account he gratefully acknowledged the help of the commonwealth of Massachusetts:

Though involved in all the calamities and distresses of a severe war, the Government discovered all the attention and readiness to promote the cause of science, which could have been expected in the most peaceable and prosperous times; and passed a resolve, directing the Board of War to fit out the Lincoln Galley to convey me to Penobscot, or any other part at the eastward, with such assistants as I should judge necessary.

His assistants, all volunteers, included Professor Stephen Sewall, first Hancock Professor of Hebrew and other Oriental languages but interested in astronomy as well, James Winthrop the librarian (son of John), and six students. Among them, interestingly enough, was John Davis, later a Fellow and Treasurer of Harvard College and collaborator of John Quincy Adams in promoting the Observatory. The instruments lent by Harvard consisted of a clock, a quadrant, some telescopes, and other apparatus, but, despite all the permissions, aides, and equipment, either because of British restrictions or owing to faulty maps, the site chosen for the party was not in the path of totality.

Still, not all was lost. Fifty-six years before the English observer Francis Baily noted the phenomenon (1836), Williams saw, described, and made a drawing of what has become known as "Baily's beads," but which in fairness should bear the name of the Harvard observer. These "beads" are formed by the light from the edge of the sun shining between the mountains on the moon. Williams' account of this feature, characteristic of every total eclipse just before and just after totality, is unmistakable:

> The sun's limb became so small as to appear like a circular thread or rather like a very fine horn. Both the ends lost their acuteness, and seemed to break off in the form of small drops or stars some of which were round and others of an oblong figure. They would separate to a small distance, some would appear to run together again, and others diminish until they wholly disappeared.

Far from being blamed for partial failure, Williams in July 1781, perhaps as a mark of recognition for faithful service, was voted the use of a small silver tankard formerly belonging to Professor Winthrop.

Unfortunately, his career soon took an unhappy turn. Although, as Quincy writes, Williams had "entered with activity and devotion on the duties of his office," had published "astronomical observations and notices of extraordinary natural phenomena," had lectured effectively and even received extra pay of £30 for work not part of his usual Hollis duties, he "did not possess the wisdom to keep his expenditures within his income." In May 1788 the record states that "whereas sundry reports greatly to the disadvantage of Professor Williams have been circulated through the state, which, if true, will reflect great dishonor on the University," the Corporation appointed a committee to "make inquiry into the truth of said report, more especially to a settlement of accounts between him & the administrator of the Revd Joshua Paine deceased." When the report failed to exonerate him, on June 23, 1788, he sent in his resignation, a pathetic document:

> I could have wished to have spent the remainder of my days in those studies and employments, which I have so ardently pursued since I have been at the University. But if innumerable reports are kept up against me by the public and I cannot have a trial by the laws of my country, I must give up the prospect of future peace and usefulness here.

He concluded his letter by committing himself, his character, and prospects in life "to that Being whose words & works it has been the great business of my life to study and explain." The Corporation, nettled by his reference to a trial unjustly denied, took care to state that all the facts in the case had been shown him but that any legal proceedings were beyond their province. Williams had tried desperately to save his position. "If I leave my office," he wrote the treasurer on June 2, "I have no other prospects but those of misery, poverty, and calamity,"[33] but he failed to move the indignant authorities, who once again had to dismiss a Hollis professor.

Two weeks later, on July 8, 1788, the President and Fellows met to consider a suitable successor, but "there being no prospect of [the professorship's] being immediately filled, and it being of importance that the students be instructed in the usual branches in that department till the place can be supplied," they voted "that Mr. Webber, the mathematics tutor be desired to give the students private mathematics instruction . . . [and to] the senior class, after commencement, such a course of Astronomical Lectures, as those of that standing have been used to attend in the fall of the year several years past." They also stipulated that "the President be desired to give Mr. Webber such assistance in the Astronomical Lectures as he may find convenient."[34]

Samuel Webber, born in Byfield, Massachusetts, in 1759, was too poor to enter Harvard until he was nearly twenty-one. As a student he showed special aptitude for mathematics but followed the usual pattern of going into the ministry after graduation. He preached only two years, however, before becoming head of the Dummer Academy in New Hampshire, and then in 1787 returned to Harvard as tutor. Despite the clear implication that he was merely a temporary substitute who would require the assistance of President Joseph Willard (well known for his knowledge of astronomy and his correspondence with eminent men of science in Europe, including the Astronomer Royal), Webber performed well enough to be named Hollis Professor on June 5, 1789.

In the preceding October the Corporation had restated the duties of the post.[35] In addition to lectures in mathematics to all four classes (including fluxions if any student wanted the subject), weekly public lectures, "a complete course of Experimental Philosophy" for "resident Bachelors and the two Senior Classes," with explanation of the "principles and construction of the various machines" used, the professor must offer a course of lectures in astronomy every fall to seniors. Under solar astronomy, he must "explain the Precession of the Equinoxes, the Nutation of the Earth's axis, and the motion of the Apogee. Under Lunar Astronomy, the Moon's Libration and the motion of her Apogee and Nodes: And under Sydereal Astronomy, the Aberration of Light." He was expected also to keep up with advances in philosophy and astronomy and acquaint his students with them, but unlike his predecessor, who seems not to have benefited financially by the extra £30, Webber would receive no fee or special allowance beyond his usual compensation. And lest science too largely overshadow the Source of it All, the professor is "directed, while he is delivering his Philosophical and Astronomical Lectures, to make such incidental reflections upon the Being, Perfections and Providence of God, as may arise from the subjects, and may tend seriously to impress the minds of youth." On top of everything, if students demanded it, he must give them private tutoring; he should even encourage them to request it!

Webber held the chair for seventeen years, until 1806, when he was elected president of the University. Quincy writes that Webber "quitted his professorship . . . with reluctance" and praises his combination of "patience and facility with a thorough acquaintance with his subject," demonstrated not only in his courses but in the writing of a widely used two-volume text in mathematics. Webber also served on a United States boundary commission in 1796, he was active in the American Academy, and he apparently carried on the tradition of the mathematical theses. Some of them were prepared under his tenure by men later closely associated in one capacity or another with Harvard, for example, John Farrar, who succeeded him as Hollis Professor in 1806–07.

Quincy's high estimate of Webber is not altogether shared by some, who thought him undistinguished in science. Professor Morison describes him as perhaps "the most colorless president in Harvard history," whose "great ambition was to establish a Harvard Astronomical Observatory." He dismisses Webber in a brief paragraph with the remark that after a four-year term of office he was "remembered only by an 'erect declining sun-dial' that he constructed for Massachusetts Hall."[36] Historians of the Observatory Webber hoped to establish, however, must memorialize him somewhat differently.

One of the last charges placed on him before he became president was the act of the Corporation on January 13, 1806:

Before the Observatory

Professor Webber, to whom sundry papers received from Europe relative to the construction of an observatory had been committed made a written report on that subject, which was read. Thereupon, voted, that Professor Webber be requested to obtain such farther information, as he may think requisite, relative to the most approved method of constructing observatories in Europe and for that purpose to correspond with experienced artists & scientific men, & to procure accurate & satisfactory drawings & if in his judgment the plans and drawings, which have been or may probably be obtained will not give information on the subject sufficiently precise, that he be authorized to procure a model or models of one—or more—of the most approved European observatories. And that Mr. Webber be requested to avail himself of the assistance of John Lowell, Esq.[37] while he may remain in Europe in the accomplishment of the aforesaid objects; Mr. Lowell having generously offered & afforded his aid relative to this business.

The sundry papers referred to contained information that Lowell, in Europe for his health, had gathered chiefly from the noted astronomer Delambre. Since Webber was the professor most interested in astronomy at Harvard, presumably the initiative for the move came from him. The discouraging costs of an establishment such as governments in Europe had long supported doubtless killed his hopes of implementing in the foreseeable future a plan based on the "precise" information obtained. His death on July 17, 1810, left no one with equal enthusiasm for astronomy or of sufficient authority to pursue the matter further, and Harvard lost its opportunity to establish the first college observatory in the country.

II

Ten years after John Lowell's initial report on the requirements of an observatory he met with President Kirkland and the other Fellows on May 10, 1815, to take up the matter again. They voted that the President, Treasurer, Mr. Lowell, with Professor Farrar and Mr. Bowditch, "be a Committee to consider upon the subject of an Observatory, and report to the Corporation their opinion upon the most eligible plan for the same, and the site."[38] This vote and that of 1806 may properly be termed the first corporate acts in the United States for the establishment of an observatory. On the basis of it the subcommittee composed of Professor John Farrar and Nathaniel Bowditch outlined in a letter minute instructions for a fact-finding visit to the Greenwich Observatory

27

and other establishments to be made by William Cranch Bond,[39] a twenty-six-year-old instrument maker and self-taught astronomer. Though never privileged to attend the lectures of any Hollis Professor, twenty-four years later he became the first director of the Harvard College Observatory. Bond, about to sail for England, primarily in hopes of realizing a legacy but also for business reasons, was not altogether an accidental choice for the mission to Greenwich. By this time both Farrar and Bowditch had reason to know of his passion for astronomy, which had begun at the age of seventeen when he had watched the unusual solar eclipse of 1806 from a rooftop in Boston. In September 1811 Farrar and Bowditch had observed and described in the *Memoirs of the American Academy*[40] the comet of that year, but in his article Farrar graciously acknowledged that it had been seen

in the early part of the year . . . and recognized by Mr. William Bond, Jun., an ingenious mechanic of Dorchester, Massachusetts, who has obligingly favored me with the following notices: "I remarked on the 21st of April a faint whitish light near the constellation Canis Major, projecting a tail about one degree in length, and set down its place as follows: Right ascension 108°, declination 9°S. April 24th 9 o'clock P.M. Right ascension, 108°, declination, 7° or 8°S. Its motion and the situation of its tail convinced me that it was a comet. I noticed it several times in May, and supposed that its motion was toward the western part of the constellation Leo."

Bond in turn attributed his determined pursuit of astronomy to the encouragement that Farrar, and especially Bowditch, had given him. Although in his paper Farrar had described the young amateur astronomer as "an ingenious mechanic of Dorchester," there was no reflection on the respectability of the Bonds. They belonged on the maternal side, at least, to a distinguished family, the Cranches, who had gained prominence in Massachusetts early in the eighteenth century. One of its best-known members was Richard Cranch,[41] member of the Massachusetts Legislature, judge of the court of common pleas, and brother-in-law of John Adams.

William Bond was born on September 8, 1789, in Portland (formerly Falmouth), Maine, where his father had attempted to establish the business of cutting and exporting ship timber. When that enterprise failed, he moved his family to Boston and in 1793 set up a clock and silversmith shop. Here at a very early age the boy became his father's apprentice, and, although he left school after the elementary grades, he developed the native mechanical skill that later served him well. Among other ingenious devices he built the first American ship's chronometer in 1812. It was used successfully on a voyage from Boston to Sumatra and compared favorably with similar instruments made much later. George Bond describes his father's first *astronomical* apparatus as "a sundial and pieces of string held at arm's length, with which he plotted the stars and comets, 'after the fashion of Ferguson'." He also observed lunar culminations and developed the marked interest in meteorology and magnetism that earned him a government appointment (see Chapter II). It was his performance in this post that drew President Quincy's attention to him as a suitable director of the Observatory in 1839. In 1815, for this young man without any academic standing to be chosen as an official emissary of the College was no small tribute to his abilities.

Farrar's letter of June 23, 1815, reveals both the serious purpose of the committee and his exact knowledge of what to look for in the establishment of an observatory. No one at Harvard was better equipped to guide Bond on his tour of inspection. Described as "one of the most inspired teachers and lecturers ever to grace a Harvard platform," Farrar is likewise generally credited with having introduced into the curriculum, some years before Agassiz's arrival, the most advanced continental science through the astronomical and mathematical literature of France. As Hollis Professor he had originally taught both mathematics and astronomy, but in 1833, when Benjamin Pierce was appointed to a separate professorship in mathematics, Farrar concentrated entirely on astronomy. He translated and adapted a number of French works, among others Arago's *Tract on Comets* (1832), and prepared *An Elementary Treatise on Astronomy* (1834) for the "Fourth Part of a Course in Natural History, Compiled for the Use of the Students at the University," with selections from Biot's *Elementaria d'Astronomie Physique,* "adapted to the latest discoveries and improvements." For comets and fixed stars he turned to John Herschel's *Astronomy.* The contents listed such sections as General Phenomena, Theory of the Sun, Theory of the Moon, Theory of the Planets and Comets, and Fixed Stars.

Farrar instructed Bond to observe at Greenwich the exact location of the Observatory, the construction of the piers, the kinds of instruments—telescopes, clocks, meridian circles, and so forth—the method

29

of rotating the dome, the prices of the various apparatus, and the length of time needed to produce them:

I would observe further, that, with regard to the sort of information which we wish you to bring with you, in order to answer our purpose, it must be such as to enable you or some other person, to superintend and direct in the erection of an Observatory . . . in short . . . whatever is peculiar to the building as an Observatory . . . will require special attention. You will be able to judge, therefore, for yourself, when it will be necessary to take plans and sketches, and when a mere statement of the form and dimensions will be sufficient.[42]

Armed with his letter and others,[43] Bond met with a cordial reception at Greenwich, where he followed instructions to make measurements and to study every detail essential for carrying out his commission:

From Edward Troughton, one of the leading instrument makers in England, whose firm later served the Observatory with great skill and integrity, he learned of the proper foundations for support of the telescope. The information he never forgot, for on the basis of it he designed the great stone pier still to be seen today in the Harvard Observatory. He obtained floor plans of other establishments at Glasgow and Edinburgh and at private observatories. No visit left a more cherished memory than that at Slough, where Miss Caroline Herschel, as he recalled in his diary many years afterward,[44] gave him a personal tour and a thrilling sight of the first great reflector of her illustrious brother William. Leading astronomers, including the Astronomer Royal, offered valuable advice, all of which Bond incorporated in the report he made to the committee on his return.

Not one hint, however, did he give of the desperate hardships he had undergone during his prolonged stay abroad. Inadequately supplied with funds and unable to reach Harvard's absent London agent, he had wandered the streets of the city without a shilling in his pocket, finally, in despair about where to turn, falling into an exhausted sleep on the steps of St. Paul's Church. (The college reimbursed him in part for his expenses when he came back.) He continued hopefully to work on the project by supervising the construction of a model dome which, like the foundation, he later incorporated into the design for the observatory, but it soon appeared that nothing would come of all his efforts. Bond attributed the failure to lack of money, but President Quincy mentioned the "difficulties from the European artist, whom they desired to employ." Perhaps that, too, was money.

Seven years later the question of an observatory again arose. At a meeting of the Corporation on March 13, 1822, following a report of the "Committee on the Purchase of Land of Edward Dana," they voted to accept the first offer of 2 or 2½ acres, "provided that Mr. Lowell, Mr. Bowditch and Professor Farrar shall be of opinion that it will answer the purposes of the Observatory, and if they shall think otherwise, or shall not agree on the subject, that they be requested to report a statement of facts."[45] Nothing in the record shows any further action that year. In 1823, however, a new plea came from John Quincy Adams (Fig. 8), whose conception of his role as Secretary of State included the promotion of learning, in which astronomy was central. And to advance it an observatory worthy of the country and his college must be established. On September 15, 1823, Adams wrote a long letter to the member of the Harvard Corporation known to share his own passion for the project, Judge John Davis, who as a student had accompanied Professor Williams on the eclipse expedition of 1780:

I have conversed with President Kirkland, with your brother, with F. C. Gray, Col. T. H. Perkins & Sheriff Hall upon the subject of undertaking immediately the erection of an astronomical observatory in connection with a Professorship of Astronomy, & find them all heartily disposed to favour the project. The President & Col. Perkins are willing that the bequest of Mr. James Perkins[46] of 20,000 dollars payable on the decease of his widow should be applied to the institution of the Professorship. Whether the sum can be realized so that the two establishments may be accomplished at once & together will be subject to further consultation but I conceive the first and most urgent object is the erection of the building; & this I hope may be undertaken without waiting for the Instruments to be imported.[47]

The president of the University (Fig. 9) had mentioned several possible sites, but to Adams only one seemed desirable. For the sake of students and the professor of astronomy it should be near the college. He thought there "would be a peculiar propriety in giving the name of Perkins to the Professorship," but the income would have to be supplemented by the subscription he proposed launching. "Should any person be inclined to take a preeminent charge in the expense of erecting the building the other subscribers may be willing that it should bear his name."

31

Fig. 8. John Quincy Adams, from a painting by William Page. (Harvard University Portrait Collection.)

I propose to subscribe myself, one thousand dollars, a sum more suited to my circumstances & means than to my inclination. Aware that it gives me only the right of asking the aid of friends with whom I can take liberties to an object interesting to the reputation of our country, to the credit of our University, & to the progress of Human Sciences, I wish not the appearance of taking the lead in an undertaking to which I can at this time contribute so little of that effectual support which alone can realize it. I desire therefore that my intention to subscribe this sum may be reserved to your own knowledge & that of F. C. Gray to whom

FIG. 9. President John Kirkland, from a copy of the original painting by Gilbert Stuart. (Harvard University Portrait Collection.)

I have mentioned it until it shall be ascertained that an adequate sum to the completion of the plan is actually raised. If it shall be accomplished the honor of having accomplished it will be justly due to those who shall most liberally furnish the means by the devotion whether of their time, their talents or their money. If my taking a warm interest in the object can be of service to the end freely use my name. But use it for that purpose only & only to those who will not suspect a motive of making myself conspicuous at the expense of others, in the ostensible appearance of zeal for a public end.

33

The Harvard College Observatory

His next paragraph shows how shrewdly he foresaw the possibilities of friction unless a "system of regulations should be prepared prescribing the duties of the professor & providing for the purposes of the Institution." This advice he was compelled to repeat almost word for word some twenty years later as chairman of the Visiting Committee of the Observatory.[48] In the beginning, however, the question might "perhaps be referred entirely to the Corporation," but, since some of the subscribers might wish to be consulted, he suggested a meeting as soon as a sum of $30,000 should be subscribed.

To his son George Adams he wrote on both October 4 and 5 appointing him agent, contributing a subscription of $1000 and giving instructions to confer with Judge Davis. For quite special reasons Adams worded his letter to President Kirkland of December 15, 1823, somewhat differently:

The whole of this astronomical establishment appeared to me to be necessary to secure to our country the useful and honourable results which I was sure could not fail to flow from it, if undertaken and accomplished with that liberal, persevering and practical spirit by which alone it could be effected. The sum [$1000] thus appropriated as my portion of the subscription, was limited, not by my inclinations, but by the necessary comparison of my means with other and indispensable calls upon them . . . I authorize you to place my name wherever you may think it will be most effective to promote the subscription—It is my earnest desire that the object may be undertaken and effected—My first wish was that it might be obtained without encroaching upon any other fund of munificence to the University, not even upon the Perkins donation—I should be sorry to see any of the general property of the University made tributary to it; and I earnestly hope that the beneficence of the Legislature as now enjoyed will not be withdrawn—The learning and Science of a Nation are its immortality upon earth, and its approach to Heaven—Our University, is our glory and our defence—To adorn and improve it is to honour our Country and instruct future ages—The Legislature of a Sister State, under the auspices of our Patriarch Jefferson, is pouring forth with a profuse, which in such a case can never be a lavish hand, resources for the advancement of Literature, and the Institution of a new University.

With such an example before it, could our Legislature

select this moment to *withdraw* their wonted bounty from ours? Forbid it Shades of Hollis and of Harvard!—Forbid it, Spirit of Virtuous emulation! —Forbid it, Genius of our Constitution! Forbid it, Reverence to the memory of our Pilgrim Forefathers![49]

34

This letter obviously had a dual purpose: to press for an observatory at Harvard, of course, but of even greater urgency just then to plead with all possible force for a continuance of the state support of $10,000 annually that the University had enjoyed since 1814. But this appeal, like the subscription, failed. Harvard lost the subsidy for good, and an eloquent circular sent out on January 28, 1824, doubtless as a direct consequence of Adams's letters to Davis and to Kirkland, left unmoved the heirs of those who had rallied so beautifully after 1764. Not the call to emulate Europe, not even the usually effective utilitarian appeal of commercial benefit, served. The president, although entirely sympathetic, was faced with the problem of somehow meeting the financial crisis created by the withdrawal of the state subsidy. Even his own salary was docked and his secretary dismissed. In such circumstances he could hardly press the Corporation for anything so costly as an observatory.

The undaunted Adams stubbornly persisted nevertheless. On October 16, 1825, he again wrote to Judge Davis: **1700075**

During my short visit to my father and friends from which I am returning, the subject of the establishment of an Astronomical Professorship at Harvard University and the erection of an Observatory with provision for making a constant succession of observations for publication at suitable periodical times, was brought again to my mind by several interesting incidents, among which was the appearance of the Comet which at this very time is lighting our starry nights *unobserved,* for want of such an institution. I had reason to hope from conversations with some of our friends, that the moment was now favorable for a new effort to effect the purpose. When two years since, I made the offer of subscribing one thousand dollars towards raising the sum which would be required for the object proposed, I thought it necessary to limit the time within which I should hold myself responsible for the payment and that time being now past, I now write chiefly to renew the offer, limiting the time to two years from the first of January next.

Nor did he appeal to Harvard alone. On December 6, 1825, in his first annual message to Congress, President John Quincy Adams again urged the duty and right of government to promote learning and emphasized as a major aspect of it an astronomical observatory. On that occasion he used for the first time the imaginative metaphor his enemies ever after made their favorite political target:

Connected with the establishment of an University, or separate from it, might be undertaken the erection of an Astronomical Observatory, with provision for the support of an Astronomer, to be in constant attendance of observation upon the phenomena of the heavens: and for the periodical publication of his observations. It is with no feeling of pride, as an American, that the remark may be made, that, on the comparatively small territorial surface of Europe, there are existing upward of one hundred and thirty of these *light-houses of the skies;* while throughout the whole American hemisphere, there is not one. If we reflect a moment upon the discoveries, which, in the last four centuries, have been made in the physical constitution of the universe, by the means of these buildings, and of observers stationed in them, shall we doubt of their usefulness to every nation? And while scarcely a year passes over our heads without bringing some new astronomical discovery to light, which we must fain receive at second hand from Europe, are we not cutting ourselves off from the means of returning light for light, while we have neither observatory nor observer upon our half of the globe, and the earth revolves in perpetual darkness to our unsearching eyes?

Unfortunately, Adams's noble image of a grand program for science and an observatory under national patronage met the fate of his impassioned plea to the General Court of Massachusetts in 1823 for continued support of Harvard College. Indeed, so persistent was the antagonism to him and his proposals that even in 1832, when Congress was legislating the establishment and operation of the Coast Survey, it expressly declared that "nothing in this act, or the act hereby revived, shall be construed to authorize the construction or maintenance of a permanent astronomical observatory." Nevertheless, his long battle for astronomy between 1835 and 1846 as Chairman of the Congressional Committee to determine the disposition of James Smithson's bequest to this country,[50] his famous journey in 1843 at the age of seventy-seven to lay the cornerstone of the Cincinnati Observatory,[51] and his unwavering devotion to the cause at Harvard, where he served as Chairman of the Observatory's Visiting Committee until his death, give him a central role in the history of American astronomy.[52]

III

Despite the strenuous efforts of John Quincy Adams and others who preceded and followed him, Harvard lagged, apparently in the hope of finding the means to create a large establishment comparable to those its representatives had carefully inspected and reported on in Europe. Other colleges, however, were beginning to move on a less ambitious scale.[53] The importance of systematic astronomical observation was being increasingly urged in this country, but perhaps the predicted return of Halley's comet in 1835, which was widely publicized, sparked the demand for superior instruments and appropriate plans for observing the event. George Airy, when Plumian Professor of Astronomy and Director of the Cambridge University Observatory, had reported in 1832 to the British Association for the Advancement of Science, "I am not aware that there is any public observatory in America, though there are some able observers."[54] By 1850 he could have counted ten, four in New England alone.

The University of North Carolina claims credit[55] for the first observatory building, erected in 1831. It had begun to order instruments in 1824, but until the small structure was ready these had been scattered in various spots on the campus and even at one time mounted on the roof of the president's house. Among them were a meridian transit, a Dollond refractor, a Hadley's quadrant, a zenith telescope by the English instrument maker William Simms, and several other pieces of equipment by well-known European firms.

Yale received $1200 in 1828 for a Dollond telescope, the gift of Sheldon Clark, a farmer from Oxford, Connecticut. It was a refractor of 5-inch aperture, the largest in the country at the time, though the housing for it, a remodeled tower, proved inadequate for the best observation. Even so, however, it performed well under the first two directors, Elias Loomis (1831) and Denison Olmsted (1836), whose observation of Halley's comet several weeks before news of its reappearance had arrived from Europe led Professor Farrar to lament that Harvard had nothing to match Yale's facilities.[56]

In 1836, under the direction of Professor Albert Hopkins, Williams College acquired a modest building for its 7½-inch equatorial built by Alvan Clark[57] of Cambridgeport and the 3¾-inch transit from Troughton & Simms. In 1837 Elias Loomis left Yale for the Hudson Observatory of Western Reserve College, Ohio, and acquired for it in Europe a number of excellent instruments and books. (Several years later he began to put together material for his book, *The Recent Progress of Astronomy, Especially in the United States,* the first, and in some respects the only, over-all survey of American astronomy ever attempted in one volume.) West Point's first observatory, erected in 1839 primarily for training cadets, possessed a 9¾-inch equatorial by the American maker Henry Fitz, a mural circle by Troughton & Simms, and several other instruments.

The Philadelphia High School, unique as the only secondary school in the United States to provide first-rate astronomical equipment and buildings, at a time when most institutions of higher learning lacked them, received its instruments in 1840. They included a 6-inch equatorial by Merz and Mahler, mounted on the model of the Dorpat (Russia) telescope, and other instruments also from the Munich firm, which became the chief supplier of several American observatories, including Harvard. One of the High School's first directors, Sears Walker, later established close ties with the Harvard Observatory in its cooperation with the Coast Survey.

Perhaps the most famous observatory of the early 1840's was Cincinnati's, built by public subscription as a result of Ormsby MacKnight Mitchel's series of brilliant lectures before huge audiences there and elsewhere. The building, for which John Quincy Adams at the age of 77 laid the cornerstone in 1843, was an imposing classical structure of 80 by 30 feet. It housed in 1845 an excellent 12-inch telescope made in Munich, at the time the most powerful in the United States. Among other establishments founded in this decade were the observatories at Tuscaloosa, Alabama (1843), Georgetown (1843), Amherst (1847), and Shelby College (1847), of special interest because of its later association with Harvard through its third director, Joseph Winlock. The building originally authorized by Congress in 1842 as a "depot of charts and instruments for the navy," which eventually became the Naval Obsevatory, was completed in 1844, but, as a governmental rather than a strictly academic institution, it belongs to another category.

At the Harvard bicentenary celebration in 1836, undaunted by its failure to equal its younger sister colleges in the power of its telescopes or even in a place to observe with those it possessed, Edward Everett, the orator of the occasion, nevertheless spoke as if Harvard had already realized some early imagined prophecy: "From the towers of our Academy, the optic tube, lately contrived by the Florentine philosopher, shall search out the yet undiscovered secrets of the deepest heavens."[58] His fanciful flight had to be brought down to earth, however, by President Kirkland's sober reminder that such a happy prediction still awaited fulfillment:

Before the Observatory

Taking part in the celebration of the lapse of two hundred years from the foundation of this society, we have to reflect on what we owe to God and man for its gradual enlargement; to all the aids, public and private, of past times. The donations and bequests of recent benefactors have done much to increase its means. Let it not be supposed, however, that all which is desirable is accomplished. The disposition to enhance the literary apparatus of our University, and extend the sphere of her usefulness, has ample scope. She has wants which demand additional patronage. We should be able on this side the water to look at the heavens through our own eyes and instruments; and for this purpose we require an observatory.[59]

It is to the everlasting honor of President Quincy that three years later he boldly took the step which, after many faltering efforts, enabled Harvard to look at the heavens with its own eyes and instruments.

II

THE TWO BONDS

1839–1865

When William Cranch Bond (Fig. 10) in 1839 "was drawn to Cambridge by the strong hand of President Quincy" (Fig. 11), the College could offer him little beyond some instruments once used by John Winthrop, a house, and promises. No salary went with the post, and none was paid until seven years later. Quincy's official account of the arrangement is somewhat ingenuous:

> In October, 1839, the Corporation were informed that Mr. William Cranch Bond was engaged under an appointment and contract with the government of the United States, with a well-adapted apparatus, in a series of observations on "meteorology, magnetism, and moon-culminations, as also upon all the eclipses of the sun and moon and Jupiter's satellites," in connexion with those which should be made by the officers of the expedition to the South Sea, commenced in 1838, under the authority of Congress, for the determination of longitude and other scientific purposes. Being also apprized of the reputation sustained by Mr. Bond as a skilful, accurate, and attentive observer, they made arrangements with him, with the consent of the government of the United States, for the transfer of his whole apparatus to Cambridge, appointed him Astronomical Observer to the University, and took measures to raise by subscription a sufficient sum to erect such buildings as were immediately required.[1]

Neither in this passage nor in the more detailed Appendix in the *History* does Quincy reveal the actual sequence of events or his own personal initiative in seizing upon the one man in the area best qualified to help Harvard launch its observatory. Only much later did he admit having gone to Bond before asking the sanction of the Corporation to propose transferring his private observatory and instruments from Dorchester to Cambridge, where, "by his labors, character, and influence," he could "draw the attention of the students and the public more forcibly to astronomical science, and create a general interest in the community on the subject, and perhaps form a nucleus for an efficient institution."[2]

Much to his surprise, Quincy reported, Bond at first demurred. He doubted his ability to fill a "public" office, he feared that the equipment would fail to satisfy possibly too high expectations, and above all he preferred "independence in obscurity to responsibility in an elevated position." Quincy's persuasive powers, however, prevailed over these and other objections, and on October 26, 1839, the Corporation duly voted that "the President, Mr. Lowell and the Treasurer form a committee to make any arrangement that they may deem expedient with Mr. Bond for the transfer . . . to Cambridge, and for the establishment of an Observatory there and also to receive any donations for that object on such terms as they deem it for the interest of the College to accept." On November 30 they confirmed the agreement, and by December 31, one week after Bond's last Dorchester observation on Christmas Day, he recorded his first transit data in Cambridge.

Bond's reluctance stemmed not alone, as Quincy had generously implied, from innate modesty. After unspeakable hardships in youth and early manhood, he had now achieved a certain success in the firm of William Bond and Sons, watch, clock, and chronometer makers. He owned a comfortable house in Dorchester, and in the private observatory he had built in 1823 and gradually equipped with superior instruments he had pursued astronomy every moment snatched from sleep or business. His years of experience in rating (and building) chronometers had much to do with his appointment in 1838 to make longitudinal and other determinations for the Exploring Expedition sent out by the United States in 1838–1842 under the command of Charles Wilkes.[3] Bond had a four-year contract that would supplement his other income, even though in his zeal to perform well for the Navy he had spent a ruinous sum on additional equipment. The support of a large family and elderly parents rested on him alone, and to gamble on the unpaid Harvard position would merely add to his burdens. In the end his temporary sacrifice proved of the highest service to the College and brought renown to the name of Bond.

FIG. 10. William Cranch Bond, first director of the Observatory (1839–1859), from the portrait (1849) by Cephas G. Thompson. (Harvard College Observatory.)

The Dana house (Fig. 12) became the first Harvard Observatory. Until moved across the street in recent times to make way for the Lamont Library and to serve as the University's official guest house, it stood on the land originally favored by John Quincy Adams in 1822. In 1839, somewhat less favorable for the purpose, owing to the new structures and freshly laid out streets, this site nevertheless offered the only

FIG. 11. President Josiah Quincy, founder of the Observatory, from a painting by William Page. (Harvard University Portrait Collection.)

immediate possibility. On February 12, 1840, John Quincy Adams's committee, appointed to raise money for altering the house, reported that $3000 had been subscribed by "thirty gentlemen of Boston and vicinity" and that remodeling had already begun. Bond has left in Volume 1 of the *Annals of the Observatory of Harvard College* (now generally referred to as the *Annals*) an admirable account of the ingenious

The Harvard College Observatory

Fig. 12. The Dana house, which served as the Observatory from 1839 to 1844, showing the cupola constructed to accommodate the telescope. Originally on the site of the Lamont Library, it now stands on Quincy Street and serves to house the University's guests.

makeshifts necessary to convert a private professorial dwelling into an observatory. The most important of these was a cupola with a revolving roof to accommodate a telescope, which was to be mounted on a platform extending to the framework of the structure.

The instruments, such as they were, would be arranged in the various rooms according to their use. Even supplemented by those that Bond had transferred from Dorchester, the apparatus was not very promising, since it had remained much as it was after the fire. Professor Lovering in a letter to President Quincy dated June 9, 1845, stated categorically that before Bond came in 1839 "the College did not possess a single instrument which was adapted to making an astronomical observation which would have any scientific value." He listed an unreliable astronomical clock, a small transit instrument "so far below the average of such instruments" as to be quite useless, and three telescopes that might "answer decently well" for observing the moon, Jupiter's satellites, Saturn's rings, and perhaps other bodies, but very imperfect for any "nice observation." Even if improved they would hardly do for "a large class of astronomical observations" since they lacked an equatorial mounting and proper clockwork. In addition, there were two quadrants and two sextants. In short, "the University possessed no instrument of much value for determining either time or position, the two great elements which the theoretical astronomer wants, and which the Observatory is expected to furnish."[4]

Nevertheless, plans moved ahead. A small wooden building housed the Lloyd magnetic apparatus, which had been given to the Observatory from the Rumford Fund of the American Academy of Arts and Sciences. Its purpose was to promote cooperation with the far-flung magnetic stations at Greenwich, Toronto, St. Helena, Cape of Good Hope, Madras, Singapore, and Van Diemen's Island (Tasmania), and with the Russian Government and German Magnetic Association. In carrying out this program simultaneously with his observations for the Wilkes Expedition, Bond embarked on the course of worldwide collaboration with other astronomers and institutions that has characterized the Observatory's history ever since. The magnetic observations were made for three years (1840–1843) on the system proposed by a committee of the Royal Society, but in the absence of sufficient funds Cambridge had to rely for the latter part of the period on volunteers, students calling themselves "The Meteorological Society of Harvard." One leading member of it, Bond's elder son William, a most promising astronomer, unfortunately died in 1842.

Although Bond toiled heroically to meet the exacting demands of these programs, it soon became clear that the present quarters and instruments would never fulfill Harvard's ambitions for a worthy astronomical establishment. Accordingly, on September 4, 1841, the Corporation appointed a committee to purchase a suitable site, and within three weeks it was able to report having contracted for eleven acres of the Craigie Estate, then known as "Summer-House Hill." It was unprophetically described as excellent for "its unobstructed prospect, and its freedom from all liability of having its range of vision obstructed in the future." Additional land bought shortly afterward to give the Observatory command of the entire hill brought the total to twelve acres at a cost of $8300, of which six, valued at $4100, would belong to the Observatory. It is ironic to recall, in view of the vast expansion of the University, that in 1841 the treasurer consoled himself for this enormous outlay by his certainty that some of the area could be sold off to private persons without loss!

Fig. 13. Professor Benjamin Peirce, an early promoter of the Observatory; from a painting by Daniel Huntington. (Harvard University Portrait Collection.)

Another year passed before further efforts seemed feasible. In July and August 1842 Bond and Benjamin Peirce (Fig. 13), who had just become Perkins Professor of Astronomy and Mathematics and who already had succeeded in getting subscriptions, began their campaign for a new telescope. The Corporation willingly authorized the appointment of Joseph Cranch, Bond's brother-in-law and cousin, as an agent to investigate costs and makers on the Continent and in England and to

46

consult leading astronomers there. One piece of information the instigators of the move already had: that the best telescope in the world belonged to the Russian Observatory at Dorpat, for which it had been made by Merz and Mahler, the former Fraunhofer establishment in Munich. Accordingly, on August 12 Peirce drafted a letter in which he advised the agent to go to Germany and draw up a contract for a telescope, the chief features of which must be

in every respect equal or superior to the Dorpat telescope . . . The contract should be founded on an exact description of the Dorpat telescope and of the excellence of its performance . . . equal in all the following particulars—the perfection of its achromatism, its power of giving definite images of the larger stars, its power of separating double stars and of measuring their distance asunder, its power of exhibiting the colours of the stars, the firmness of its mounting and the accuracy of its clock work; the eye pieces should be of the highest degree of excellence and the delicacy and accuracy of the micrometers should be capable of standing the severest tests.[5]

The agreement should specify also time of completion, terms of payment, safe delivery, insurance, and final cost—which should be no more than $6000. Later, this proved to be an unrealistic figure for the quality desired. Bond happily wrote his close confidant William Mitchell,[6] "You will rejoice with us that official instructions have been forwarded to Mr. Cranch to search Europe for the best telescope and parallactic mounting for the Cambridge observatory."

Promptly on September 2 Cranch replied that though he had already written to Munich he would postpone going there until he had received the best advice from such trustworthy astronomers in England as George Biddell Airy, James Challis, and others. These authorities, Cranch wrote, "gave pleasing encouragement to your negotiations," manifesting "the best feeling towards rendering every facility in promoting the object of American undertakings in regard to science." As an Englishman Cranch himself harbored "an excusable partiality and a very warrantable one to prefer the good word of an English house of credit to all the uncertainty and doubts of foreign." He added that one firm in London had promised, "We can do anything after we find the object glass, like the receipt in the Cookery book for roasting Hare—first catch the Hare."

Cranch eventually caught the hare in Munich, as planned, but not until a dramatic astronomical event opened up a wholly new prospect for the Cambridge observatory.

II

On February 28, 1843, a bright comet, like a sudden blazing dagger of light, pierced the sky alarmingly close to the sun. "Never before, within astronomical memory," writes Agnes Clerke, "had our system been traversed by a body pursuing such an adventurous career."[7] Its dazzling brilliance even in midday, its rapid flight (a rate of 360 miles a second), its phenomenally long tail, estimated at some 200 million miles, its swift angular motion that at perihelion took it from one side of the sun to the other in little more than two hours, the rapid changes in the length and form of the tail, like "a torch agitated by the wind"—all created the most intense excitement and in many quarters deep apprehension. Despite advances in astronomical knowledge, sudden heavenly apparitions still had not lost their power to stir latent fears of imminent doom. Comets especially had a long and evil reputation as ominous portents, figuring prominently, in the late Professor Perry Miller's words, "in the panorama of conflagration."[8] To Increase Mather, writes Miller, they augured caterpillars, tempests, inundations, and sickness. Even as late as 1832 Arago had found it necessary to publish his elaborate treatise on comets to reassure a frightened public that a forthcoming visitor would not collide with the earth or any other planet. No comet, he insisted, had ever been known to fall into the sun, and though every conceivable disaster had been laid to its appearance—earthquakes, volcanic eruptions, violent storms, pestilence, plagues of flies and locusts, even the biblical Deluge—"by thus making out for each year a complete catalogue of all the miseries of this lower world, any one might foresee that a comet would never approach the earth, without finding a part of its inhabitants suffering under some calamity or other."[9] (Lest we pride ourselves on twentieth-century immunity to such fears, we may note that in 1968, when the asteroid Icarus came within 4 million miles of the earth, a frightened group fled to the high mountains to take refuge as far as possible from the ocean, in case a crash produced a great tidal wave.)

It so happened that the comet of 1843 coincided exactly with the widely publicized prediction of the Millerite sect[10] that the year would see the coming of the Last Judgment. Even those inclined to scoff at such religious fanaticism might well have wondered whether there was something in the dire prophecy. New England newspapers added to the general anxiety by printing reports of worldwide panic. A Liverpool journal voiced relief that "there will be no recurrence of this awful visitor in the days of the existing generation of men. This comet made the most direct dart at the sun of any one on record." In far away Turkey it was taken as a sign of the last day, predicting fearful disasters. The population of Malta was surprised and alarmed. Others saw it as a "thrilling warning to *all* the inhabitants of this precarious and transitory earth," one of the most "remarkable manifestations of Divine Power yet seen."

On March 22 Professor Benjamin Peirce delivered a lecture at the Odeon Theater in Boston to an audience of a thousand.[11] He began with some general remarks about the hypotheses of ancient writers and the superstitions attached to comets, such as deaths in high places, diseases, and other catastrophes. He quoted Arago on the prevalence of calamities regardless of comets and declared that present fears were groundless. But if a comet, he jested, seemed to bring extreme heat in Europe, severe cold in America, earthquakes and influenza in the West Indies, fires and neuralgia everywhere, and to be "the prophet of the end of all things to all of us," in Massachusetts, "the generous spirits of Boston consenting," it would bring a telescope.

And so it did. The Harvard Observatory, pressed for information about the comet, could reply only that it had no adequate instruments to calculate precisely the behavior of this spectacular apparition. To be sure, Bond had observed the comet for six nights and had been the first to detect the nucleus, even with nothing but a small telescope. He had also measured its distance from the sun as well as his instruments permitted. The comet he described as having the appearance of a small, indistinct star receding rapidly from the sun, but he could not account for the extraordinary length of its tail.

On March 29, greatly disturbed by Harvard's inability to satisfy a public demand for explanation, the officers of the American Academy of Arts and Sciences called a meeting of "gentlemen interested in Nautical and Astronomical Science . . . to consider the want felt in this country of a Telescope suitable for Astronomical observations." Held at the hall of the Marine Society on State Street, Boston, the meeting was attended by "several scientific gentlemen, members of the Academy, together with some of the principal Merchants and Presidents of the Insurances Offices." After calling the session to order, John Pickering, president of the Academy, turned the chair over to Abbott Lawrence and immediately offered the resolution: "That the reputation of a great commercial city, and the interests of commerce and science equally require, that means be provided for making more exact astronomical observations in this vicinity than can now be made; and that a Committee be appointed to procure subscriptions for a Telescope of the first class for this purpose."

Professor Benjamin Peirce followed Pickering. He agreed heartily that a good telescope was essential—in fact, efforts were already being made in that direction—but he wanted to make clear at the outset that Bond's observations should not be slighted. He considered them of great value, perhaps "as perfect as any relating to comets in the present century," though owing to inadequate instruments Bond had taken longer to calculate them. Peirce was canny enough to add, however, that the Philadelphia High School had surpassed Harvard in the number of its observations. Other remarks were made by Josiah Quincy and Samuel Eliot, both of whom, speaking for Harvard, assured the company that if the best instruments were procured they would find the most capable observers there. J. Ingersoll Bowditch then moved for the appointment of a committee to secure subscriptions and to bring in a report.

On April 7 the committee, consisting of John Pickering, Francis Gray, Jonathan Phillips, William Appleton, and Israel Lombard, met again. They reported "the expediency of procuring a Telescope of the first class for an astronomical observatory," entered into a general discussion of telescopes, their history and uses, and heard Bowditch lament that if the painful absence of anything worthy of the name of an observatory continued, this country would sadly lack trained astronomers. The needs of both navigation and education demanded action. William Bond, who was present, wrote later, "I was called out to receive the pleasant news that David Sears[12] offered five thousand dollars to be devoted to the building of an Observatory Tower, provided twenty thousand dollars should be subscribed for the instruments."

A week later Sears wrote to Quincy that he was somewhat disturbed to find that in the published report of the Boston meeting nothing had been said about Harvard "as a fit depository for their astronomical purchases." Lest anyone be in doubt about his intention, he insisted it should be clearly understood that "the erection of the Observatory must depend upon the condition, that the astronomical instruments are to be furnished by the College, and are to be under its care, and patronage." Reassured by Quincy, and informed that subscriptions were coming in, Sears was satisfied. Within the remarkably short span of six weeks the zealous committee announced complete success—eighty-two individuals, seven firms (mostly insurance companies), the American Academy ($3000), the Massachusetts Humane Society ($500), and the Society for the Diffusion of Useful Knowledge ($4000) had met Sears's handsome offer by contributing the total of $20,000. "I rejoice at the early accomplishment of an object important to science, from which much good may be expected," Sears wrote Quincy, "and I am gratified to find that your call has been promptly answered, and the necessary funds provided." He shortly afterward supplemented his first gift with an additional subscription and in 1845 gave another $5000 to be the nucleus of a capital fund for the permanent endowment of the Observatory. The massive Sears Tower still today commemorates the first donor to bring a long-deferred hope to fruition.[13]

On July 31, 1843, Samuel Eliot, treasurer of Harvard, officially recommissioned Cranch:

> You are, no doubt, aware that a number of gentlemen here, determined that our Cambridge College should not longer be without a good telescope, have contributed the sum necessary to purchase one of Merz and Mahler's best, viz., 42,000 florins, and it is desirable that some one should visit their establishment at Munich, to make the contract with them on behalf of our college. Mr. Bond has assured us of your ability to undertake such an office, and of your probable willingness to do us the favor. I write, therefore, to request you to act for us in the matter, to go to Munich, and to make the best bargain you can for us . . . either for an object glass already made, or for one to be made and mounted by them in the best manner . . . Our purpose is, as you will see from the enclosed letter from Mr. Bond, to procure an instrument fully equal to the best they have ever made, and even superior, if that is possible.

Although Eliot had left technical specifications to Bond, in stipulating "an instrument fully equal to the best" the firm had made, or even superior to it, he no longer meant the 9½-inch telescope at Dorpat (Tartu). Harvard wanted one that would match or surpass the great 15-inch refractor at Pulkovo, the Imperial Russian Observatory near St. Petersburg (now Leningrad), built in 1835 at the express decree of the Czar. Determined to make it the "astronomical capital of the world," the Czar had transferred the German astronomer Friedrich

Struve[14] from Dorpat to direct it, and already it had shown brilliant fulfillment of that ambition.

Cranch's previous inquiries had prepared him well for the new investigation. After further consultation with astronomers in England, he journeyed to Munich to arrange with Merz and Mahler the construction of two glasses from which to choose. If both proved unsatisfactory they could be rejected. For the mounting he turned to the well-known English firm of Troughton and Simms, which had given Bond much help in 1815.

The next few months saw plans drawn up for the building and the director's new house, the exact site of the pier foundations staked out, and actual groundbreaking on August 15. Meanwhile, now free of his Navy contract, Bond concentrated as well as possible on magnetic and meteorological observations, lunar culminations, occultations, eclipses, meteors, auroras, and so on, but more important than these was his deep concern for a proper architectural plan. He had carefully retained the information given him in England long ago. George Bond later wrote of his father's anxiety, "His own views of what was required . . . differed so widely from those of the architect as to occasion a good deal of disquietude, and he would never admit his responsibility for its external appearance, nor for its internal arrangements, excepting in the piers, the plans for the dome, and the machinery of meridian openings, observing chair, etc."[15]

The dome was a particularly sore point. Eliot had chosen as architect Isaiah Rogers, well known in Boston and elsewhere for the design of theaters, banks, and hotels, but totally without experience in planning any structure intended for scientific work. Fancying himself something of an inventor, Rogers had persuaded Eliot that his method of rotating the dome was superior to that which Bond had devised in 1815. Rogers' original device Bond described not very clearly in his diary entry for June 27, 1846, as "chain rollers having peripheries of combined conic sections so-called." His own idea was to rotate the dome "on the simple spherical form" of cannon balls. For months, while the construction was going on, he argued against the chain rollers, analyzing in detail their inefficiency and inconvenience. "Whether you are convinced by the above arguments or not," he wrote Eliot, "I shall feel satisfaction arising from a performance of a duty in entering my protest against the present proceedings."

Unmoved, Eliot replied, "I trust you will not regard my defence of the plan adopted as a pretense, on my part, of superior science," but having weighed all the arguments, he still felt that his responsibility to the College could not allow him to say "build as you please, and put myself blindfolded into your hands." If Bond is dissatisfied with the structure, "I can only regret it, but I shall certainly not reproach myself with having omitted to do anything of which I am capable, for its convenience and security."

In the end, Rogers' design failed, as Bond had stoutly maintained it would, but for many months, until Eliot yielded in favor of Bond's method,[16] the disagreement clouded their relation. The despondency of that period Bond felt freest to express to his confidant, William Mitchell. In a letter to Mitchell from Dana House in May 1844 Bond wrote, "In regard to the principal features of the observatory, I suppose we must say it is getting on as well as can be expected. Merz is progressing with the second object glass, the carpenters promise to have the new house ready by July [Bond doubted them], but for two years we must be prisoners of hope." He confided to Mitchell also that the years 1840–1843 he looked back upon, "in common with the rest of the family, as the most unpleasant in my whole life."

In September 1844 the Bonds and all apparatus, even the small wooden buildings housing some of it, left their home of nearly five years for Garden Street, then on the outskirts of Cambridge, and by November Bond could report genuine progress on the new Observatory. The stone pier, a massive cone-shaped structure, rested on a firm foundation at least twenty-six feet below ground, to satisfy Bond's almost obsessive concern for rigid support. "The tremor of an instrument," his son wrote, "would annoy and fret him as a harsh discord does the cultivated ear of the musician." A College appropriation of $1000 would assure space for meridian and transit rooms, and additional apparatus would include various transit instruments, a sidereal clock, and two refracting telescopes of relatively small aperture but useful pending the arrival of the Great Refractor. In December, with a transit instrument borrowed from the government, Bond determined the latitude of the Observatory, declared by Benjamin Peirce to be the "only one in the country [measured] with the requisite precision for the higher problems of astronomy. It was ascertained with the Prime-Vertical, and may be depended upon to a second of arc." Bond also continued collection of data for finding

53

the difference of longitude between the Cambridge and Greenwich observatories, and with one of the small refracting telescopes mounted equatorially in a separate building he observed sunspots (one of his major preoccupations), solar eclipses, a transit of Mercury, and comets.

In all of these activities George Phillips Bond joined his father—a collaboration of such closeness and harmony that it has never been entirely possible or necessary to distinguish their roles during the time they worked together. The wife of Richard Bond, William Bond's youngest son, whose mechanical skill greatly assisted in perfecting some of the recording devices and in chronometric ratings, has charmingly recalled their method:

> When I was a little girl I used to spend my August vacations at the observatory and for several years in succession Professor W. C. Bond and G. P. Bond were engaged in taking observations and drawings of [sun] spots . . . One observer, with a sharp pencil, traced the spots as they were reflected on the paper, while the other wrote down any notes or observations, of time, or peculiar appearances, or explanatory of the drawings . . . But both of them, besides being gifted with extraordinarily keen vision, had eye and hand and mind so thoroughly trained, that even to children it was fascinating to watch the certainty and accuracy of every touch, their enthusiasm and delight in the work, and the quick response and recognition of either to a remark or suggestion of the other.[17]

George Bond, William's third son, was born in 1825, and after completing a preparatory course at the Hopkins Classical School, entered Harvard in 1841. He was graduated in 1845, but even as a student, though not at first attracted to astronomy, he had regularly assisted his father since his older brother's death in 1842. In 1846, shortly before the completion of the new Observatory building, he received his official appointment as assistant observer.

From that year, in fact, one may date the effective operation of the present Harvard Observatory. It now had its own permanent home. An active Visiting Committee, headed first by John Quincy Adams, and consisting of President Quincy, David Sears, Abbott Lawrence, J. Ingersoll Bowditch, Robert Treat Paine, and (briefly) John P. Cushing, showed zealous concern for the new establishment and its observer. In June, William Simms wrote from London that he and Joseph Cranch had now made the final selection of the object glass, which they pronounced unequivocally to be superior to that at Pulkovo. To make cer-

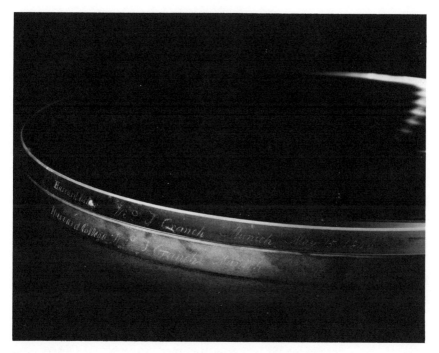

Fig. 14. The objective lens of the Great Refractor, showing the identifying inscription "Harvard College W.S. J. Cranch Munich May 15, 1846." (Photograph courtesy of Nathan Hazen.)

tain that the right lens would be sent they had had engraved on the edge with the point of a diamond the identification, "Harvard College W.S. J. Cranch Munich May 15, 1846" (Fig. 14). In September Bond presented his own first official report to the Visiting Committee. Besides announcing the pleasant news that the lens had turned out perfectly and that a transit circle ordered from Troughton and Simms was nearly completed, Bond emphasized his satisfaction with the changes in the mechanism of the dome. His test in June had entirely vindicated his insistence on his own rotary method. As one witness of the trial reported, the dome revolved almost at the flick of a finger. His specially designed chair was also under construction. Altogether, he held out to the committee a pleasing prospect that everything would be ready for "the reception of the two great instruments before winter sets in." Routine duties meanwhile included the usual observations, computations by Professor Peirce and George Bond on the orbits of comets, daily meteorological records for the *American Almanac,* and data on the temperature of the earth gathered from two separate depths. The time-consuming magnetic observations for the Coast Survey, however, he could no longer fit into his busy schedule.

In August 1845 President Quincy, replying to a number of pointed queries by John Quincy Adams (Fig. 15), had made an eloquent plea for a more businesslike and just arrangement with the Bonds:

As soon as [the great telescope] arrives, Mr. Bond will necessarily be required to devote his whole time to the Observatory, and ought to be enabled to give up his business. . . . This cannot be even asked, unless a sufficient . . . salary be provided for him. And this can only be obtained by applying to the same liberal and patriotic spirit which has so nobly laid the foundations of this institution. Without a scientific observer, all that has been already contributed will be lost to science, and useless. To secure such an Observer, an adequate salary is absolutely necessary, now, and in all future times.[18]

Not until April 1846, however, did the "same liberal and patriotic spirit" prevail, stirred perhaps, as often happens, by information that another institution had bid for Bond's services. George Bancroft, Secretary of the Navy, searching for an astronomer at the Naval Observatory in Washington, had offered Bond the post on terms far more favorable than those at Harvard—the prestige of a national office, a generous salary, and freedom from constant anxiety about maintenance and equipment. Since 1839 Bond not only had served without pay but had even met from his own modest purse various minor needs of the Observatory. Though gratified by the recognition, he declined in a letter which admirably expressed the ideal characterizing his entire career: "An astronomical observer to be useful in his vocation should give up the world, he must have a good eye, a delicate touch, and above all, entire devotion to the pursuit." On April 11, 1846, from funds subscribed by "citizens of Boston," the Corporation voted salaries for the first time, to William Bond $1500 annually for two years and to George Bond $640, the pay of a tutor in the College. The house would remain free of rent.

The more formal annual reports from 1846 to 1855, along with other documents relating to the early history of the Observatory, occupy the first half of Volume 1 of the *Annals.* A far more intimate view of the first three years at the new site the Bonds have left in their joint diary,[19] dating from April 15, 1846, through December 1849. Entries in the months before and at the time of the arrival and mounting of the Great Refractor not only reveal an incessant round of observations, conferences, computations, and correspondence, but touch on contemporary events and comment on personalities and situations with a sharpness and candor not intended for other eyes. William Bond rejoices in the election of President Everett (Fig. 16) and describes vividly the inauguration on Thursday, April 30, 1846, with its dinner and fireworks afterward: "a happy day for Cambridge . . . a day of as pure and unalloyed enjoyment as perhaps the world ever gave its votaries." (That Everett himself, who had somewhat reluctantly accepted the post, shared this joy seems doubtful, for Professor Morison calls the new president's three years in office the "most wretched" of his life.)[20] On May 16 Bond visits the Treasurer and "is glad to find him much better disposed to follow reasonable advice than he was some time ago." Eliot had been "in great error in listening to Rogers who has all along shown himself utterly ignorant of the wants of an observatory." He is a "wretched mechanic." And as for architectural taste, "nothing could be more discordant with the true principals of the art, than the whole affair." Lucky for the enterprise that Rogers had little to do with the pier! "We have now, I hope, done with him."

The diary follows George Bond's avid hunt for comets, in which he was greatly aided by two excellent comet seekers, the gifts of President Quincy and J. Ingersoll Bowditch. In the course of a few years George independently discovered at least eleven comets, in some instances before their observation was announced elsewhere. Besides comets, the planets Saturn and Jupiter attracted the steady attention of both father and son. In September 1846 they noted particularly the double rings of Saturn, one of the first in a long series of observations of this planet that led ultimately to their discovery of its eighth satellite, Hyperion. But perhaps two entries a week apart are of particular interest in view of the long-continued speculation about a fundamental problem in astronomy and the twentieth-century solution of it: *Where was the center of the Universe?*

On September 26, 1846, George Bond wrote, "Saw the numbers of the Astr. Nach. up to August 3d they contain Mädler's new theory of the motion of the sidereal systems." On October 3, at the home of Professor Peirce, where George had gone every day from 11 to 2 to "compute tables of Parallax in AR and Dec for this observatory,"

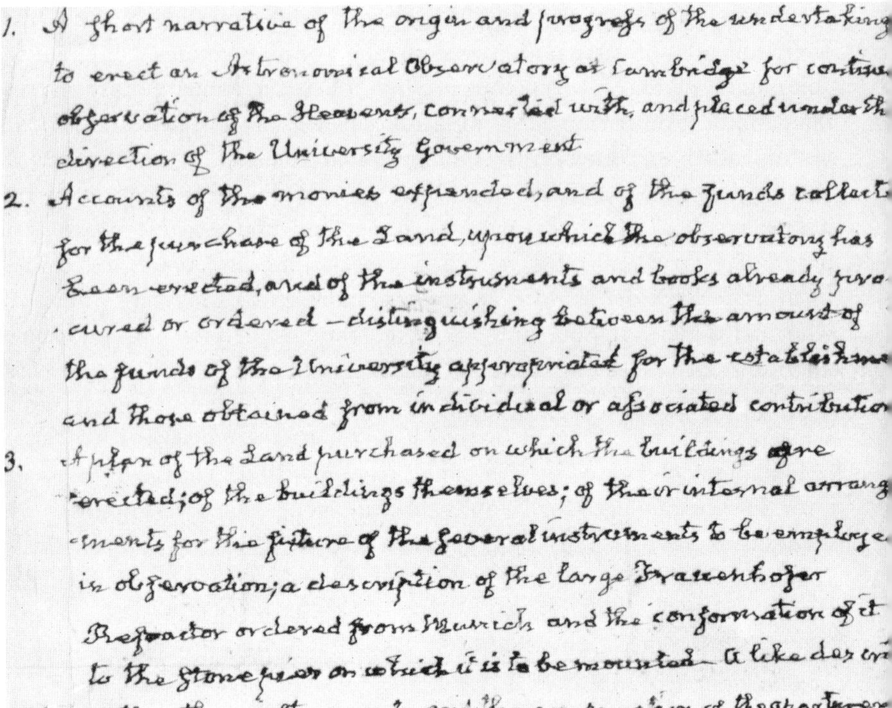

1. A short narrative of the origin and progress of the undertaking to erect an Astronomical Observatory at Cambridge for continual observation of the Heavens, connected with, and placed under the direction of the University Government

2. Accounts of the monies expended, and of the funds collected for the purchase of the Land, upon which the observatory has been erected, and of the instruments and books already procured or ordered — distinguishing between the amount of the funds of the University appropriated for the establishment and those obtained from individual or associated contributions

3. A plan of the Land purchased on which the buildings are erected; of the buildings themselves; of their internal arrangements for the fixture of the several instruments to be employed in observation; a description of the large Fraunhofer Refractor ordered from Munich and the conformation of it to the stone pier on which it is to be mounted — a like description of the other instruments and the conformation of the apartments in the building to their fixture for observation.

4. The appointment of an Observer, and one Assistant; a measure indispensable to make the whole establishment effective for the purpose of continuous observation. In the first instance it is very important and desirable that these Offices should be conferred upon Mr Bond, and his son: but as a permanent institution it seems that provision should be made for a regular succession to these Offices. That the mode of their appointment, the occasions of vacancy the tenure of their Offices, their right of occupation and custody of the buildings, both dwelling house and observatory, and of the adjoining grounds, should be regulated and prescribed

Fig. 15. Pages of a letter (1845) from John Quincy Adams, chairman of the Visiting Committee, to President Josiah Quincy, outlining his views on the administration of the Observatory

58

That the line of division between the duties, of the observer and those of the Perkins professor of Astronomy and Mathematics should be accurately drawn — That the extent to which the Professor shall have a right to the use of the Instruments, and of access to the observatory, for the purposes of instruction in his department to the Students at the University, should be clearly defined. Whether some liberty of occasionally assisting the observer in making observations may be indulged to Students whose inclinations may take a special direction to the study of physical astronomy, and whether the teaching of the use of the instruments, to all the Students of the higher classes may not be included among the joint duties of the Professor and of the Observer.

— The regular compensation to be allowed to the Observer, and to the Assistant, in addition to the occupation of the dwelling house and the grounds — The duties to be required of them, and the remedy for the possible neglect or inability to perform them. The definition of the duties, and the exercise of the power of removal are points of peculiar delicacy, to be drawn up with care, without reference to any present circumstances or persons. The fund to be raised by contributions of individuals — its amount — The amount which it will be in the power of the Corporation to add to it, to constitute the whole fund to be duly invested and from the income of which the Salaries of the Observer and Assistant may be permanently provided for and other expenses to no considerable amount but occasionally necessary may be defrayed.

¹ Letter found among the M.S.S. of President Quincy and presented to the Observatory by Eliza Susan Quincy. — May 6. 1868.

The original, found among the Quincy papers, was given to the Observatory by Miss Quincy.

FIG. 16. President Edward Everett; from a painting by Bass Otis. (Harvard University Portrait Collection.)

he noted, "Have been reading over Mädler's theory of the motions of the fixed stars round Alcyone [the central star of the Pleiades] along with Prof. Peirce." The author of this arresting new theory was Johann Heinrich Mädler, who had succeeded Struve at Dorpat—giving rise, no doubt, to the acid comment of a traveler in Russia that "the Germans owned the Russian skies." In his paper, "Die Centralsonne," just published in *Astronomische Nachrichten* (Nos. 566–567), he had categorically concluded: "I therefore pronounce the Pleiades to be the central group of the entire system of fixed stars limited by the great stratum of the Milky Way, and *Alcyone* as the individual star which combines the greatest amount of probability of being the true *central* sun."[21]

George Bond, intrigued but somewhat dubious, wrote:

In the first place his [Mädler's] arguments for setting aside other theories are not of much weight yet are such as ought to be considered. This theory raises so many questions to be decided that the consideration of it is exceedingly interesting, for to answer them does not require data beyond our reach, but such as is already at our command or as we may reasonably hope to obtain. If it be true only as the rudest outline even it will be the first of a grand series of discoveries in the sidereal systems which must soon follow.

Pursuing the inquiry, he suggests,

will give direction at once to our else haphazard rambles among the stars. On what points are we to concentrate observation and What are the phenomena to be expected that we may apply the most proper means for discovering them? What stars are we to select as giving the most probable annual parallax? What view are we to take of the magnitudes & number of the stars of our system? And the length of time which will be probably requisite to obtain some definite knowledge of this system?

To such inquiries as these this theory if true will return satisfactory answers and of many others of no less interest it will lead to most probable solutions.

If at the moment the theory seemed to rest "on a most frail & uncertain basis," all that was needed to test it would be "such a general or statistical knowledge of the stars as we may fairly consider within our reach"—their number and distribution, accurate comparisons of them with nebulae, and the parallax of neighboring suns. We may then "confidently await a very speedy settlement" of the truth or error in the theory. "What a magnificent triumph of Science" that would be! And yet, and yet, "what a pity if a loose screw in Bradley's Quadrant should be its final upshot!"[22]

George Bond's excitement about the implications of Mädler's pronouncement was widely shared. Ormsby MacKnight Mitchel, director of the Cincinnati Observatory and publisher of the short-lived *Sidereal Messenger,* considered the theory of such moment that he published Mädler's articles on the "central Sun" in three successive issues of that journal (vol. 1, nos. 3, 4, 5, September and October 1846), with the explanation that the results arrived at represented the "most laborious research" of many years and that, if the discovery should be verified by observation, it would give to the subject of the proper motion of the fixed stars "a new character, and a deeper interest."

Bond's optimism about a "speedy settlement" was hardly shared by Sir John Herschel,[23] who concluded from his own study of Mädler's calculations that, although the concept advanced was "by no means to be discouraged as a forerunner of something more decisive," as it stood it was highly improbable and would require observations over thirty or forty years to settle its validity. In fact, it took somewhat longer.

In 1943 the Cincinnati Observatory observed its centenary. Invited to give the leading address on that occasion, Harlow Shapley, then director of the Harvard Observatory, paid appropriate tribute to the founder and his service to American astronomers in having published Mädler's work in the *Sidereal Messenger.* Taking as his topic "The Problem of Being Central,"[24] Shapley could bring to it the results of his own long preoccupation with the question that so deeply stirred Mitchel. Shapley placed the center of our Galaxy not in the Pleiades but in the direction of Sagittarius; yet, he went on:

Satisfied as we are that we now know accurately the direction to the galactic center and know approximately its distance, we are not certain that we know where lies the gravitational center of this part of the universe of galaxies, or the center of the total system of galaxies which we call the Metagalaxy . . . In fact, we seek in vain for a metagalactic center.

For George Bond, however fascinated by Mädler's view of a central sun, the more immediate center of interest was the "Great Equatorial." On October 21, 1846, he records: "Got news that the object glass . . . is likely to arrive in November." The same day he made the following entry:

This Evg received intelligence of the discovery by Dr. Galle of the planet predicted by Leverrier. He found it the very day that he received a letter from M. Leverrier directing him where to search. The fulfillment of the prediction is more a matter of admiration than of surprise. The name of Leverrier will hereafter take a high rank among the greatest of astronomers.[25]

The existence of this planet, subsequently given the name of Neptune, had been suspected for some years because of anomalous perturbations in the orbital motion of Uranus, the planet discovered by Sir William Herschel in 1781, which he had at first named Georgium Sidus. (It was labeled "the Georgian" for some time in the English *Nautical Almanac*.) In 1840 Bessel had stated that "we have here to do with disturbances whose explanation can only be found in a new physical discovery. Further attempts to explain them must be based upon the endeavor to discover an orbit and a mass for some unknown planet." In 1843 young John Couch Adams, of Cambridge University, began to apply himself to the problem, and by 1845 he had obtained an approximate orbit of a disturbing planet, but he made no attempt to publish it. In France, at the suggestion of Arago, Urban Leverrier independently tackled the identical problem and in August 1846 announced to the French Academy of Sciences the results he had obtained. The actual discovery on September 23, 1846, set off an unpleasant international controversy, one for which the Bonds' diary, beginning with the entry for Wednesday, October 21, besides noting their own observations of the planet, provides invaluable contemporary information:

We immediately set to work to find the stranger having found an approximate place from observations of Sept. 23–24 (the first is the date of the discovery) we fixed upon the star 7648 B. A. Cat. On directing the Telescope to this Star there were to be seen in its immediate vicinity 8 or 10 stars, perhaps more, of from the 7th to the 10th magnitude—we observed six of them with the annular micrometer. We fixed on no. 3 as the planet & so it proved to be. We saw it again on the 22nd . . . It is of the 7–8 magnitude & precisely like a star. Its brilliancy is unaccountable. Father was the first to see it. People seem to take a good deal of interest in the new planets.

The Bonds felt especially triumphant that with their small telescope they had observed the new body, having "received on the 21st the earliest & most important astronomical intelligence we have ever yet had . . . *only 28 days*" after its discovery, and they followed with mingled amusement and disdain the unlovely feud between the French and the English about both priorities and the name of the planet. On November 21, Bond notes,

the evening was not favourable for observing, but we got a set of observations on the planet, which is to be called 'Leverrier.' The name was given by Arago, to whom Leverrier delegated his right. This course does not give entire satisfaction, as the French would not allow the designation of 'Herschell.' M. L[everrier]'s modesty is a little Frenchified, but after all he may easily be excused for allowing a new world to be called after him.

On December 5 Bond gives an accurate summary of the events and describes the heightening tension between the two countries:

It seems that matters are taking a pretty decided aspect between the French & English astronomers, respecting the priority in the discovery of this new Planet.

Airy, Herschell & Challis have laid claim on the part of Mr. Adams for a fair division of the honours.

They say that as early as June last Adams sent to Airy a position of this new Planet nearer to the true than the latest determination of M. Leverrier giving besides an inclination, which M. L. did not.

Airy neglected to search for it probably from a press of other business. Adams then desired Challis to look for it. He accordingly *observed* it *twice in August*, more than a month before it's discovery by Galle, but neglecting to compare the places,[26] unfortunately did not recognize it.

The publication of these facts in the London Athenaeum (I think) touches the honour of 'La grande nation'.

A proposition at a sitting of the French Academy to consider the claim of Mr. Adams is hissed down, & Mr. Airy is proclaimed a charlatan: a discovery the undisputed honours of which belong to M. Arago.

The new planet attracted scores of visitors to the Observatory—the president of Harvard, professors, and friends. Everett, always eager to publicize the alertness of Harvard observers, promptly forwarded to Leverrier the Bonds' notes on his triumph.

Meanwhile, in the midst of the excitement about Neptune, the Bonds learned to their dismay that the London bankers, ignoring explicit orders to ship directly to Boston, had allowed the precious object glass of the new telescope to go out from London bound for New York:

Now it runs the risk of the tariff of December (20 pr ct). It is to run the gauntlet at its unshipment at New York without a friendly hand

The Two Bonds

to smooth its passage into this Western World. & next come all the dangers of its passage from New York to Cambridge which is far more dangerous than it's journey across the Atlantic.

But their fears were groundless. The glass reached New York on the *Margaret Evans* November 28 after possibly the fastest passage then on record. A day later would have put it "in the power of the consignees to delay her entry until the operation of the new tariff." A week later the glass

arrived on Friday being provided with a private car for its exclusive accommodation, it arrived in the forenoon looking very much like a bale of old rags. We began unpacking with Prof. Peirce between 3 & 4. There were first two tow bags with straw between, the whole bundle being perhaps 5 ft. through, next was a square deal box of 3 ft. packed in straw, next a smaller box of about 2 ft. & inside of this the object glass without any padding part of the box is lined with velvet. from the color of the cloth it appeared jet black, and the "coup d-oeil" quite exceeded expectation. It looked very large & there seems no chance for fault finding before mounting it. We were delighted to find that it exceeds 15 inches in diameter. We succeeded in deciphering the signatures on its edge, "Harvard College, W. S.–J. Cranch May 15th 1846" [Fig. 14].

An age that has long taken for granted the great 200-inch at Palomar, that plans even larger telescopes, and that has hardly kept up with the names of rockets prior to the last Apollo can only with the greatest difficulty recapture some of the excitement aroused in this country by the acquisition of the most perfect glass of its size yet to be produced. It is true that Cincinnati had received its excellent 12-inch in 1845, but here was the first real rival of the Pulkovo telescope; indeed, according to the maker, it was actually superior in quality and even by a fraction in size. In another six months the arrival of the mounting would end all suspense about its powers.

The diary for July 5, 1847, the first entry since February, after noting a few routine activities, describes in minutest detail every operation involved in the reception of the long-awaited tube (Fig. 17). It reached Cambridge late on a cloudy Friday afternoon, June 11, in 13 huge cases weighing about 6 tons and had to be unloaded with the greatest speed by a dozen men in order to beat the shower expected at any moment. All the next day the workmen opened crates and hoisted them to the dome floor, where they unpacked, in order, tools for mounting

Fig. 17. The Great Refractor, bought from funds subscribed by citizens of Boston. The largest telescope in the United States at the time, it was a twin of the instrument at the Russian Imperial Observatory, Pulkovo.

the clockwork, eyepieces, micrometers, hour circles, friction rollers, and the rest. Every part was eagerly examined; every one proved in size and workmanship the skill of the maker. Ten days of stone work on the pier were followed by two strenuous days of anxious placement, during which the slightest slip or misstep might have landed the whole apparatus in the cellar. At 4 P.M. on the second day "the great object glass was fastened in & we soon got a sight of the moon." In the evening the Bonds looked again, "but the eye could scarcely bear its light, with the low powers which [because of the incomplete staging] we were obliged to use."

The stars disappointed them at first, but they consoled themselves by laying the blame on the atmosphere. The weeks following required a number of delicate, even hazardous, adjustments, but on September 22:

At 3½ AM Turned the Great Telescope upon the Nebula in Orion—with a power of 180—the atmosphere was in the most favourable state that it has been since we had the Grand Refractor—the revelation was sublime, the first appearance was like bright clouds—the fifth star in the Trapezium was conspicuous, many stars were seen among the clouds of light, & about the borders. The stars came out in this way—first the fifth star of Struve bright and clear—then four stars above the Trapezium distant about 40″ or 30″ from its centre lying in nearly one parallel of declination, about 30″ and 20″ apart—these were steadily seen, with numerous glimpse stars in their neighbourhood—then a faint star between the trapezium and what is call'd the fishes mouth . . . the whole appearance of this nebula was altogether different from the representations given in Books—it is much more extensive, appears, as our milky way does of a very clear night when the moon is absent, sprinkled with stars.

On that same day in September 1847 William Bond wrote to President Everett:

You will rejoice with me that the great nebula in Orion has yielded to the powers of our incomparable telescope!

This morning, the atmosphere being in a favorable condition, at about 3 o'clock the Telescope was set upon the Trapezium in the great nebula in Orion. Under a power of 200 the 5th star was immediately conspicuous; but our attention was very soon absorbed with the splendid revelations made in its immediate vicinity. This part of the nebula was resolved into bright points of light. The number of stars was too great to attempt counting them;—many however were readily located and mapped. The remarkable character of the star θ in the Trapezium was readily recognized with a power of 600. The companion, "Struve's sixth star," was distinctly separated from its primary, and certain of the stars composing the nebula were seen as double stars under this power.

It should be borne in mind, that this nebula and that of Andromeda have been the last strong-holds of the nebular theory; that is, the idea first suggested by the elder Herschel of masses of nebulous matter in process of condensation into systems. The nebula in Orion yielded not to the unrivalled skill of both the Herschels, armed with their excellent reflectors. It even defied the powers of Lord Rosse's three-foot mirror, giving "*not the slightest trace of resolvability*": by which term is understood the discerning *singly* a number of sparkling points.

As if aware of improper boasting, Bond added:

I feel deeply sensible of the odiousness of comparison, but innumerable applications have been made to me, for evidence of the excellence of the instrument, and I can see no other way in which the public are to be made acquainted with its merits.[27]

Its merits were unmistakable, for his second examination of the nebula on October 7 further convinced him of the telescope's superior powers. "It is delightful," he wrote in his diary that day, "to see the stars brought out which have been hid in mysterious light from the human eye, since the creation. There is a grandure, an almost overpowering sublimity in the scene that no language can fully express." And he gives additional details of the stars he had seen in the nebula—"the sixth of Struve plainly," the "5th star of the Trapezium with any power it being more distant from its principal . . . a second class of stars, four of which . . . are seen steadily when the atmosphere is favourable," three stars "below the opening called the Fishes mouth, *and three others,* forming altogether a miniature representation of the Constelation of Orion, excepting the star χ which in Orion is a multiple star. If we should make out a cluster here, the resemblance will be most remarkable."

In his enthusiasm Bond had made the very natural mistake of supposing that because, owing to the high magnification and the brilliance of the image, he could distinguish faint stars within the nebula he had "resolved" the whole. Again and again in the diary he returns to the observation of this fascinating nebula. On October 11 he notes that he could see the sixth star with a power of 300, that the "Nebula has a bright clouded appearance in the neighborhood of the Trapezium and border," that four stars above it "were quite plain," that many others were seen in glimpses, and that "sometimes a cluster would come out plain for a few seconds." To him the appearance was altogether very different from the description given by Sir John Herschel in *Results of Astronomical Observations . . . at the Cape of Good Hope,* a copy of which the author, much to Bond's pleasure, had sent him. He projected a map "of the Telescopic stars" in the nebula, and in December 1847 when he turned the telescope in the direction of Orion, where it was very clear, "the stars and nebula were beautifully defined, the impression was irresistible that the nebula was resolved." He caught "momentary glimpses" of stars "in great numbers."

It so happens that the Orion region lies in what astronomers call Gould's Belt, an extension of the Milky Way that is rich in faint stars. In all probability, this background of faint stars, viewed through the nebulous haze, gave the impression that the nebula itself was indeed a cloud of stars, and therefore completely resolvable. Actually, as Sir William Huggins showed spectroscopically some twenty years after Bond's observations, the nebula is a cloud of gas. It glows only because of the bright stars embedded in it.[28]

News of the final mounting of the telescope and its success lured crowds to the Observatory eager to enjoy the celestial wonders the Great Refractor revealed. On a Saturday evening in October, for example, the number of viewers was so large that for an hour they lined up to catch a mere 30-second glimpse of Saturn. Important visitors came also—on October 6 ex-President Quincy and President Everett, with John Quincy Adams, who was enjoying his first sight of the establishment since his return from Washington. He was "very cheerful, and interested as ever in the prosperity of the Observatory." Others included members of the Corporation, Professor Longfellow and Lady, Captain Wilkes, and many more. Strangers from far away turned up: a "gentleman and Lady" arrived on a busy day and expressed such disappointment at not being admitted—they would be leaving for their Michigan home the next day—that Bond invited them in and gave them what they called "the most gratifying sight of Boston . . . We could form no idea, they declared, of the interest taken in these matters at the West. They had . . . read in the papers our account of the Telescopes performance."

The Harvard College Observatory

Fame "at the West" delighted the Bonds, but the crowds at home, regarding the Observatory as a rival to Barnum, soon became intolerable, "a perfect Babel." William Bond wrote Everett despairingly that to 400 persons crowding into a 30-foot room a minute at the telescope could hardly prove instructive, not to speak of the possible damage to the instrument from dust raised by so many tramping feet. The condition of the dome room, he sharply notes in his diary, was something "aweful," very much "like those portions of the Common in Boston which have been the most crowded, on the morning after a fourth of July." Within a few weeks relief came when the President and Corporation voted to close the Observatory except on special occasions or by invitation. Their promptness in sympathetically understanding the problem fortunately spared the Bonds the kind of disruptions that all but destroyed the Dudley Observatory in Albany, in the famous controversy between the trustees and the director, Benjamin Apthorp Gould.[29] Indeed, the Bonds were generally fortunate in their relations with Harvard presidents, donors, and exceptionally generous and knowledgeable Visiting Committees, who seemed always anxious to promote in every possible way the work of the Observatory. Their wise attitude left the director and his assistant free to carry on their appointed astronomical tasks of observing sunspots, nebulae, planets, and comets, in a setting Bond lyrically described in two pages of manuscript omitted from his account published in 1856. It is worth quoting in part for its view of a scene long since vanished. From the balconies of the Observatory one could see

the unrivalled beauty of landscape . . . eastward to the thriving town of Somerville with its beautiful hills and thickly congregating rural mansions, the retreat of the wearied citizen from the dust and noise of the metropolis; beyond . . . the heights of Charlestown, the memorable Bunker Hill of revolutionary memory, with its towering monument—and below . . . the tall masts of the ships of war indicating the position of the government navy yard. Southerly . . . the aspiring chimneys of the extensive manufactories of east Cambridge and the northern parts of Boston and Roxbury . . . spread like a panorama before the eye . . . the nearest view embraces the Academic Halls of Harvard, beyond [are] the heights of Dorchester where Washington presented to the astonished eyes of the British commander who then occupied Boston, the spectacle of a fortress raised as if by magic, and so situated as to leave him no alternative but an immediate retreat.

Turning to the south the fine outline of Blue Hills of Milton are presented to the view . . . the beautiful towns of Dorchester, Milton and Brookline, admirably diversified with hill and dale and thickly studded with pleasant dwellings intervening. Westerly the range of beautifully moulded and highly cultivated hills of Brookline, Brighton, and Newton . . . the more elevated hills of Waltham and Watertown the chosen residence of some of the more wealthy families . . . northward we see the hills of West Cambridge and Lexington, again reminding us of the "time that tried men's souls." The wooded heights of Woburn and Medford bound the view to the north till finally again returning to Somerville the most conspicuous object that meets the eye, is the recently erected building called Tufts College, built by members of the society of universalists from funds provided by Mr. Tufts of Somerville.

How pleasing the whole prospect!

Now, as Peirce had promised, the great comet had indeed brought to Massachusetts a telescope worthy of the liberal spirit that made it possible. But to repeat the usual romantic explanation that "the heavens themselves" had provided Harvard with an Observatory is to distort history. Four years before the comet and seven before the arrival of the Great Refractor, Josiah Quincy had boldly taken over the Dana house and installed William Bond with his instruments in it. Perhaps in fact one may say that the true genesis of the Observatory dates from 1671, when John Winthrop sent to Harvard its first "tube"—"3 foote and a half with a concave ey-glasse."

III

Of all the entries in the diary for 1847, perhaps one of the most significant is that of October 23, in William Bond's writing:

It clear'd off afternoon. I came home early to prepare for the reception of the Public. Mr. Whipple[30] has brought out his apparatus for the purpose of attempting to take a Daguerrotype of the Moon & Sun, by aid of the great Telescope, we turn'd it on the Sun my arm happening to be just close to the eye glass when I caught the solar focus—it instantly burnt a hole in my coat sleve, I felt it on my arm, the heat was so intense that we consider'd it most prudent to refrain for the present, and adjourned to the small observatory, where on trial Mr. Whipple was satisfied that he should be able to accomplish the object.

Eventual accomplishment of that object represents one of the triumphs of mid-nineteenth-century astronomy. The fruition of it came later, nowhere more richly than at Harvard. This first daguerreotype experiment of 1847 was made a year earlier than the date usually given for the beginning of astronomical photography here. Although it failed, it strikingly proves the Bonds' early awareness of the enormous potential photography offered—a tool second only to the telescope itself in the advancement of astronomy. In attempting to apply the new technique, this observatory was first in the United States, preceded only by the observatory in Paris, where Fizeau and Foucault made the earliest images of the sun.

New inventions often raise questions of chronology and priorities, and photography is no exception. For present purposes, however, it is practical to go back no further than January 7, 1839. On that date Arago made his historic announcement to the French Academy of Sciences that Daguerre had succeeded in fixing images with the camera obscura ("Communication sur la découverte de M. Daguerre concernant la fixation des images qui se forment au foyer de la chambre obscure"). He withheld details, but of significance here is his immediate declaration that the new process would offer invaluable aid to science, especially in physics and astronomy. In fact, he reported, Daguerre had already attempted an image of the moon, which, if not altogether successful, had at least shown the "first chemical impression of the moon's light."[31]

Eight months later the public, whose curiosity had been whetted by rumor and speculation, received specific information about the miraculous invention. Among the persons who had seen some of the results in private exhibits was Sir John Herschel.[32] He had himself experimented as early as 1819 with problems of a chemical fixing agent and in 1839 with the use of sensitized paper. He was greatly impressed. "This is a miracle," he is quoted as exclaiming, adding that attempts in England by himself and others "are childish amusements in comparison." Also during the interval prior to Arago's final disclosure of Daguerre's technique, American papers had picked up from the French and English press stories of the still mysterious process. In fact one Boston paper, the *Mercantile Journal,* actually compared it with the work of the English pioneer, H. Fox Talbot,[33] whose experiments in the mid 1830's evidently were known here. Another widely circulated account appeared in the New York *Observer* a month later. This was a copy of a letter that Samuel F. B. Morse, in Paris at the time, wrote his brother describing a visit to Daguerre's Diorama. He had seen examples, he declared, that would open up "a new kingdom to explore."

The Two Bonds

Finally, on August 19, 1839, at a joint meeting of the Academies of Sciences and of Fine Arts, Arago, fully sensing the drama of the occasion, told the world the secret of Daguerre's invention. In his oration he deliberately ignored the earlier work of Fox Talbot, but again, it is important to note, he took care to emphasize the importance of the new technique to astronomy. We may hope, he declared, to be able to make photographic images of our satellite ("des cartes photographiques de notre satellite"), and that in minutes we can carry out one of the most time-consuming, detailed, and delicate tasks in astronomy.[34] Besides the moon, the planets, stars, and sun, too, could be captured in this way. Photometry, comparisons of the intensity of the light of the sun, moon, and stars, and eventually a practical means of enlargement—all would be possible.

Five days later, English newspapers carried full reports of Arago's remarks and predictions, and on September 20, 1839, when the *British Queen* docked in New York with copies of the London *Globe*, "practical photography began in America." Within a few days, writes Professor Taft, "the nature of the method of obtaining Daguerreotypes was known in most of the principal cities . . . and while not detailed, the accounts were sufficient to start America's first amateurs to work."[35]

Though hardly an amateur, one of the first and most successful to take up the new process was Dr. John Draper, who had already experimented with the Fox Talbot method. A trained chemist and an investigator of the properties of light, he had an early lead over others because he understood how to increase the sensitivity of Daguerre's plate. As a result, by March 1840 he could announce that from a 20–30-minute exposure he had achieved a "representation of the moon's image," 1 inch in diameter, the first to reveal any noticeable lunar detail. His claim to have produced the first photographic portrait in this country, that of his sister Catherine, seems to be doubted, but his scientific achievements in the medium cannot be questioned. His biographer has written:

Draper's work in photographing the diffraction spectrum took him to the edge of a whole universe of study: the mapping of the spectrum . . . It seems that Draper was the first to take with any precision a photograph in the infrared region, and the first to describe three great Fraunhofer lines there . . . He also photographed lines in the ultra-violet at about the same time as Edmond Becquerel . . . Draper may fairly be described

73

as one of the chief, if not the chief, of the pioneers in pushing with the techniques of photography beyond both extremes of the visible spectrum.[36]

The relevance of his accomplishments to this history is closer than at first may appear. As the father of Henry Draper,[37] whom he inspired to pursue these fundamental researches much further, and in whose memory Anna Draper endowed one of the Observatory's greatest sustained programs, he can at least be regarded as the father-in-law of the stellar-spectra work at Harvard.

Boston was quick to exploit the new invention. Among the first to take up daguerreotypy was John Adams Whipple, who with his partner J. Wallace Black[38] owned an elaborate establishment that in time employed forty assistants. Whipple not only devised a number of mechanical improvements that vastly enhanced the quality of his products, but taught others the secret of his great specialty, paper photography. As early as 1844 he began to experiment with glass negatives and ingeniously devised his own steam-driven method for cleaning and polishing his plates. He was also among the first to recognize the merits of collodion. In 1853 his "chrystalotypes" for making likenesses on glass that could be reproduced readily—hailed as "the sign of a new era in photography"—won top prize at the World's Fair in New York.

It was this able American professional, who saw the possibilities of extending daguerreotypy to scientific fields rather than merely to family portraits, that the Bonds were fortunate enough to secure for their experiments in celestial photography. Just four months after the mounting of the Great Refractor in 1847, with Whipple's aid the Bonds led the movement in this country that revolutionized the whole study of astronomy. After the first abortive try on October 23, 1847, the diary records no immediate experiments, but evidently in 1848 new attempts were made. In his Annual Report for that year, submitted November 15, William Bond wrote:

The physical condition of the sun's disc has been attended to whenever the state of the atmosphere has admitted of distinct delineation. Some experiments made with the Daguerreotype and Talbotype processes, for the purpose of obtaining impressions from the image formed by the telescope, have not been attended with complete success. The application of these processes to astronomical purposes is met by a serious difficulty in the variable refraction of the atmosphere. However, we do not despair of ultimate success, when our time and means are adequate to the requisite expenditure.[39]

Since the next report, that for 1849, was submitted on November 7, Bond could hardly have included mention of experiments resumed in December. For that information one must turn to the Observatory notebook, where on December 18 it is recorded: "On the evening of the 18th just as we were commencing observations on Mars, Messrs Whipple and Jones came to take a Daguerreotype of the Moon." The diary omits all reference to this effort, probably because the entries for the month of December appear to have been made entirely by William Bond, whose interest in photography was high but his day-to-day involvement somewhat less. He reports a number of visits to Boston to receive chronometers from abroad and to attend to Observatory business matters, and, if it is disappointing to find nothing about daguerreotypy, he does provide one of those vivid glimpses into the contemporary scene that add to the fascination of this document. Going into the city on the first of December he was horrified to learn of the arrest of John White Webster,[40] Harvard's chemistry professor, for the murder of the eccentric George Parkman. The whole town was engrossed in the excitement, "to the exclusion of every Topic." Bond was personally affected: "We who are intimately acquainted with Dr. Webster cannot harbour a suspicion of the kind for an instant." Everything in the man's life "is utterly at variance with such an idea."

Moreover, Dr. Parkman was a familiar figure in the William Bond and Son clock shop. He had called there on November 23, the very day of his disappearance, to pay part of a bill, promising to return in the afternoon. But he was never seen again after he entered the Massachusetts General Hospital about two o'clock. On December 14 Bond writes that the coroner's jury "brought in Poor Dr. Webster as guilty," and, though the family believed in his innocence, "Public Reports are in the utmost degree dark and horrible against him." It is hardly surprising that, with this shadow overhanging the community and colleagues of Dr. Webster, Bond should fail to note an experiment that may or may not have amounted to much. He spent December 18, the day of the photographic experiment, at the Observatory comparing chronometers and merely jotted down for that evening, "Mars and Transits PM." Not a word appears on December 19 about the interruption of the evening before. The Annual Report for 1850, dated December 4, does, however, make up for the omission. Two entries on the subject appear:

The smaller equatorial has occasionally been used for Daguerreotype experiments [Fig. 18]; and it will answer a most valuable purpose when placed, as is intended, in the dome of the new building, where, in connection with the comet-searcher and the requisite apparatus for photographic operations, it can be used more efficiently than in its present location.

He leaves the subject at this point to deal with other matters—the state of the instruments, the arrival of a new student assistant named Charles W. Tuttle, the disappointing meagerness of library funds, chronometer ratings, observations on Mars by which he plans to determine the solar parallax, various measurements of the positions of stars in the cluster of Hercules—and then, in the middle of a paragraph that starts off on other topics, he comes almost casually to the really significant item in the Report, regarded by one writer as "the official beginning of stellar photography with the 15-inch refractor," because it produced "the memorable first portrait of a star (other than the sun)":

With the assistance of Mr. Whipple, daguerreotypist, we have obtained several impressions of the star Vega (α Lyrae). We have reason to believe this to be the first successful experiment of the kind ever made, either in this country or abroad. From the facility with which these were executed, with the aid of our great equatorial, we were encouraged to hope that the way is opening for further progress. If it should prove successful when applied to stars of less brilliance than α Lyrae, so as to give us correct pictures of double and multiple stars, the advantages would be incalculable.[41]

This image of the star was made on the night of July 16, 1850, as reported in the notebook: "Daguerreotyped α Lyrae." It is worth noting that, although the earlier statement in the Report assigns to the "smaller equatorial" the occasional experiments, in this instance, at George Bond's suggestion, the 15-inch was used, even though it involved the inconvenience of removing the micrometer. More informative than the bare statements in the formal Report, however, was a long, somewhat technical explanation sent in to the Boston *Advertiser*, and apparently considered of sufficient public interest to be copied by another (unidentified) paper, a clipping from which also appears in the Observatory Scrapbook. The style is George Bond's. Should anyone question the purpose of stellar photography, he writes, "one of the first direct applications of it would be the measurement of the angles of position and distance of double stars. It is interesting to be assured of the fact that the light emanating from the stars possesses the requisite chemical properties to produce effects similar to certain of the solar rays, and that these properties retain their efficacy after traversing the vast distance which separates us from the stellar regions." The experiments would also tell something of the "nature of the light emitted from the stars." Two conditions hamper completely satisfactory measurements. One of these is "the variable nature of the atmospheric refraction, when influenced by sudden changes in temperature," and it should be correctible by increasing the sensitiveness of the plates. More serious was the "irregular motion of the machinery which carries the telescope." It needs "uniform sidereal motion, in order that the successive rays from the star may fall on precisely the same part of the plate which is to receive the impression." This improvement the Bonds expected to achieve with a clock drive then under construction, for they definitely intended "to pursue the subject of daguerreotyping the stars, proceeding step by step from the brighter to those of lesser magnitude. We do not despair of obtaining ultimately, faithful pictures of clusters of stars and even nebulae."[42]

Progress in 1851 provided a cheerful note:

Not the least interesting part of our employment has been the assisting of Mr. J. A. Whipple in taking daguerreotypes of the moon and several of the stars. With the moon he has been particularly successful, and the improvement on last year's impressions is very decided. Some of these, taken by the aid of our great object-glass, excited the admiration of eminent men in Europe, to whom Mr. G. P. Bond gave the specimens. For these Mr. Whipple has been awarded a prize medal at the great exhibition in Hyde Park.[43]

These daguerreotypes were made between March 12 and 14, when, Bond reported, he and Whipple, experimenting with both exposure time and focus, obtained "a better representation of the Lunar surface than any engraving" he had seen. Whipple's award was one of three given to Americans at the great British Exhibition of 1851, "with highest commendation"; it was "perhaps one of the most satisfactory attempts

One of 4 daceurreotypes of the partial eclipse of the Sun 28 July, 1851, made at HCO by Whipple

(b)

(a)

(c)

(d)

Fig. 18. Early experiments at the Observatory in solar and lunar photography: (a) daguerreotype of the moon, made in 1850; (b) daguerreotype of the partial solar eclipse of July 28, 1851, made by Whipple with the 15-inch refractor; (c) collodion-process photograph of the moon, made by Whipple with the 15-inch refractor, May 8, 1857; (d) stereoscopic collodion-process photographs of the moon made by Whipple on February 7 and April 6, 1860, with the 15-inch refractor.

that has yet been made to realize by a photographic process, the telescopic appearance of a heavenly body, and must be regarded as indicating the commencement of a new era in astronomical representation." One other significant result of Whipple's exhibit was its impact on Warren de la Rue, who in England carried the process to its greatest heights in this period, and who consistently acknowledged his great indebtedness to the Bonds and Whipple for having inspired his own efforts. In addressing the British Association for the Advancement of Science, he credited Bond with having made the first photographic picture of a celestial body, a daguerreotype of "our satellite . . . I remember seeing one of those pictures in the Exhibition of 1851, and was so charmed with it that I determined to try and do the like at the first opportunity . . . It is almost needless to say that Professor Bond, of America, continued to follow up his researches in celestial photography" and measured "the distances and angles of positions of double stars and determined their magnitude."[44]

Bond and Whipple did continue their efforts with some regularity for a brief period. On March 22, 1851, following the excellent daguerreotype of the moon, Bond recorded in his notebook a discovery that led to interesting results later: "Succeeded in Daguerreotyping Jupiter." They took six plates and "could distinguish the two principal equatorial belts—Time about as long as the Moon required or not much longer." Several years after this observation he returned to it in a significant contribution to the American Academy,[45] communicated September 11, 1860, "On the Results of Photometric Experiments upon the Light of the Moon and of the Planet Jupiter." This paper opens with the paragraph:

On the 22d of March, 1851, several daguerreotypes of Jupiter were obtained on plates exposed at the focus of the great refractor of the Observatory of Harvard College. The belts were faintly indicated; but the most interesting fact in connection with the experiment, apart from its having been, as is believed, the first instance of a photographic impression obtained from a planet, was the shortness of the time of exposure, which was nearly the same as for the Moon, whereas, considering the relative distance of the two bodies from the Sun, it was to have been expected that the light of the Moon would have had twenty-seven times more intensity than that of Jupiter, supposing equal capacities for reflection. The experiments were repeated on the 8th and 9th of October, 1857, by Mr. Whipple, using the collodion process, with a like result.

Aside from Bond's detailed comparison of the light of the two bodies, too technical to discuss here, several interesting facts emerge from his study. He gives an exact list of the dates and exposure times of all the plates taken between March 12, 1851, and April 28, 1860, and shows that on April 4, 1857, he had abandoned the daguerreotype process for collodion. In the first quarter of 1852 Bond and Whipple had photographed the moon on February 26, March 3, and April 24. The exposure times apparently varied from 6 to 20 seconds. By 1857, in addition to the greater advantage collodion offered, the mechanical difficulties that had hampered earlier efforts in 1851 had been overcome. A driving clock on the principle of the "spring governor" (replacing the original Munich clock) had been built by the Clarks and, adapted to the large telescope, now provided the "most perfect regulation of rotary motion yet devised."

These new facilities encouraged Bond to invite Whipple and Black to return to the Observatory for a whole new series of experiments, primarily concerned with the fixed stars. He communicated a memoir to the American Academy on May 12, 1857, "Results of an Examination of the Photographs of the Star Mizar with its Companion and Neighbouring Star Alcor,"[46] in which he stated that "with singular and unexpected precision" they had obtained "images of the group of stars composed of Mizar of the second magnitude, its companion of the fourth, and Alcor of the fifth." He predicted that as soon as even greater sensitiveness could be achieved "it would scarcely be possible to overrate the importance [of photography] to a science of stellar astronomy." These impressions, including some of the double star Castor, he claimed as "the first, and till very recently, the only known instances, of application of photography to the delineation of fixed stars." Bond was apparently distinguishing between the earlier daguerreotype of Vega and the plates made with the collodion process, as well as calling attention to the initiation of double-star photography.

The gap of three years between his first successes with collodion and his resumption of experiments on comparison of the light of the moon and of Jupiter represents lack of both time and money. In June 1857 Bond had called J. Ingersoll Bowditch's attention to the fact that the small allotment for photography from the Phillips endowment was exhausted, and, as abandonment of the work would be discreditable from every point of view, he hoped Bowditch would mention the matter to

The Harvard College Observatory

"some of those gentlemen who are desirous that the observatory should maintain its standing among similar institutions." Still, even with no foreseeable resources, Bond determined to carry on as well as possible. In July his father wrote indulgently to Maria Mitchell of this passion for photography: "George is, and has been for months, almost hidden from the ken of us mortals in the clouds of Photography. I think he has been astonishingly successful in developing stellar photography. You must come see for yourself."[47]

Miss Mitchell did see for herself, and, as she was soon to go to Europe, she took with her examples of George's work to present to the Astronomer Royal upon her arrival in London. They were exhibited at the Royal Astronomical Society and highly praised:

> The Astronomer Royal exhibited a collodion photograph of the double star ζ Ursae Majoris, with its companion g Ursae Majoris, formed by the magnificent refractor at Harvard College . . . which has been placed in his hands by Mr. Bond . . . Upon the same plate of glass there are, in fact, two complete photographs of the entire system . . . The representation of the stars is very beautiful . . . It bears good testimony to the excellence of the telescope and the accuracy of the movement of its clockwork . . . The Astronomer Royal expressed his feeling that a step of very great importance had been made, of which, either as regards the self-delineation of clusters of stars, nebulae, and planets, or as regards the self-registration of observations, it is impossible at present to estimate the value. The most cordial thanks of astronomers are due to Mr. Bond and to . . . Messrs. Whipple and Black, by whose perseverance this object has been obtained.[48]

In one of George Bond's best articles, "Celestial Photography," published in 1859 in the *American Almanac,*[49] for which during a number of years he took the responsibility of providing all current astronomical information, he wrote:

> It is a remarkable fact in the history of astronomy, that, if we except the very recent application of electro-magnetism and of the art of photography to the purposes of celestial observation, the changes which have taken place within the last one hundred and fifty years in the methods of determining the positions of the heavenly bodies have been almost entirely confined to the improvement and extension of processes earlier suggested and employed for the same object. Scarcely a single instance of note can be recalled of the substitution of others based upon new and original conceptions.

The Two Bonds

He pointed out that the clock, the telescope, the transit, the equatorial, the meridian circle, the micrometer, and other instruments were all known by the first part of the eighteenth century, all of them were soundly conceived, and all were still in use. Where, therefore, could astronomers now look for advancement? For the present, at least, only the two mentioned offered the best hope—"new elements of power waiting only for intelligent direction to become of great and lasting benefit." As yet photography had barely obtained a foothold, but "the evidence that the new method already ranks among the first in . . . precision has since been strengthened by the most satisfactory proof." He then listed the advantages later astronomers have taken for granted: rapidity of measurement, consistency of results, permanence of the plates, freedom from personal error, greater scope for capturing simultaneously groups of stars or clusters, enormous economy in time spent on observation and measurement, opportunity to compare images taken at different times, means of determining relative magnitudes of stars. Bond saw them all:

We have indeed yet to wait for the complete fulfillment of most of its promises and suggestions, but enough has been accomplished with inadequate means, and by hands oppressed by other duties, to encourage further efforts, and to warrant the expectation of a rich return. The processes, when fully perfected, will be of a character to be best conducted on an extensive scale.

Telescopes of the largest size would of course be necessary, but even more important were "localities where the atmosphere will be found to be peculiarly favorable to the chemical action of light," for even the highest optical powers would not accomplish much "in ordinary climates and states of atmosphere . . . The surface of the globe must be explored for favored spots where a perfectly tranquil sky will afford the desired field for celestial exploration." If such places were established and improved, "the fruits of the enterprise would be beyond all computation rich and interesting." What more admirable means, he asks in conclusion, "can be imagined for the resolution of the great problems of sidereal astronomy? . . . A single night can be made to yield the results of months of labor, the data can be kept indefinitely and examined critically when convenient, analyzed and collated and corrected where necessary." Although he reiterated these arguments in much of his cor-

The Harvard College Observatory

respondence and a number of articles, notably three on "Stellar Photography" in *Astronomische Nachrichten*,[50] which are regarded as classical statements on the subject, this one essay alone mapped the promised lands that later astronomers brilliantly conquered and occupied.

George Bond's experiments in photography and his articles and memoirs would alone justify the title conferred on him of the "father of celestial photography," but no account of his remarks on the subject would be complete without reference to private correspondence which reveals his strongest personal convictions. On July 6, 1857, he wrote his famous letter to William Mitchell,[51] reviewing the Observatory's experience and successes from the beginning to that very moment, when Whipple and Black night after night were opening up "new vistas requiring exploration":

> The field for experiment is too vast to be at once occupied, even if we were provided with unlimited means. But the results already obtained in the disconnected attempts we have thus far been enabled to make, are of the highest interest, and suggest possibilities in the future which one can scarcely trust himself to speculate upon. Could another step in advance be taken equal to that gained since 1850, the consequences could not fail of being of incalculable importance in astronomy.

In 1850 α Lyrae had required an exposure of 100 seconds, and even then the image was not perfect, whereas now it could be "photographed *instantaneously* with a symmetrical disc perfectly fit for exact micrometer measurement . . . Now we can take [all the stars] that are visible to the naked eye." But much more must be done, for which "we need at least for a year to come the services of excellent artists who have hitherto literally given us their assistance, expensive materials and instruments." They should be liberally paid, free to give nights to it without working during the day so that they could come to the task unfatigued.

> Could we but press this matter on, we should soon be able to say what we can and what we cannot accomplish by stellar photography—the latter limits we certainly have not as yet reached. At present the chief object of attention must be to improve the sensitiveness of the plates, to which I am assured by high authorities in chemistry there is scarcely any limit to be put in point of theory. Suppose we are able finally to obtain pictures of seventh magnitude stars. It is reasonable to suppose that on some lofty mountain and in a purer atmosphere we might, with the same tele-

scope, include the eighth magnitude . . . To increase the size of the tele-scope threefold in aperture . . . would increase the brightness of the stellar images, say eightfold, and we should be able to photograph all the stars to the tenth and eleventh magnitude inclusive. There is nothing, then, so extravagant in predicting a future application of photography to stellar astronomy on a most magnificent scale.

In a postscript he suggests further advantages—that "the intensity and size of the images taken in connection with the length of time during which the plate has been exposed measures the relative magnitudes of the stars," and that "the measurements of distances and angles of posi-tion of the double stars from the plates" are as "exact as in the best micrometric work."

Bond's vision is all the more remarkable in view of his inability to foresee the time when astronomers would no longer have to make their own plates or when the ability to photograph stars of the eleventh magni-tude would be enormously surpassed. Astronomers now record stars of the twenty-first and twenty-second magnitude, objects more than a hun-dred thousand times fainter than some he referred to. He did consider the size of telescopes—that they must be the largest possible—and, al-though he specifically stated that much could be accomplished with his present instrument, if it were taken now to the kind of high altitude he recommended, that same 15-inch would be able to photograph stars at least of the 16th magnitude, 10,000 times fainter than those visible to the eye alone.

To the Astronomer Royal he wrote on September 28, 1857, of the "grand impulse" that could be given to astronomy by taking advantage of photography—that indeed it would not only aid the eye but actually substitute for it; most desirable would be a mountain site. His emphasis on a purer atmosphere occurs again in a letter to Henry Draper in 1864:

I believe that we shall never know how much may be acccomplished in astronomy by this beautiful art, until some one inbued like yourself with zeal and the knowledge which comes only from practical experience, shall transfer his apparatus to more favored skies, whose atmospheric disturbances shall be less annoying than here.

In our climate the case is absolutely hopeless. Through seventeen years, during which I have constantly used a large telescope, it has never afforded me a glimpse of a celestial object not sensibly deteriorated by undulations in our atmosphere.[52]

The Harvard College Observatory

In an eloquent plea addressed to Bowditch for support of the Observatory and what it could accomplish in this medium, Bond outlined a five-year program that would emphasize photography at an annual cost of $10,000 for a full-time artist and materials and an experienced astronomer with good equipment to explore different parts of the world, perhaps to settle on California! He pointed out the lavishness of the Russian government's recent appropriation of $50,000 for an expedition to Persia for the sole purpose of testing the atmosphere—not even for photography. "Why," he asked,

should we always have to wait for the example of governments in Europe in the encouragement of scientific enterprises? If our Observatory had possessed the means, we should have sent off an expedition of this kind, years ago; it was actually proposed, but of course nothing could be accomplished without money. We might now, with equal means, get our expedition to the interior of California, secure the best of the results, and get back before the Russians have started.[53]

Can there be a more startling proof of the long history of rivalry in the skies?

In summing up George Bond's immense contribution to and influence on the development of celestial photography, a recent writer[54] on the subject points to at least four areas in which he led: by proving "that the stretching of the photographic emulsion during drying was negligible," Bond greatly extended "the scope of positional astronomy, at the same time decreasing the labor and increasing the accuracy of each determination of position." He helped lay the foundation of "accurate stellar photographic photometry" by showing the relation between the size of the photographic image of a star and its magnitude; in noting the difference between photographic and visual magnitudes he suggested "the use of color indices"; and he repeatedly insisted that optimum seeing could come only from observations in clearer atmospheres, hence at higher altitudes.

On this last point he was remarkably prophetic, for if in his letter to Henry Draper his view of the hampering atmospheric conditions of Cambridge from the earliest days of the Observatory seems extreme, his emphasis on elevated sites as a major desideratum has been justified by the placing of nearly every major observatory since his day on

86

a mountain. The Lick Observatory on Mount Hamilton and the Carnegie Solar Observatory on Mount Wilson, to name only two, have set the model for others even more recent.

IV

Important as the Observatory's pioneering photographic efforts proved to be, they were necessarily only a small part of the work during these years. George Bond used no idle phrase therefore when he wrote of "hands oppressed by other duties." It described succinctly the burden he and his father carried, especially between 1847, when they first began observations with the new telescope, and 1859, the year of William Bond's death and George's appointment as his successor. A survey of their performance, only a portion of which can be suggested here, reveals the truly staggering load they assumed and sustained with unremitting devotion. Always severely limited in means, dependent on a few transient volunteers or wretchedly paid assistants, under constant pressure to impress the public with new discoveries of their own or instant awareness of others' investigations, they year after year maintained a program that would have taxed a far larger staff and that doubtless shortened George Bond's life. They assiduously observed comets, moon culminations, occultations, nebulae, clusters, sunspots, planets, double stars, eclipses. Anticipating Samuel Pierpont Langley's arrangement with the Pennsylvania Railroad in 1869, they initiated time service without charge to roads in Massachusetts and a number of New England cities outside Boston—a service that Joseph Winlock more fully developed as a much-needed source of income when he became third director of the Observatory in 1867.

The Bonds prepared three zone catalogues of stars, served the government in surveys for railroad routes and other Western explorations, collaborated with the Coast Survey in establishing boundaries and determining longitudes, and directed a long series of chronometric expeditions in cooperation with England. They supplied magnetic variations to settle disputes over titles of land surveyed by compass, furnished information to manufacturers on rainfall to determine the effect on water power, and kept up an extensive personal and professional correspondence with most of the great astronomers of the world—Sir George Airy, Sir John Herschel, Leverrier, Struve, Galle, to name only a few. Much of their

87

exchange was technical, but without exception it indicated the most agreeable relations, which were not always characteristic of those nearer home. Writing to John Russell Hind in 1850 on the naming of the minor planet *Victoria,* which he had discovered (the Bonds were the first in this country to report Hind's achievement, in the *Astronomical Journal*), William Bond expressed his gratification and encouragement from "the very friendly tone which the veterans of Europe have assumed towards the astronomical 'Pioneers' of America. May the day be far distant when local prejudice and personal considerations" hold sway. A close reading, in fact, of the letters to and from the Bonds reveals a great deal about the intellectual climate of the day, the support of science here and elsewhere, and the perhaps natural mixture of admiration and envy accorded them by other less prominent "pioneers" on the American scene.

The Bonds possessed unusual skill in designing and producing efficient mechanical devices. To William Bond belongs chief credit for the successful method of rotating the dome and the original design of both the observing chair and the chronograph controlled by a spring governor. This mechanism, which became the most widely adopted instrument for recording electrically the time of a star's transit, was referred to as the "American method." The unjust attribution[55] of the invention to others was for a time a source of great bitterness to him, but he stated his case publicly in the Boston papers, and his claim is convincingly documented in Volume 1 of the *Annals.* Not content, however, to restrict their activity to the merely practical, the Bonds kept abreast of new theory and, as the fifteen-page bibliography of their works in Holden's *Memorials* reveals, contributed regularly not only to the publications of the American Academy, but to leading astronomical media—among others the *Astronomische Nachrichten,* the *Monthly Notices* of the Royal Astronomical Society, and, until their relations with Benjamin Apthorp Gould grew less friendly, the *Astronomical Journal.*[56] George Bond, always drawn to the mathematics of astronomy and highly competent in it, had consistently supplemented his observations with numerous tables, computations on the orbits of comets, and several papers of outstanding originality. One of these, "On Some Applications of the Method of Mechanical Quadratures,"[57] published in *Memoirs of the American Academy of Arts and Sciences* in 1849, anticipated by three years a similar discussion by the noted German mathematician Johann Encke, who graciously acknowledged the fact in a private letter to Bond and to the Berlin Academy. The copy of an essay written in 1856, "On the Use of Equivalent Factors in the Method of Least Squares,"[58] sent to Sir George Airy, drew the enthusiastic reply, "Your paper has delighted me; it is the most reasonable treatise on the subject I have seen for a long time."

In addition to such activities, the Bonds had to prepare innumerable reports, illustrate their own observations (which they often did with drawings of marked artistic quality), personally care for and clean instruments, compile data for publication, beg for funds, regularly confer with the president and the treasurer of Harvard and the Observatory's Visiting Committee, arrange to admit students at stated times, acquire a working library and list acquisitions, answer innumerable queries, keep the newspapers informed, and prepare copy for the *Annals* long before funds were available for their publication.

Nor did they refuse to cooperate with other enterprises. In 1849, responding to a German proposal that a new determination of the sun's distance should be undertaken, the United States authorized a naval expedition to Chile under the command of James Gilliss to observe Mars and Venus for this purpose. In order to obtain simultaneous observations in the Northern Hemisphere Gilliss enlisted the aid of a number of observatories. Harvard was in a peculiarly favorable position to serve, not only because of its excellent facilities, but because the Bonds had been giving close attention to Mars for prime purposes of their own. In his report for 1850 William Bond had explained, "We have a great many unreduced observations on Mars, taken near the opposition. They are intended to be used as data for determining the parallax of the sun from the diurnal parallax of the planet observed east and west of the meridian, and also for comparison with the observations made on the western coast of South America by Lieutenant Gilliss under orders of the United States Government." Clearly, though fulfilling their obligation to Gilliss, between 1849 and 1850 the Bonds were preparing to advance their own theory, ten years before it was adopted elsewhere, that differential observations of Mars would provide a more convenient method of determining the parallax than would a costly and uncertain expedition to another hemisphere. The distance could be determined by the same person, with the same instruments, and under similar conditions throughout.

The Harvard College Observatory

The disappointing results of the Gilliss venture and the resulting criticisms of northern observatories caused considerable ill feeling among astronomers participating, but the Bonds had the satisfaction, as William Bond boasted in a letter to William Mitchell, of endorsement by no less an authority than the Astronomer Royal, Sir George Airy. He had strongly recommended the method to astronomers "as the best hitherto devised, and has had a list of stars proper for observing with Mars prepared for the next opposition of 1860, which will be just ten years after us."

The demanding routine followed by the Bonds obviously allowed little time for recreation, but they somehow managed to find a few moments of leisure for such lighter duties as receiving noted guests unconnected with astronomy. In September 1850, during her Boston engagement, Bond invited Jenny Lind to "find a little relief in her arduous duties from a quiet visit to the Observatory." He carefully kept her coming a secret to give her leisure and privacy for her "examination of the heavens." She was rewarded by the sight of a fireball.

Another guest wrote in his journal for July 9, 1851:

Visited the Observatory. Bond said they were cataloguing the stars at Washington [?] or trying to. They do not at Cambridge; of no use with their force. Have not force enough now to make mag. obs. When I asked if an observer with the small telescope could find employment, he said Oh yes, there was employment enough for observation with the naked eye, observing the changes in the brilliancy of the stars, etc., if they could only get some observers.

Who but Henry Thoreau would have added the comment, "One is glad that the naked eye still retains some importance in the estimation of astronomers"?

Good observers should have been available at this time, for between 1848 and 1855 the Observatory was a part of the newly established Lawrence Scientific School. The story of that not altogether successful enterprise belongs rather to the over-all history of Harvard than to that of the Observatory, but, since for about seven years there was at least a nominal connection, a brief account of it is due.[59] In 1845 a growing movement to establish a school of science as a separate department had the support not only of President Everett, who advocated it in his inaugural address, but of Benjamin Peirce, who had great designs for strengthening the scientific program of Harvard. Peirce had made no secret of his ambition. "It has been Peirce's darling wish for a long time past," Henry Rogers wrote his brother, William, in March 1846, "to reorganize the Scientific Corps of the Faculty; but this they cannot do with———in the way, and he fears that if the Rumford Chair is filled without other changes being made, a golden occasion will be lost of the more thorough reform which he desires."

As a result of the ensuing discussions about the proposed "reform," the Corporation late in 1846 appointed a committee of three—John Amory Lowell, James Walker, and President Everett—to consider the matter. In their report, submitted and accepted the following February 13, they expressed the conviction that the time had arrived for such an experiment and outlined a specific plan. It provided for an advanced school of instruction in the theoretical and practical sciences and in the usual branches of academic learning, to be called the "Scientific School of the University at Cambridge." (President Everett, fond of English nomenclature, always insisted on referring to Harvard in this style!) Instruction would be given by professors of the regular faculty, aided by such additions as might be needed. Graduates of Harvard or other colleges would be eligible, as well as any qualified young man at least eighteen years old. Students would pay $100 a year tuition, receive a diploma, and attend lectures in the College classrooms until a special building should be provided.

Through the efforts of President Everett and the treasurer, Samuel A. Eliot, and the personal influence of Charles Storrow, who was closely associated with the extensive Lawrence enterprises, Abbott Lawrence on June 7, 1847, offered Harvard $50,000 for a school of science. In his long letter to the treasurer, Lawrence outlined the proposed program with remarkable precision. Disturbed by what seemed to him the serious neglect of the practical sciences in the American educational system, he urged concentration on three important branches—engineering, mining, and the invention and manufacture of machinery. These subjects must of course embrace kindred sciences—mathematics, chemistry, geology, metallurgy, and the various skills related to architecture. His gift would make it unnecessary, he carefully pointed out, to encroach on other commitments to science, particularly astronomy, "which has already engaged the public sympathy . . . I cherish a wish to see the Observatory, the telescope, and every instrument required to prosecute the heavenly science ready for use . . . I wish to see all these branches of science prosecuted with vigor, and moving forward in perfect harmony at Cambridge."

That excellent ideal did not altogether materialize. To Eben Horsford, earlier named Rumford Professor of Chemistry, was assigned the responsibility of implementing the plan. In February 1848 he issued the official announcement of the School, named by the Corporation in honor of the donor. In addition to Horsford and in accordance with Lawrence's original suggestion, there should be immediately appointed a professor of engineering and a professor of geology. The former proved to be less readily found, but for geology the candidate whom Lawrence, Lowell, and Everett were determined to capture was at hand. Louis Agassiz, who had arrived in the United States in 1846 to give the Lowell Lectures and who had charmed everyone from the hardest-headed capitalist of State Street to the planters and their ladies in South Carolina, was the great prize. Other professors rounding out the program included Asa Gray in botany, Joseph Lovering in experimental philosophy (physics), and Jeffries Wyman in anatomy and physiology. The two Bonds and the Observatory were drawn in for astronomy. Significantly enough, no professor of engineering was found until 1849, when Henry Lawrence Eustis, formerly professor of civil engineering at West Point, was called to Cambridge. The Rogers brothers, who had been following the course of events because of their own eagerness to see a school of science established, exchanged vigorous views on both the choice of Eustis and the outlook for the new Harvard program. "Military engineering is hardly wanted in this community," Henry Rogers wrote William, in November 1849, "and something more should be given in the Scientific School of the applications of physical science, than even civil engineering." And William had predicted, "The Lawrence School can never succeed on its present plan in accomplishment of what was intended."

But what was intended itself underwent a change. Although in the beginning a number of able professors stood ready to teach well-motivated and mature students of science, clearly, whatever else emerged from the School in this stage of its history, it was not likely that Agassiz, who became its dominant spirit, would supply the practical engineers and manufacturers of machinery that Lawrence had originally envisioned. Agassiz's biographer has written:

In sponsoring Agassiz for the position, Lawrence effectively transformed the character of the original scheme he had proposed . . . Agassiz was interested in applied science only to the degree that it might have a practical effect upon his future. His experience as a geologist had been limited to glacial studies, and these hardly filled the utilitarian requirements set forth in the original bequest.[60]

Nor did his subsequent concentration on natural history and the establishment of a museum fulfill the scientific goals of other departments. Some departments, finding themselves unsuited to the School, withdrew. The Observatory, for different reasons, did so in 1855. As members of the School's faculty the Bonds were assigned the subject of astronomy, instruction in which had languished since the early vigorous periods of Winthrop and Farrar. Indeed, an examination of the catalogues for nearly a half century after the arrival of the Great Refractor reveals the curious paradox that, while intense effort went into the establishment of a great observatory, astronomy as a subject or department in the College enjoyed no independent status. The Observatory, it is true, both during its affiliation with the Lawrence Scientific School and afterward, when it received separate listing, had offered "Practical Astronomy and the Use of Astronomical Instruments" directed by the Bonds. The apparatus, mainly for magnetic observations, would be "entrusted to a class of students desirous of taking charge," but they would have neither lectures nor classes and would be expected to consult appropriate texts on their own. The fee was $50 a term. Usually not more than one student turned up, and sometimes none. For lectures relating to astronomy any undergraduates interested must go to the mathematics professor, Benjamin Peirce ("Celestial Mechanics"), or to Professor Joseph Lovering ("Celestial Mechanics and the Undulatory Theory of Light").

This situation was hardly satisfactory to anyone. In 1853, at the close of his administration, President Sparks attempted to rectify Harvard's poor offerings by suggesting increased instruction at the Observatory, but William Bond's response anticipated that of later directors, whom the question continued to haunt.[61] He wrote Sparks on January 6, 1853, that the Observatory lacked room, it was unsafe to entrust expensive instruments to novices, and both day and night observations would suffer seriously from interruptions for teaching or demonstration. Nor were there funds or extra assistants available. Nevertheless, Bond said, "Those really interested and willing to work we are always glad to have with

Fɪɢ. 19. President James Walker, from a painting by William Morris Hunt. (Harvard University Portrait Collection.)

us." Shortly after James Walker (Fig. 19) became president, the Corporation voted $500 to prepare a room and equipment for students. "Public interest," Walker wrote the Bonds, required receiving students who wished to avail themselves of such facilities and setting aside one or two evenings for members of the senior class who may want to "behold the heavens through your great telescope." Since this was hardly a profound approach to instruction in astronomy, it resolved nothing. The result was that for nearly four decades the subject remained one of the serious weaknesses of scientific studies at Harvard.

The formal separation of the Observatory from the Lawrence Scientific School was voted by the Corporation on July 28, 1855. The Observatory with its officers would be constituted "a distinct department of the University, under the name of the Observatory of Harvard College," and the Visiting Committee in its report for 1854–55 expressed "peculiar satisfaction . . . that the Observatory has once more an independent existence as a separate department of the University." No longer would it be "classed as a mere branch of the school."[62]

Yet if the relatively brief association of the two failed to benefit either the Bonds or astronomy, at least one great debt the Observatory owes to the School. The School provided as fourth director of the Observatory one of its most distinguished graduates, Edward C. Pickering, who in 1877 began the administration that for forty-two years determined the course of astronomical research here and far beyond the confines of Harvard.

The Bonds were thus relieved of extra responsibility, but at no time during the connection with the School do their astronomical activities seem to have been hampered. Aside from the significant photographic experiments already described, one particular triumph was their earlier discovery of Hyperion, Saturn's eighth satellite, on September 16, 1848. Sears Walker hailed it enthusiastically as the greatest telescopic achievement of the century, and Josiah Quincy, chairman of the Visiting Committee, called it "the only addition to the solar system ever made on the continent of America." By "a singular coincidence," he added, "the satellite had been observed by the British astronomer William Lassell the same day," but this was no reflection "on the merit of Mr. Bond." In marked contrast to the acrimonious debate over the discovery of Neptune, British astronomers amiably proposed that in Great Britain the satellite would be referred to as "the discovery of Mr. Bond and Mr. Lassell, and in America the discovery of Mr. Lassell and Mr. Bond."[63] William Bond remarked in the diary for November 25, "the thing passes off in good humour, which is saying a good deal in these days." (In fact, Bond's discovery preceded Lassell's by two days.)

In 1848, among other written works, William Bond published his memoir on the nebula in Orion, and George communicated to the

Fig. 20. Edward Bromfield Phillips, whose bequest in 1848 established the Phillips Library and the Phillips Professorship of Astronomy. From a painting at the Observatory.

American Academy the first results of his study on the nebula in Andromeda.[64] Each paper represented prolonged observations, and each was illustrated by exquisite drawings.

In 1848 the munificent bequest of Edward Bromfield Phillips (Fig. 20) solved the vexed problem of salaries and made possible the establishment of the Phillips Library and Professorship. Phillips, a classmate of George Bond's and a kinsman of Josiah Quincy's, took his own life at the age of 25, it is said because of an unhappy love affair. In his will, because of his friendship for George Bond, he left $100,000 to the Observatory, and although, like subsequent bequests to Harvard, it was threatened with a contest by dissatisfied heirs, the matter was amicably settled. The Observatory received the full amount, providing an income of $5000 a year, its first endowment not subject to restriction in the immediate use of funds. Partly as a result of this windfall, but for an even more compelling reason, the Bonds and the Corporation discussed new statutes[65] for the Observatory which, under the sympathetic leadership of President Jared Sparks (Fig. 21), were finally established. Subject only to the superior authority of the Corporation, the director would have "full liberty to order the course of proceedings in regard to the kind of observations and manner of making them," as well as all publication of results. The purpose and hope of the new statutes were to clarify once and for all the relation of the director to the Perkins Professor, Benjamin Peirce. Having been very influential in establishing and promoting the Observatory, Peirce had assumed a proprietary attitude toward it and fell into the annoying habit of regarding himself as a public interpreter of investigations made there. Even so, the new rules failed to prevent one particularly flagrant example of his practice, with lasting consequences that proved unhappy to all parties.

On the night of November 11, 1850, George Bond noticed "the filling up of light inside the inner edge of the inner ring of Saturn," and "below the edge . . . a dark band. Am very confident," he jotted down in his notebook, "of having seen to-night a second division of the ring near the inner edge of the inner ring." He promptly drew a diagram of it. On November 16, examining the planet at 7:30 in the evening, he observed what he described as a "dusky ring" (designated by modern astronomers variously as the "gauze" or "crape" ring). In his notebook he wrote, "Where the dusky ring crosses Saturn, it appears a little wider at the outside of the ball than in the middle. Where the new ring crosses Saturn it appears not so dark as the shadow of the ring below on the body of the planet." A half hour later he noted, "The best definition of Saturn's ring we have ever had . . . there can be no doubt that it exists; its inner edge is sharply defined."

Several weeks later the Reverend William B. Dawes and William Lassell, who on the same date (November 11) had independently found the ring with a small refractor at Dawes's observatory in Wateringbury,

FIG. 21. President Jared Sparks, from a painting by Rembrandt Peale, presented to Harvard by Lizzie Sparks Pickering, wife of the fourth director of the Observatory.

The Two Bonds

England, confirmed George's observation. Bond then began a detailed mathematical analysis of the ring's composition. After reviewing all previous observations and the theories of some of the most eminent authorities who had speculated on the problem—La Plata, Plana, Sir William Herschel, and others—he concluded that, contrary to accepted views, the rings were not solid, but fluid. His reasoning can be studied in Volume 2, Part 1, of the *Annals,* where he gives the entire history of his observations and discoveries relating to Saturn from 1847 to 1857. Earlier, however, on April 15, 1851, he communicated his discovery and his conclusions to the American Academy. Other papers on the subject he sent to the *American Journal of Science* and to Gould's *Astronomical Journal.*

George's discovery and discussion at the Academy immediately attracted Benjamin Peirce. Not waiting for Bond to follow up his first communication with a complete description and analysis of his own work to the public (pending a proper engraving of the planet and its rings), Peirce journeyed to the Cincinnati meeting of the American Association for the Advancement of Science, and on May 12 he took as his topic "The Constitution of Saturn's Rings." Newspapers[66] hailed the paper as "by far the most important communication yet presented to the Association . . . and the most important contribution to astronomical science made since the discovery of the planet Neptune." Peirce did not claim the discovery of the ring himself—he duly credited "the young Bond" with that. But while praising him for "his bold and ingenious theory . . . clearly sustained by his own simple and novel computations," Peirce, "by entirely different reasoning, independent of observation, shows that Mr. B's positions are true of any planet existing or conceivable; that no solid ring can encircle a planet. Saturn's rings are a fluid, rather denser than water." Whether or not Peirce meant to admonish young men not to explore too deeply into the mysteries of what sustains a ring around a planet, he wound up with the moral:

But in approaching the forbidden limits of human knowledge, it is becoming to tread with caution and circumspection. Man's speculations should be subdued from all rashness and extravagance in the immediate presence of the Creator. And a wise philosophy will beware lest it strengthen the arms of atheism, by venturing too boldly into so remote and obscure a field of speculation as that of the mode which was adopted by the Divine Geometer.

William Bond, whose indignation was aroused as never before, on May 18 wrote to Alexander Dallas Bache, president of the Association, enclosing a copy of George's paper, not, he declared, to involve Bache in any attempt "to correct the monstrous wrong which has been done to George," but to insist that the paper be published and due recognition given for his discovery and careful mathematical analysis. Not content with that, the outraged father saw to it that the case should reach a wider public. He sent to the Boston *Traveller,* which published it on June 16, a complete record of the affair, with exact quotations from George's notebook and his diagram drawn the night of his observation. Despite that publicity, however, in his tenth Lowell Lecture, February 13, 1852, Benjamin Apthorp Gould, Peirce's disciple and advocate, slurred over George's part in the discovery, coupled it with another's, and then said:

> There had been of late observations of a ring within the outer bright rings of Saturn . . . Mr. Kirkwood, and the young Mr. Bond . . . both suggested that the rings might be fluid. Professor Peirce took up the question in earnest, and less than a year ago demonstrated by analytical reasoning, that a ring about a central body must, in order to exist permanently, without falling upon the central body, be fluid. After an investigation of the subject, he arrived at this result, at present the last discovery in astronomical science.[67]

In 1857, however, James Clerk Maxwell in his Adams Prize essay, "On the Stability of the Motion of Saturn's Rings,"[68] corrected George Bond, Peirce, and Gould by showing that the rings could be neither solid nor fluid, but were composed of myriads of small particles like tiny Saturnian moons revolving in their own orbits around the planet. That theory was confirmed in 1895 by James Keeler, then at Allegheny Observatory, in his paper, "A Spectroscopic Proof of the Meteoric Constitution of Saturn's Rings," but these later investigations detract nothing from the credit due George Bond. In successfully questioning earlier, greater authorities, he inspired further inquiry that led to a more acceptable theory. It is probable, however, that Peirce's appropriation of Bond's observation, and Gould's support of Peirce, even if the analysis differed somewhat, created the strained relations and eventual enmity that clouded George Bond's directorship of the Observatory from 1859 to 1865. Holden suggests that from the time of the episode Peirce in particular "gave something of his support to covert reflections against the observatory," a rather mild view to take of the sharp criticisms and implacable hostility emanating from these sources.

In 1851 George Bond made his first journey to Europe.[69] Although only twenty-six years old and a mere assistant in an Observatory that was, scientifically speaking, in its infancy, he was received by the leading astronomers of the world without the slightest trace of condescension. In London, visiting the Great Exhibition, where the Bond chronograph was awarded a medal, he found at the Crystal Palace "the most luxurious kind of sightseeing ever imagined"; nothing in the Arabian nights could compete with the "realities" of modern creation. Later, he addressed a meeting of the British Association for the Advancement of Science on the subject of the Bond chronograph. He emphasized its wholly American origin, partly to refute a vague allegation of German or other influence, but chiefly to point out its revolutionary effect on practical astronomy. At the reception that followed, Prince Albert paid him the honor of inquiring very closely into the principle of the invention.

His visit to Greenwich introduced him to such noted astronomers as Lord Rosse, with whom his father had corresponded about the nebula in Orion, John Couch Adams of Neptune fame, Sir John Herschel, and the visiting Belgian astronomer-mathematician, Lambert Quételet. The same evening he dined with forty eminent men of science. In Paris Leverrier received him cordially, though he made no secret of his intense bitterness at the refusal of certain Americans to accept the validity of his calculations leading to the discovery of Neptune. He directed his ire particularly at Benjamin Peirce, who had stoutly maintained in several communications to the American Academy that the discovery was a happy accident. (Bond rightly noted that Leverrier had confused Peirce's remarks with those of Ormsby Mitchel in the *Sidereal Messenger,* which Peirce explicitly disclaimed.[70]) Leverrier accompanied George to the Paris Observatory and arranged for his attendance at a meeting of the Academy of Sciences, where the daguerreotypes of the moon by Whipple "produced a sensation" uncommon in that august body. In France also he met, among others, Jean Foucault and Jean Biot, who spoke with admiration of the Harvard Observatory and its work.

In Germany the astronomers were gathering for an expedition to Sweden to observe the eclipse of July 28, among them Ruemker of Hamburg, Petersen of Altona, Repsold, and Bond's valued English correspondents, Dawes, Carrington, and Hind. So exceptional an opportunity Bond resolved not to miss, especially since he could view the eclipse on his way to Russia. With other observers he went to Lilla Edet, a Swedish village, "a wild, remote place, surrounded with rocky hills and woods, where the beasts and wolves still roam," but in the central line of totality. His description, sent to his father from Copenhagen, was published almost in entirety by the Boston *Traveller*. It loses nothing of scientific value by its lyric appreciation of the glorious sight, "in comparison with which," he wrote, "all the trouble I have been at to get here seems as nothing." The weather at first was discouraging. Three hours before the eclipse the sun "was still hidden in an impervious mantle of gloomy cloud." Then, "as the time drew near when the Moon was to completely cover the Sun, the clouds dispersed entirely, leaving a thin veil of cirrus, which, without question, often accompanies the most perfect vision . . . The definition of the edges of the Sun and Moon was remarkably excellent . . . The awful, unearthly aspect of a few cumuli around the horizon" was a fitting precursor of what was to follow:

The outline of the lunar mountains was finely projected in profile upon the sun; and these, as the crescent of light narrowed, produced a singular appearance on the southern cusp. Presently, as I watched intently, the edge of light was broken up into beads, moving towards the point. The complete covering of the Sun's disc by the Moon succeeded instantaneously. Up to this point I may say that I was in a great measure prepared for all that I saw. I had seen other eclipses, and this was like one of them on a great scale. The change which takes place in less than a tenth of a second, so entirely alters the scene, that the second which precedes the instant of total obscuration gives one no idea of what is to follow. Some seconds elapsed before I had my thoughts sufficiently about me to remove the screen from the eye-piece. What I then saw, it is utterly beyond my power of language adequately to express [Fig. 22]. The corona of white light which encircled the dark body of the Moon, resembled the aureola, or glory, by which painters designate the person of the Savior, its radiance extending from the circumference to a distance equal to about half of the sun's diameter . . . How shall I attempt to describe those other wonders . . . these flame-like protuberances projecting from the inner edge of the corona? . . . Before the end of totality, the left hand flame disappeared entirely . . . Those on the side which the moon was first to leave, increased, until the moment before the edge of the sun appeared; when the rosy light was suffused over the limb of the moon, near where the sun-light first broke forth, and then all vanished as quickly as it had at first appeared.

Fig. 22. The solar eclipse of July 28, 1851, as drawn by George Bond at Lilla Edet, Sweden.

The scene, he concluded, "was surpassingly beautiful—the most sublime of all that we are permitted to see of the material creation."

(At home William Bond reported to the Boston *Traveller,* "To the unwearied and skillful exertions of J. A. Whipple, we are indebted for a series of dag[uerreotype] impressions of the progress of the eclipse, taken in the small dome of the west wing of the observatory." He also mentioned some of the observers with him: Robert Treat Paine,[71] that staunch friend of the Observatory and able amateur astronomer in his own right; Sears Walker, of the Coast Survey; Charles Tuttle,[72] like George Bond an avid comet hunter; and the fifteen-year-old mathematical prodigy Henry Truman Safford (Fig. 23). That young man, as

FIG. 23. Truman Henry Safford, mathematical prodigy, at the age of about nine. From a portrait by an itinerant painter. (Harvard University Portrait Collection.)

a result of widespread publicity about his prowess as a "lightning calculator," in 1846 had been brought to Cambridge with his whole family from a Royalton, Vermont, farm, to prepare him for Harvard, and his passion for astronomy had led him straight to the Observatory then and eventually, as will appear, to temporary direction of it.)

Fig. 24. Front view of the Imperial Russian Observatory at Pulkovo.

At the other end of the world, meanwhile, George Bond, still in the glow of the brilliant spectacle he had witnessed, moved on to Russia, pleased and flattered to be met at Kronstadt by a special emissary to conduct him to Pulkovo (Fig. 24). He was "domiciled at the Observatory with his excellency Struve," William Bond wrote proudly to his friend William Mitchell, "as kind and communicative as if George had been his own son." Struve (Fig. 25) spent hours in the great dome with his young visitor, who eagerly compared the famous refractor with its twin at home. Bond noted with envy the amount of assistance the Struves enjoyed at the expense of the Czarist government—two men in the dome, one soldier-servant just to turn it, another to keep the record. The "grand works now in progress"—a fundamental catalogue of 500 stars based on prolonged observations—impressed him deeply, as did the rich library, with a unique collection of Chinese astronomical works and all other important publications in astronomy, which gave the director incalculable advantages for research and comparison of results. "In a quarter of an hour," George noted, to settle some matter

105

Fig. 25. Friedrich G. W. Struve, first director of the Pulkovo Observatory and founder of a distinguished line of astronomers.

on the planet Saturn, "we were in possession of a great mass of evidence on the subject."

After the Russian visit he went to Berlin, where he visited Galle and Encke, and, armed with a letter of introduction from Agassiz, called on von Humboldt, who had apparently followed with great interest astronomical progress in America, and who knew of the work at Cambridge. At Leipzig George hunted up d'Arrest, whom he described as "a true man of science," who loved astronomy enough to work for

106

forty pounds a year! Others met on this triumphal tour included Gauss, Argelander, Challis, Lord Cavendish, Whewell, Lassell—a heady company for this modest young man. In a letter to his father about his reception he wrote:

I should do injustice to my own feelings, which prompt me to the warmest expressions of gratitude, were I to neglect to notice the numberless instances of kindness and attention extended to me . . . At the same time, I could not be insensible that the marks of consideration, though gratifying in no common degree, and exceedingly gratifying to one's personal vanity, may easily be misconstrued as to lead to very false notions of importance and are really a dangerous trial of character . . . A young man perhaps just commencing a scientific career . . . is thrown into immediate contact with the great masters of science—men of worldwide reputation, whose acquaintance and personal friendship may well be accounted a high honour—men, too, often as distinguished for courteous and affable manners, as for their intellectual powers. By them he is treated with marked attention. He is a stranger. The very distance he has travelled for the sake of being benefitted by their society conveys a compliment.

He comes too from America, in a scientific capacity . . . he will return there to disseminate in a distant region his impressions of the reputation, character, and position of those he has associated with. These are reasons enough and more might be adduced, if necessary, to explain what is really a distinguished reception. A meeting on terms almost of equality with those every way his superiors—invitations to their select clubs and societies and perhaps public notice of his presence—all this and much more without adding in the slightest degree to the evidence that he is any way more deserving than thousands of his countrymen less fortunate, because farther off. In all likelihood, for a long time to come, some months or a year spent in Europe will be regarded in the United States as a very important qualification to the man of science, and so it may be, but it will be well it is not made a means of building up a reputation of the first order, upon second-rate abilities.

He closes this remarkable evaluation of himself, vis à vis the great men who had welcomed him, with a plea for recognition and support of science by public, permanent establishments and resources such as European governments granted Kew, Brussels, and Pulkovo.

Among other purposes of the European trip the purchase of books for the Observatory library had been considered by the Corporation of sufficient importance to appropriate as much as $3000 from available funds. While in Germany, Bond heard of the sale of the library once belonging to the late mathematician Karl Jacobi. He succeeded in buying over 500 books and 400 pamphlets at 40 percent less than the printed catalogue prices and for only half the sum he could spend. The collection greatly enriched the as yet meager Phillips Library and included works that are today very rare.[73]

A serious illness followed Bond's return, from the effects of which he never fully recovered, but when he was able to resume work he and his father began to prepare material for their long-desired first published report. For lack of funds it was not issued, however, until 1856.

In 1855 Josiah Quincy announced to the Corporation that he intended applying the fund his father had left to Harvard toward "the permanent interests of science," a $10,000 endowment, the whole income of which would go to publication of Observatory *Annals*.[74] It enabled the Bonds for the first time to plan on gathering together for printing a large mass of observations, computations, and memoirs, which when published bore on the title page "Printed from Funds Resulting from the Will of Josiah Quincy, Jr., Who Died in April, 1775, Leaving a Name Inseparably Connected with the History of the American Revolution."

In September 1856 a flattering communication addressed to George Bond from the Department of State informed him that the President of the United States, with the advice and consent of the Senate, had appointed him chief astronomer of the agency delegated under the Act of Congress of August 11 "to carry into effect articles of the Treaty between the United States and Her Britannic Majesty concluded on June 15, 1846." The treaty concerned the Oregon boundary question.[75] The commission accompanied the letter. William Bond, reporting the offer to President Walker, at the same time called attention to the poor pay George still received—$1250 after ten years of service! The offer had not been discussed; it was a tribute solely to George's ability, without any solicitation whatever, and in the interests of his family he might have to accept it. If he left, however, the consequences to the Observatory would be disastrous. A few days later the Corporation voted to increase George's salary to $1500, and on October 14, after gravely considering the appointment, he wrote President Franklin Pierce that, though honored, he must decline:

My association with the institution with which I am at present connected dates back to its infancy and to my own earliest years. Its past and its

The Two Bonds

future are identified with my own and I cannot desert it while anything which my labors may contribute remain wanting to its permanent prosperity. If my present position were less agreeable or its claims less pressing, I know of no situation in your gift which would be more desirable than that you have tendered nor any upon the duties of which I should enter with more pleasure. I cannot help regarding this mark of confidence as an act of personal kindness to myself for which I beg leave to express my deepest obligations.

For the next three years, therefore, the devotion of the Bonds to one another and their deep loyalty to the Observatory continued to enhance its prestige at home and abroad. Only with the death of William Bond on January 29, 1859, at the age of 70, did this remarkable partnership come to a close, a relationship that can be matched in the annals of astronomy perhaps only by those of the Herschels and the Struves,[76] with whose families the Harvard Observatory and its directors maintained ties for over a century. George Bond wrote William Mitchell that his father's last hours had "passed in calm resignation and undisturbed tranquillity and cheerfulness." Condolences poured in from everywhere, genuinely deploring the loss to the scientific world and especially to that science which "he was one of the first to plant and cultivate in this country," the first American, chosen in 1849, as one of fifty honorary associates of the Royal Astronomical Society. William Bond was also a member of the American Philosophical Society and a corresponding member of the Institute of France and of the Accademia dei Lincei of Rome. At the American Academy, following eulogies by Josiah Quincy, Benjamin Peirce, Robert Winthrop, and others, the members resolved "that we are grateful for the long and valuable services of William Cranch Bond, who has proved that an American mechanic can accomplish one of the highest positions in science, and whose astronomical discoveries have illustrated his country and his observatory and stamped his own name honorably and indelibly upon the records of history."

The Visiting Committee, "privileged to enjoy his counsel and friendship for over thirty years," paid tribute to him at two successive meetings in February 1860. On the 20th they said:

In taking their accustomed survey of the premises, the instruments, and their appurtenances, they were impressively reminded of the absence of the brightest ornament of the Institution, one whose life, talents, and virtues were identified with the origin and history of the Observatory. Many of the objects of their inspection and scrutiny were the conceptions of his genius and the handiwork of his skill. These may perish, but the light which his labors have shed upon science will be a more enduring monument,—and the memory of William Cranch Bond, the first Director of the Observatory, will be long and affectionately cherished, not less for the manifest excellence of his character and deportment through life, than for his eminent services to the Observatory and to the science of our country; and that firmament which he loved so well to contemplate will perform many annual circuits before his influence and the impetus which he gave to the enterprise will be forgotten or essentially impaired.

Others memorialized him also, among them President Everett in one of the Mount Vernon Papers[77] he was writing to promote the preservation of George Washington's home. "I am not satisfied," he wrote George, "that full justice has been or will be done to him by those best able to do it." But the one best able to do it was George himself, who wrote to Charles P. Curtis, an early benefactor of the Observatory:

There is something to my mind actually appalling in the contemplation of the amount of my father's labors from the time when he was first enabled to indulge freely his passion for observation. The accumulated volumes, filled with manuscript records, give me a shudder at the thought of the weary frame, and straining eye, the exposure, and the long sleepless nights which they recall. Beyond any doubt the fatigue of these arduous, though in a great measure self-imposed duties shortened his term of years,—for his gain doubtless. He no longer sees "through a glass darkly," but, as we devoutly believe, is brought "face to face" with the great truths of nature.

v

On February 26, less than a month after his father's death, the Corporation elected George Bond Director of the Observatory and Phillips Professor of Astronomy. In acting with promptness it put an end to much unpleasant speculation that had added to his afflictions during those sad weeks. President Walker wrote him immediately, but the appointment had to await official approval by the Board of Overseers, and final notification of their unanimous consent reached Bond only on March 10. Although any other choice would have been unthinkable, at least two other contenders yearned for the post and obviously had powerful friends to speak for them. One of the most hopeful for a brief time was Benjamin Apthorp Gould, who in January 1859 had been summarily ousted from the directorship of the Dudley Observatory after a scandalous controversy with his trustees, and who had been encouraged to believe that he could and should now succeed William Bond. Unfortunately, on February 5 he learned that his chief rival (discounting the claim of George Bond) was his dearest friend, Benjamin Peirce. In a letter to Peirce, February 8, he wrote that although until the 5th no word had come from Cambridge about the vacancy at the Observatory, "many friends . . . from the most different parts of the country have written me on the subject, offering their good wishes and their efforts; and I was beginning to look upon myself as a probable candidate." Such an appointment would vindicate his course at Dudley and would also bring him back to Cambridge and "at last some haven near my home and kindred, and friends of a lifetime. And I supposed," he went on,

that my persecutions and sufferings borne on behalf of science would not improbably balance any supposed right of inheritance on George Bond's part—so that the corporation might feel free to decide, upon grounds of attainment and competency alone. But all this presupposed that no other person would be thought of for the place; and least of all did it occur to me that the position would be to *your* taste. But at the same time with your letter of last eveg came another which mentions that *you* would be gratified at receiving the appointment—I write therefore to beg you not to regard me as in any way an obstacle, for under such circumstances I shall not and will not be a candidate.

He would so inform other friends who were anxious to exert themselves on his behalf, but to Peirce he insisted that it would be

a great source of distress . . . that in any way or shape my wishes could for a moment seem to run counter to yours; and I only wish that I knew how to insure your immediate success. The idea of having *any* competent person there, would be a glorious one. The Lord only knows who could take your place in the College. I only know that it is certain no graduate of Harvard could.[78]

Some ten days later he asked Peirce, "What is the great argument in behalf of the 'biggest boy'—Natural claim by right of inheritance:

111

or because he was a class mate of Phillips? or because the best person for the place?"

Rumors of maneuvering for the post evidently spread far beyond Cambridge, for Ormsby Mitchel, who was to succeed Gould at Albany, in a letter of condolence to George Bond added that he deeply regretted to see agitation about his father's successor in the public journals: "I had hoped that as a just tribute to his memory the friends of anyone aspiring to fill his exalted place would have restrained their zeal and impatience within respectful limits." He trusted that the reports were entirely without foundation, for he could not believe that "the appointing powers of the University have in any way consented to perpetrate so gross a wrong."

George Bond had all along been quite well aware of those "who have done what they could to influence the election here adversely to my interests." His friend C. H. F. Peters had warned him especially on February 10 that the ex-director of the Dudley Observatory would now be looking for another place and "you may have still some trouble." As for Peirce, Holden writes, "Directly after the death of the older Bond he went to the Observatory to announce to his [Bond's] son that he himself was a candidate for the vacant directorship. He was not elected to it, and became an open enemy."[79] Nevertheless, as George Bond wrote Peters the day of his official election, he saw little use in keeping up hostilities. "This eternal strife is a very great evil," and he intended to do whatever he could to overcome it. Ex-President Everett had written him on March 14:

> I hope it is not too late for me to congratulate you on being chosen in your father's place. I took a great interest in the election which at one time seemed to be threatened with opposition, and in common with other friends exercised what influence I possessed to have a right result, which, however, I must say, was brought about by the merit and undoubted qualification of the successful candidate.[80]

A postscript to this letter, omitted by Holden, reads:

> Mr. Ingersoll Bowditch told me he should offer you some friendly suggestion, as to a communication to be made by you to Professor Peirce, tendering aid and co-operation. I fully concurred with Mr. Bowditch in hoping you would take some such friendly step. It will put you on strong ground.

The Two Bonds

Interestingly enough, George had not waited for the advice. On March 10 he wrote William Mitchell:

Now that my position at the observatory is settled, I wish to compose matters, if possible with Professors Peirce and Bache, so far as can be without a compromise of independence. This contention and perpetual hostility of interests is a miserable occupation for men who are capable of better things.[81]

Two days later he addressed to Peirce a conciliatory but by no means abject letter:

Dear Sir:
You do not need to be reminded that the mutual confidence and friendship which once subsisted between us has for some time past been disturbed, if not wholly interrupted.

There can be little use in dwelling upon the circumstances which have occasioned this unfortunate difference; perhaps it would be better if they were forgotten.

My object now is to propose and to open the way for a return to a better state of feeling.

He emphasized the advantage to the Observatory of cooperation from one whose position and attainments commanded such respect and offered to assist Peirce's scientific investigations in every possible way, by furnishing results of observations or any other means; but "only by a mutual consent to give up past differences" could any such "cordial and earnest" collaboration be achieved. "On my part," he promised, "nothing shall be wanting to accomplish that end." Holden believed that Peirce sent no reply, since none has been found in the Bond correspondence; however, the Bond folder among the voluminous Peirce papers contains a rough draft of one, though it may never have been sent. It is undated and not altogether decipherable, and its insertions and excisions indicate some uncertainty about what attitude to take:

Dear Sir:
I agree with you that the time is come when the dead [?] past should be buried in oblivion, when the director of the observatory should take the Perkins Professor by the hand, and when an effective cooperation should be established between them. Whatever plan you may propose in this direc-

113

tion, I rejoice to accede to, and shall [be happy *canceled*] desire to confirm [?] by investigations & any system of observations [*undecipherable*] from the observatory which will tend to elucidate to mankind the laws which the Creator has written upon the firmament.[82]

The same Bond folder includes some curious bits of doggerel scribbled over the page in Peirce's execrable handwriting. In their obsessive anger toward George Bond they hardly indicate a promise of reconciliation: "The poet mourneth to the fixed stars the unexplained irregularities in the motions of this Georgium sidus"; "Another's woes she weeps, nor feels her George's predilections"; "In Bond age better far to die, than live old Harry's son."

If Peirce did send his letter, it apparently changed nothing. The fact is that George had never been admitted into what he sardonically referred to as "the sacred brotherhood," otherwise known as the Lazzaroni. That band of scientific comrades, headed by Agassiz, had been organized some time around 1853 as an alliance designed, in the words of Agassiz's biographer, "to control the institutional forms of science in America."[83] In Cambridge, its original home, its direction lay with Agassiz, Peirce, and Gould; its leader in Washington was Alexander Dallas Bache. So long as the elder Bond lived, the Observatory remained by statute firmly in his hands, invulnerable to challenge by any member of the group. With his death, however, the opportunity seemed to have arrived to assert their power, and failure to do so doubtless added greatly to the animus against George. Peirce continued his criticisms, directed not only at George but at other astronomers of whose work he disapproved. Holden mentions particularly his severe strictures on Bond's study of Donati's comet, which all other authorities found unequaled. But there were other offenses.

In August 1859 a committee of the American Association for the Advancement of Science issued a proposal for a new determination of the sun's distance. To press the case for it, the framers of the statement repeated the assertion made in the formal report of the Gilliss expedition[84] of 1849–1852 that failure to achieve success had arisen from lack of proper cooperation by northern observatories, including Harvard. Peirce addressed the American Academy on the subject, apparently delivering a philippic of such sharpness as to elicit from one outraged correspondent who heard of it the complaint that "he has insulted all

North American astronomers." Asked if the speech had been published in the *Proceedings* of the Academy, the unruffled Bond replied:

If you ever saw water slide off a duck's back you know something of the effect which such outpourings of indignation are calculated to produce. You may rely upon it. He will never publish it . . . I did not hear it myself, and judge of it by the accounts of others. If it were not too much like a waste of time, it might be well to overhaul the Report of the Chili Expedition. The assertion that its failure is due to the want of northern cooperation has not the slightest foundation.

No abstract or notice of Peirce's remarks appears in the *Proceedings* of the Academy.

Controversy had little charm for George. He preferred to spend his energies on more important matters now that he had the sole responsibility for maintaining the reputation of the Observatory he had grown up with and helped establish in the astronomical world. Shortly after his appointment he had moved into the Observatory house with his two little motherless girls (his wife and an infant daughter had died shortly before his father) and busied himself at odd moments with changes made necessary by his limited means. In her "Early Recollections of Harvard College Observatory," Elizabeth Bond gives a charming impression of the Observatory surroundings when her father became director. Their nearest neighbors and good friends were the Asa Grays, the Dixwells, and the Wolcott Gibbses. William Bond, her grandfather, had been a passionate gardener, covering every conceivable spot of the grounds with flowers, shrubs, and many exotic plants given him by Gray. Great trees added to the parklike air, vines covered the front of the house, and "at the foot of the avenue ran a brook which crossed Garden Street at the junction of Linnaean," to children a "river of great importance . . . and a fine playground where we sailed boats or built mills and dams." Unable to afford either the time or the money for lavish floral display, George Bond seeded the beds with grass, and, because he thought it a fire menace as well as "detrimental to the dignity of the Observatory," he moved to the extreme end of the grounds a large barn that in his father's time had stood near the house to shelter a cow, two or three horses, some pigs, and poultry. There were many favorite visitors, William Mitchell and his famous daughter, the astronomer Maria,[85] Dr. C. H. F. Peters[86] of the Litchfield Observatory at

The Harvard College Observatory

Hamilton College (one of the Bonds' most pungent correspondents) and even the Prince of Wales, Albert Edward, to whom, much to their delight, the little girls were presented. Company was frequent, Elizabeth explains, because in the day of "small things the Harvard Observatory loomed large and together with Mt. Auburn Cemetery was the chief show-place in Cambridge."

Although his daughter believed that the jealousies and unremitting hostility of disappointed candidates for the directorship deprived her father of "repose in mind or body," in fact Bond received constant evidence of loyalty from various quarters.[87] He was on excellent terms with Joseph Henry, whom he had visited in Washington, and who wrote afterward that he had taken much pleasure in Bond's company and had found it "refreshing in a city of strife and contention to have occasionally a quiet talk with a man of scientific mind and spirit." Bond's correspondence with eminent astronomers continued without interruption; among them was the secretary of the Royal Astronomical Society, Richard C. Carrington, whose somewhat similar interests in comets lent zest to their exchange of information. One communication is of particular historical interest.[88] On June 11, 1859, Bond thanked Carrington for a striking photograph of the tail of Donati's comet, which he compared with a drawing that he himself had made of the nucleus. The letter reveals that successful photography of a comet antedated by more than two decades the year (1881) astronomers had generally accepted. It is also worth mentioning that when Carrington planned to apply for an Oxford post in astronomy he wanted a recommendation from Bond, observing that "a word from the most important Observatory in America would be a decisive influence." It proved not to be, but at least the request indicated that Bond's opinion was presumed to have weight in England.

For the daily routine of the Observatory, since no assistant observer had been appointed to take Bond's place, he relied for a time on four unusually able young men. They were praised by the Visiting Committee for their devotion to science and were compensated chiefly by a position that "afforded them instruction and practice." Bond, however, gravely doubted the propriety and permanence of such inducements. Still, two of the four eventually became well-known astronomers: Asaph Hall,[89] who, undeterred by the poor prospects, arrived at the Observatory in 1857 with a wife, twenty-five dollars in cash, and a promise of three dollars a week, and Truman Safford, the account of whose subsequent career belongs to another chapter. The others were Horace P. Tuttle, who remained long enough to win a reputation as a discoverer of comets (he is said to have observed the great comet of 1858 before Donati announced it), and to be awarded the Lalande Prize in 1858 given by the Imperial Academy of Sciences at Paris for his additions to the catalogue of known nebulae visible in smaller telescopes; and Sydney Coolidge, grandson of Thomas Jefferson, who served with great devotion from 1853 to 1860 without pay, and who, after a somewhat adventurous career, fell in the battle of Chickamauga in 1863.

The annual report for 1859, Bond's first year of sole responsibility, indicated no lessening of the usual observations. The instruments in some cases he thought deficient, but he cared less to dwell on that than on the urgent need to find money enough to publish a vast accumulation of data, "now lying neglected and useless . . . inaccessible to the scientific world, subjected by delay to a depreciation in value which can never be restored or compensated, and exposed to injury, or possibly to entire loss, by some unforeseen calamity."

The Visiting Committee, headed by William Mitchell, sympathized with Bond's plea for publication and echoed his lament that "observations of the most valuable character, made by practiced eyes, with one of the most penetrating instruments in existence," should remain unpublished, "contributing, as they unquestionably do, one step, at least, towards the development of those ordinances of Heaven which for ages have seemed to defy the researches of man." One member had offered money for some printing, but not for the heavy engraving costs. These, the Visiting Committee urged, should come from the Corporation itself. The report concluded with the statement, "It should be a matter of gratulation to all, that the mantle of the father falls so fittingly upon the son."

Bond's next year's report adds little to that of 1859 beyond the good news that generous friends of the Observatory had contributed funds for the Donati monograph and engravings (the Corporation apparently did not), and the less welcome information that Sydney Coolidge's services were no longer available. But in March of 1860, after a thorough canvass of the limitations under which he worked, Bond addressed a long letter to that most dependable friend, J. Ingersoll Bowditch, with whom he could always be candid. The crying need was money. Again he emphasized the problem of publication. He had enough material, he pointed out, to fill eight or ten quarto volumes of 300 to 400 pages each. It included zone catalogues of 500 stars, meridian transits, moon culminations, observations on solar spots, observations and drawings of the great nebulae in Orion and Andromeda, the cluster in Hercules, and Donati's comet, specimens of astronomical photography, satellites of Neptune and Saturn, meteorological and magnetic observations, eclipses and occultations. At least $20,000 would be required for these serious "arrears in printing." To cover current expenses of light, supplies, and minor repairs (the care of grounds he paid for out of his own pocket) the University granted only $200. The buildings needed paint and the installation of gas, and as there was no allowance for fuel, he had to put on his overcoat to consult books in the library. The meridian circle, defective from the beginning, should be replaced at a cost of $6000. For proper instruction of students he recommended a new telescope of the best quality and an assistant. Then there was photography, which he predicted was "bound to become a great power in the investigation of the most profound problems of astronomy." It should be carried out on a liberal scale, with the whole time of an artist and a new telescope expressly designed for the purpose. Nothing less than $80,000, he estimated, would give this Observatory the means to remain fully active and on a par with European institutions enjoying government patronage.

The onset of the Civil War further accentuated his desperate problems. Writing to William Mitchell in January 1862, he pointed out the discrepancies between his situation and that of others in civilized communities, where "astronomy is thought worthy of public recognition." He knew that it was useless at this point to expect any aid from the government. No relief would come unless from "the liberality of private individuals, and the present is not a time to make an appeal to them." Later that same year he addressed a second statement of Observatory needs and conditions to Bowditch. He took care to compare in detail the differences between his resources and those of five observatories for which he had precise figures. The Royal Observatory at Greenwich paid the Astronomer Royal $5,000 and seven assistants $7,000. Altogether for one year (1858–59) the British government contributed $30,940, not including a contingency fund of $5,000 and $10,500 appropriated the year before for a new object glass and other instruments. The Royal

The Two Bonds

Observatory at the Cape of Good Hope received $10,195 for salaries, repairs, and incidentals, and a contingency fund of $4,000. In 1861–62, the first year of the Civil War, the Congress appropriated for the Naval Observatory at Washington $115,000, $13,000 of which was for contingent expenses, $60,000 for instruments and books, and $4,000 for four aides. About $35,000 covered the salaries of the superintendent and professors. Even the Dudley Observatory at Albany, with all its troubles, had a budget prepared by its scientific council, Bache, Joseph Henry, Gould, and Peirce, of about $12,000, not including costs of publication.

Next to Washington, the most generously supported of all was the Imperial Observatory at Pulkovo. The director's salary was $6,700, with an additional allowance for house rent. Four assistants received $10,000, also with a subsidy for rent, $10,000 was budgeted for instruments, and $17,000 for contingencies. There was even an allowance of $3,000 annually for the assistants' clothing. Publication was paid for by the Imperial Academy of Sciences of St. Petersburg.

In contrast to such affluence, at Cambridge the Phillips legacy produced $5,000 annually, the $200 from the Corporation remained a fixed sum, the Sears Fund, which in 1861 amounted to $9,911.43, could not be drawn upon for forty years, and only the Quincy Fund of $10,000 provided usable income each year of $600 for publication of the *Annals*. In view of his difficulties, Bond was remarkably objective: "The Corporation," he wrote,

has always manifested the most friendly disposition towards the Observatory. Its claims from other departments naturally must first be provided for, since they are essential to the existence of the College as an institution for instruction; and moreover all of them make a return to the treasury from the fees of students, whereas whatever is expended upon the Observatory is an outlay with only the prospect of a distant and indirect return, arising from the higher character and reputation which it may contribute to the College and the favorable influence which such an institution actively engaged in prosecuting any one branch of science must have upon all others associated with it.

Any advantages accruing in this way, he added, are still "not of a nature to be reduced to an account of profit and loss," and meanwhile, since all other departments are "more or less straitened," it is understandable that the Corporation cannot contribute to the Observatory.

In all his pleas for the Observatory, however, he uttered not a word about his personal necessities and deprivations. His daughter has written of his obsession to keep the Observatory up to standard through all the hardships of the war while making every sacrifice of comfort. He gave up coffee; he took no newspaper, depending on his brother's for war news. Fuel was insufficient, and the cold, drafty rooms in which he had to work day after day and night after night doubtless quickened the pace of his decline. "We have fallen on evil times," Joseph Henry wrote Bond, gloomily predicting "the end of our glorious union"—a mood Bond had every reason to share. "What an edifying spectacle to the rest of the world do England and the United States now present," he lamented to Carrington on January 3, 1862, "the latter on the high road to military despotism and England in league with slavery to make it a sure thing." The war was taking its toll in other ways. He shortly lost his last two assistants, Tuttle to the battlefields and Hall to Washington. "As for science and its cultivators," he commented to Peters, "we have been so poorly treated by the state in its prosperity, that now in the day of adversity we have at least the consolation of having very little to lose." (Ironically, while Bond saw little hope of help from the state, it had a few years earlier set aside $100,000 from the expected sale of public lands for Agassiz's museum, and another sizable subscription list for his vast publishing scheme brought the sum to $220,000 in less than six months. And the treasurer who had written Bond in 1847 that "spending any amount upon the observatory now, however small, is like drawing teeth—it makes me shudder to anticipate it," apologetically told Agassiz, in offering him $400 a year for care and increase of his collections, "It was not without diffidence that the Corporation offered you so small a sum, fearing . . . you might esteem it inadequate.")[90]

One consolation Bond did have in the midst of the encircling gloom. In 1862, thanks to private gifts, he was able to publish in Volume 3 of the *Annals* his great monograph on Donati's comet,[91] a triumph of astronomical observation and analysis, illustrated with more than fifty plates of beautiful engravings made from Bond's own drawings (Fig. 26). Nothing the Observatory had produced so far received such praise, from astronomers as well as friends of the Observatory and Harvard. One Bostonian, in acknowledging the work, wrote humorously that he had received it with the sort of ignorant admiration an Indian might

Fig. 26. Donati's comet, as drawn by George Bond on September 29, 1858.

have for a gift of a barometer or air pump, but in such confusing times, with everything unstable, flowing, and uncertain, he envied Bond the duty and pleasure of watching "movements of those orbs of light, movements inspired and controlled by a great primordial law, which has no limitation or exceptions. Among the stars there is no radicalism, but eternal harmony, serving order, the effluence of that law whose seat is the bosom of God!" Another recipient expressed his great pride in Bond, the University, and the country; the plates "almost revived the eerie sensation the comet itself produced of a supernatural apparition as if a spirit had passed before my face." Astronomers found it a model for all subsequent works on comets, but the greatest recognition of all came almost too late.

Bond was not alone among observatory directors suffering from lack of funds. In December Peters wrote glumly from Hamilton College that there was not even enough money on hand to pay his salary, which was in arrears by many months, and it looked as if he would have to shut down his observatory altogether. In his reply George expressed sharply his own view of the public attitude toward the support of science:

What you say of the financial prospects with which you begin the new year nearly completes the list of the twenty-five observatories *started* (not *founded*) within the past 20 years in these United States & left to die of want. Now if we except the National Observatory no doubt we have here at Cambridge the one best provided for among them all—yet I can say from bitter experience that the charge of it has been from the beginning a perpetual vexation of spirit, for the want of proper means of support.

I am weary to death of this new original & undeniably American idea of cultivating science by withholding all the aids which the rest of the world had hitherto thought requisite, & then haunting the unhappy astronomer with a dismal ghost of popular reputation & a newspaper notoriety. Excepting that in this community there are those who will do what the state neglects, out of pure largeness of views, we should long ago have made [another?] one on the list of failures—have added another warning against the repetition of the folly of a new observatory.[92]

His objection to a new observatory that might suffer the fate of others reflected also his personal disappointment in having been unable to find money for an achromatic object glass of 18½ inches aperture which the Clarks had recently completed for the University of Mississippi, but for which the war now made payment impossible, though an elaborate

Fig. 27. Alvan Clark, founder of the optical firm of Alvan Clark and Sons, Cambridgeport, Massachusetts; from a portrait in the Observatory.

replica of the Pulkovo Observatory had been built to house it. During his test of the glass on January 31, 1862, Alvan Clark (Fig. 27) had made the great discovery of the companion star of Sirius,[93] which had been predicted by Bessel in 1844 but which until this moment had eluded all observers. Bond, to whom Clark immediately reported his exciting find, was eager to examine the new lens himself. In February it gave him a remarkably sharp view of the great nebula in Orion, on which he continued to work as long as he lived. He found the glass the equal of any lenses made in Europe. "All the grand features discernible with our great refractor," he declared, "were, in this view, brought out in increased brilliancy, strength, and certainty." Its powers would have added immeasurably to his resources, especially since its focal length was identical with that of the present telescope and it could be mounted without a single change of clockwork or arrangement of the dome. Elsewhere it would have to await a structure to contain it. Unfortunately, Bond could not raise the $11,187 paid for it, and it went to Chicago for the proposed Dearborn Observatory, named for the wife of the chief donor, J. Y. Scammon. In view of its limited use there (the Chicago fire of 1871 seriously affected support of the Observatory, and in 1877 the telescope was transferred to Northwestern University), Bond's bitterness had some justification. He wrote Asaph Hall on March 2, 1863, of its original removal to Chicago,

where they propose inaugurating another of those stupendous humbugs—an American Observatory of the first class—Europe is to be dazzled and America enlightened on quite an extensive scale. The harm thus far accomplished is that they have by paying down a small sum to Clark secured the large object glass which we should certainly have soon had in our dome doing good service.

Despite this loss, the year 1863 had its compensations. In January William Sturgis gave the Observatory $10,000 toward publication, a gift "of munificent liberality" that drew from Bond one of his most graceful letters:

It is not alone the language of the gift which measures its real value as an addition to the resources of the Observatory. I am persuaded that the special object to which the income has been devoted by the terms of the donation, is of far higher importance to the best interests of the institution than any extension of its operations in other directions.—A suitable provision for the publication of its current work and of the records accumulated in past years is not merely desirable because it will add to the general usefulness and reputation of the Observatory and enable it to maintain its rank among sister institutions; it is in truth as necessary to the proper fulfillment of the objects of the establishment as is the possession of its telescopes or any other essential part of its apparatus.

If the particular direction given to your bounty, so appropriate to our present circumstances, has served to enhance its value, it has been made still more grateful by the consideration that it is but one of many benefactions conferred from the same source. As such it becomes an expression of undiminished confidence & a welcome encouragement to renewed efforts on the part of all engaged in the operations of the Observatory to a fruitful discharge of their trust.

In the spring of 1863 Bond made his second journey[94] to Europe to investigate the prices and quality of a larger telescope in hopes of replacing the one lost to Chicago. Twelve years had passed since his first tour when, even as a novice, he had received such respectful attention. Now director in his own right, full of experience and recognized as an authority, he moved with ease among the eminent men of science in England and on the Continent. From Liverpool, where he visited his friends the Hartnups (and saw "spots of efflorescence" on the object glass, "just as on ours") he went to Birmingham to inspect the variety of disks made by Chance & Co., a vast establishment covering thirty acres, with a working force of 2200. In London he saw James Clerk Maxwell, who showed him his ingenious model at the Royal Society demonstrating "the motions of a ring of thirty-six satellites" about the planet Saturn. A short train journey took him to the home of Warren de la Rue, who explained his photographic processes and exhibited a striking stereoscopic view of Saturn, the shadows of the ball on the ring "singularly like ours." He met old friends, among them Airy, Hind, and Carrington, revisited Greenwich, and marveled at the fireproof manuscript room that housed the entire collection of observations and computations made from the first day of the Observatory. Not the least portion was the voluminous correspondence of Airy, all written in his own hand without benefit of secretary.

On the Continent he had pleasant sessions with Argelander at Bonn and with Schönfeld at Mannheim. Both astronomers complained about lack of money for publication! At the small town of Speyer he inspected Schwerd's photometer and tiny observatory, and in Munich he renewed his acquaintance with the Merz family and looked at a number of object glasses, prices of which Merz informed him now ranged from $15,000 to $21,000 gold, plus transportation. (At home, the Clarks, who were becoming more and more expert, had offered a glass for $8000.) The most impressive instrument, however, he saw at Steinheil's establishment in Munich, featuring a type of lens about which Bond communicated a highly technical paper to the American Academy, "On the New Form of the Achromatic Object-Glass Introduced by Steinheil,"[95] said to offer

125

improvement over the orthodox type. Yet in the state of Observatory finances Bond could place no order anywhere.

In his absence the usual work of the Observatory had been left in the capable hands of Truman Safford, who, much to Bond's satisfaction, had been appointed to fill the vacant post of assistant observer in 1862 in recognition of his "industry and singular ability." That other highly valued assistant, Sydney Coolidge, reported missing after the battle of Chickamauga, drew from Bond in his report for the year almost two pages of tribute to the many services performed in seven years entirely without pay.

Bond had sailed for Europe on April 15. The cordial reception and hospitality that noted astronomers had offered him and the stimulation of his visits to great workshops had combined to lift his spirit from the almost despairing mood in which he had departed. He returned in July to be reminded forcibly of the place assigned to a prophet in his own country. On March 3, 1863, in the dying hours of the 37th session, Congress established the National Academy of Sciences.[96] The act climaxed weeks of secret planning (not to say plotting) by members of the Scientific Lazzaroni under the leadership of Agassiz, Bache, Gould, Peirce, and Davis. Determined to assert their power beyond the confines of Cambridge, where they had been remarkably successful in carrying out their rule, they sought governmental authority for direction of the science of the country and for passing on the eligibility of anyone aspiring to have a share in it. "We have a standard for scientific excellence," Agassiz wrote Bache triumphantly. "Hereafter a man will not pass for a Mathematician or a Geologist . . . etc. because an incompetent Board of Trustees or Corporation has given an appointment. He must be acknowledged as such by his peers or aim at such an acknowledgment by his efforts and this aim must be the first aim of his prospects." Members should not worry, "for the malcontents will be set aside or die out and the institution survive, and it now remains for us to give it permanency by our own doings." He wanted the Academy "to stand as a jury in the world, in matters of Science."

The bill to establish the Academy had been introduced by Senator Henry Wilson at Agassiz's instigation (Wilson had just named him a regent of the Smithsonian Institution). The provisions called for fifty incorporators, privately and secretly chosen, with authority to perpetuate their number and, if called upon, to "investigate, examine, experiment and report on any subject or science or art." Following the example of the French Academy, members were placed in various sections— mathematics, physics, astronomy, geography, and so forth. For Section III, which included astronomy, geography, and geodesy, the leaders chose, among others, Stephen Alexander (Joseph Henry's brother-in-law), James Gilliss, Arnold Guyot, Benjamin Gould, Lewis Rutherfurd, and Alexis Caswell; for mathematics, Benjamin Peirce, Joseph Winlock, and others; for physics, Bache. Eleven of the fifty were drawn from the army or navy, hardly scientists at all, while a number were connected in one way or another with various branches of the government. The whole affair moved with such secrecy that not even Joseph Henry, closest friend of Bache and, as secretary of the Smithsonian Institution, presumably at the head of science in the country, learned of it until March 5, when, along with the rest of the chosen, he received a letter from Wilson requesting suggestions for a date of the first meeting. Conspicuously absent from the favored fifty were George Bond, John W. Draper, Spencer Fullerton Baird, and Elias Loomis, as well as others far better known than some of those included.

Writing on March 12 to John Fries Frazer, professor of chemistry and physics at the University of Pennsylvania, whom in a private joke he addressed as "Dear Grandson," Bache felt called upon to explain at least two omissions:

There are some men too mean to bring into our Academy thus slightly intimating that I so class Geo. P. Bond and Spencer F. Baird. I have had favorable opportunities for inductions upon them in parts of their lives, and have come to distinct conclusions. The ven[erable] Smithson [that is, Joseph Henry, constantly and somewhat disparagingly so referred to in the Bache–Peirce correspondence] would I think have had both in. . . .

Though the leaders of the organization expressed great delight in their success, the list aroused indignation among many excluded and even considerable criticism from some elected. Joseph Leidy thought it would turn out "a grand humbug" and wanted to have little to do with it—"a Society of the kind, that leaves out such men as Baird, Draper, Hammond [then surgeon general of the Federal forces] . . . and appropriates a number who never turned a pen or did a thing for science, certainly can't be of much value." It appeared to him rather "to be nothing more than the formation of an illiberal clique based on Plymouth Rock." Another critic, deeply troubled by the high-handedness of the small coterie, lamented

the violent hatred openly proclaimed by men of high standing whose influence should be only good and productive of a higher tone of morality and gentleness. In such disputes . . . there is no trace of dignity, not even of decency . . . We expect more of those we call great men and they are but men after all.

Perhaps the most useful contemporary estimate of the whole unlovely procedure can be followed in the letters of William Barton Rogers,[97] founder of the Massachusetts Institute of Technology, to his brother. On March 17, having been notified of his election, he wrote:

What think you of a National Academy of Arts and Sciences in the United States, incorporated at the very close of the session of Congress just ended, and of which only two or three of the men of science knew anything until . . . announced in the newspapers. George Bond, the most distinguished practical astronomer we ever had, is *omitted!* . . . Again, Cooke and Lovering are left out, though many an unknown name is placed on the roll of honour.

Following the first meeting in New York on April 22 (one week after Bond's departure for Europe), Rogers described it to his brother, in a letter dated April 28:

This [meeting], as appointed by Senator Wilson, was held in the chapel of New York University, within earshot of Professor Draper's lecture-room, and near that formerly used by Loomis, though neither of these gentlemen was admitted to the band of fifty. As Robert [Rogers] and I ascended the stairway we met Draper going the other way. I felt the incident deeply, and early in the course of preliminary proceedings, I took occasion frankly to express my surprise and mortification that in a body professing to represent the science of this country we should look in vain for Bond and Draper and Loomis and Baird. "This," said I, "is a sad error, if it be not a grievous wrong. Surely . . . there are many here who in their hearts must feel that they have no claim to be here when such men as I have named have been excluded." The shaft struck the mark, and caused a pause in the exultation and mutual glorification in which some had been indulging.[98]

When the Constitution and Rules, elaborately prepared, were read by Bache from manuscript, nobody but Rogers dissented on any important point. Among the provisions were unlimited terms of membership and life tenure for officers. Rogers let them pass until the next morning when, just as Bache was about to put the document down, Rogers rose, and

calmly called their attention to this clause, told them that to exact that would be to blast every hope of success, and so impressed them with the responsibility of such a course that they voted the term of six years instead of for life. I had much use for my backbone, but did all calmly and without personality.

With support from a few others, he succeeded in "modifying or defeating some of the most objectionable provisions, and, what is better, of having the whole open to immediate amendment at the first stated meeting to be held in Washington next January."

Bond would have scorned making any protest on his own behalf, but in July, shortly after his return from Europe, he wrote Asaph Hall:

After what has transpired of late in "scientific circles" in Cambridge, Boston and New York you may take the statement for what it may be worth, when I tell you that I had assurances as many and as strong as could be desired, of the respect and honorable position accorded to our observatory by European scientific men. They have undoubtedly fallen into this mistake by adhering too closely to the principle of judging "by deeds, not words," and will be set all right by the latest decision of the tribunal who have the scientific reputation of this country under their special keeping.

Hall replied with equal irony that the National Academy of Sciences was now taking care of the scientific interests of the country, and the fifty distinguished men chosen "of course without any previous knowledge of their own . . . like good patriots have determined not to shrink from the duties imposed on them. Thus we have a new era of American science."

Doubtless one of the unhappiest members was Joseph Henry himself. Writing to his brother-in-law a few days after the passage of the Act, he denied all responsibility either for the move or for the choice of the members, adding:

I am not well pleased with the list or the manner in which it was made. It contains a number of names which ought not to be included and leaves out a number which ought to be found in it . . . I do not think that one or two individuals have a moral right to choose for the body of scientific men in this country who shall be the members of a National Academy and then by a political ruse, obtain the sanction of a law of Congress for the act.

Still, as he wrote Asa Gray, he intended to be present at the first meeting "and do what I can to give it a proper direction." To Agassiz, however, he adopted a tone that in view of his pacific temperament was unusually stern:

Besides the objections I had presented to Professor Bache I did not approve of the method which was adopted in filling the list of members. It gave the choice to three or four persons who could not be otherwise than influenced by personal feelings . . . and who could not possibly escape the charge of being thus influenced . . . My anticipations in several particulars have been realized—and antagonism, such as I feared has been produced in the minds of those who think themselves ill used in being left out . . . The feeling also exists, to a considerable extent, that the few who organized the academy intend to govern it.

Henry had ample reason to speak his mind with such vigor. At the second meeting of the Academy, held in New Haven, Agassiz had violently fought against including Baird on the grounds that descriptive natural history was an inferior science, though before and after coming to the Smithsonian Baird had collected at great pains and sent to Agassiz many specimens for his researches on fish. Henry pointed out wisely, however, that such a slight would bring resignations from many naturalists who could not reasonably be expected to vote in a way to "disparage their own pursuit." As for himself, he insisted, he had never joined in any intrigue for personal ends and never would, but it was sometimes necessary "to have an eye on the acts of others in order to thwart their improper designs." Baird came in, and having won that battle Henry preferred to let time revise the list. By then, however, Bond, who had been slowly dying of tuberculosis, had little need of this or any other earthly recognition, especially from those who had intrigued against him.

Henry had complained to Bache that in the last dozen years the quarrels had resulted in an immense waste of time and emotion that would have been better spent on other purposes and that unless they ceased and "our Cambridge friends" were kept in check some of the best men would leave rather than be perpetually exposed to annoying disputes about the policies and control of the Academy. He had in mind among those likely to resign Arnold Guyot, who, though previously a close associate of Agassiz's, obviously differed from him in his estimate of men. Guyot wrote Henry from Princeton that, since the death of Gilliss had created a vacancy, "Bond should be called in. Our Cambridge friends had their own way in the last meeting. Let justice have its own in the next." But for Bond it was already too late.

In their report for 1863, dated January 26, 1864, the Visiting Committee made no direct reference to the slight to Harvard's and the country's leading astronomer, but their concluding paragraph can be regarded only as an outright rebuke to those who had perpetrated this wrong. They begged leave

to perform a grateful duty in bearing testimony to the diligence, fidelity, and success with which the Director of the Observatory continues to fill his arduous and responsible station. It affords them much satisfaction to know that his rank among the most distinguished living observers is fully recognized by the most eminent astronomers and scientific bodies in Europe.

Although they hardly needed further confirmation of their own judgment, it was especially gratifying to them that in 1864 two distinguished societies elected Bond to membership—the Academy of Science at Munich and the Royal Astronomical Society, which in April elevated him from honorary to associate, "as a mark of their high esteem of his astronomical researches and personal regard," as well as in gratitude for "being favored with his scientific publications." The Committee also found considerable satisfaction in quoting Galle's comments in his supplement to the latest edition of Olber's work on the orbits of comets: "The most complete collection of observations on the physical phenomena of the great comet of 1858 is to be found in Professor Bond's account of it . . . a work which stands alone in this regard in astronomical literature, and which has not yet been equalled in the marked beauty and truth of its illustrations."

Bond himself, though pleased by such firm support from abroad and from his own Committee, had little patience with what seemed to him the unhealthy state of astronomy at home. His letters to his close confidant Peters during the last half of 1864, if perhaps colored by his rapidly failing health, lost nothing of vigor in their comments on the ethics and scientific weaknesses of certain American astronomers.

Yet he found some hope among the younger men, "even within the sphere of influence" of his old enemies, the Coast Survey. A new assistant at the Observatory in particular, William Augustus Rogers,[99] he praised for "true zeal and entire honesty, traits not over prominent among American savans." That judgment of Rogers proved prophetic, for when he returned to the Observatory in 1870 as a full-time staff member he took charge of the observations assigned to Harvard by the Astronomische Gesellschaft (the German Astronomical Society) for the revision of the great star catalogue, the *Durchmusterung* (see Chapter III). Although he left in 1886 for a professorship at Colby College, he carried on his supervision of the work, which was finally published in 1896 and which occupies seven volumes of the *Annals*.

Bond's rapidly declining energies in the late summer and fall of 1864 threw major responsibility on Safford, who observed the asteroids Aglaja, Urania, Thetis, Iris, Olympia, Echo, and others, comets, and satellites of Saturn in continuation of the series he had begun in 1863. He carried on zone observations, by which he found the positions of 2000 stars, many formerly unknown, and with Rogers observed thousands of star transits. His major personal achievement, however, was his memoir on Gould's polestar position, which boldly questioned the accuracy of the place assigned to it in the Coast Survey *Catalogue*. Although very ill, Bond mustered up sufficient strength to write a vigorous introduction to Safford's paper, not only because he wished to promote Safford's work, but because, as he explained, "to American Astronomers the subject has become one of some moment in consequence of the adoption of Dr. Gould's erroneous position of this most important of all the stars by the Coast Survey and by the National Observatory and American Nautical Almanac, thus giving to it, to the extent of their influence, the character of a national standard authority." He no longer cared that his remarks would "be set down as my most grievous offence against the Coast Survey," he wrote Peters, and to Hall he expressed more concern that Safford's corrections should receive the attention they deserved than worry about a storm of indignation that the "sacred brotherhood of science" might raise over having the truth spoken out plainly. His old friend Robert Treat Paine, staunch as usual, praised the introduction as "spicy and to the point."

As the winter came on it was clear to all that only a few months remained to Bond. Members of the Committee calling on him and ob-

serving his feverish color and feeble frame urged him to lighten his
work, and the compassionate Bowditch, instead of advice, sent him "the
enclosed" with which "to find out where the best trout and the finest
deer roam." A brief stay in Maine refreshed but could not cure the
invalid, and in the end he was forced to give up everything but the
one obsessive thought that dominated the last weeks of his life—the
completion of his long work on the nebula in Orion (Fig. 28), his
final tribute to his father, whose work, once severely criticized by Struve,
he determined to vindicate. He wrote Peters shortly before his Maine
trip that he had made great progress on it, having just finished working
out the final positions of 800 to 900 stars and wanted nothing more
than the strength to finish what he had undertaken. By January 7,
however, when he wrote to Hall, he had given up hope:

My disease makes progress, and leaves me little hope of putting the
materials of my work on Orion—to which I had devoted so much labor—
into condition such that another could prepare them for press. In truth,
I am becoming resigned to the idea that most of it is destined to oblivion.
I had planned to accomplish something considerable, and this is the
end. "It is not in man that walketh to direct his steps."[100]

Even so, until the day before he died, his daughter has written, long
after he was too feeble to hold a pen, he labored at the memoir, which
was left to Safford to complete.[101] She has described touchingly those
last weeks of great suffering borne with patience, relieved at moments
by the kindness of friends. Some of them, knowing of his passion for
color (he had predicted that it would certainly be added to photography), brought their jewels for him to look at—"dying as he was,
the sight of their flashing colors seemed to afford him exquisite pleasure."
The man who delivered ice to the house, who could lend no gems but
had heard of Bond's fancy, brought him a large block of ice cut from
Fresh Pond, "as green as an emerald, perfectly clear," and sparkling
with every tint in the sun. Though too weak to walk across the room,
Bond "insisted on being wrapped in shawls and carried into an open
hall," where it was so cold the ice could not melt. He would sit there
for a long time relishing the bright colors shining from it.

On February 10, seven days before he died, he had advance word
of the great honor about to come to him, the award of the gold medal
of the Royal Astronomical Society, the first to any American, for his

FIG. 28. The nebula in Orion, as drawn by George Bond in the period 1859–1863.

magnificent memoir on the Donati comet and his other work. In presenting it, Warren de la Rue, then president of the Society, delivered what he did not know at the time was the best of all eulogies spoken or written after Bond's death. Long aware of Bond's achievements, indebted to him for the inspiration of his own astronomical photography, and remembering their pleasant meeting in 1863, de la Rue not only analyzed in detail the great work on the comet which had led chiefly to the award, but brilliantly summed up Bond's whole astronomical career. Of the volume on the comet he said:

One of the most successful workers in recording the phenomena of this Comet was Professor G. P. Bond, and the sketches and drawings made at the Observatory of Harvard College form the main contribution to the splendid graphic illustrations, which are a remarkable feature of this altogether remarkable production. The pains taken to ensure a truthful representation of this Comet of 1858, both in its eye- and telescopic-features, few are better able than myself to appreciate; and I am able, from my own practical experience, to state that the success which has attended these efforts has been deservedly won by battling with greater difficulties than would probably be imagined by the uninitiated . . . A less zealous worker would have been deterred from prosecuting to its final completion his self-imposed task, the execution of which will redound in all time to the credit of the recipient of the Medal this day.[102]

From his discussion of the comet de la Rue turned to Bond's theory of planetary perturbation, which led to his memoir on mechanical quadratures, the determinations of longitudes by father and son, the investigations of Saturn and the discovery of Hyperion and the dusky ring, the catalogue of 5500 stars near the equator, his assistance in devising the electromagnetic method of recording transits, and the photographic work that had stimulated de la Rue's own experiments. Then, handing the medal to the British Foreign Secretary for transmission to Bond, he wanted especially to have conveyed to the recipient "an adequate expression of the enthusiasm which the announcement of his name as the Medallist of the year has elicited from this assembly."

But on February 17 Bond died,[103] leaving, as his father had done before him, a standard every director thereafter strove to meet and an institution that out of small beginnings and painfully limited means the two Bonds first placed on a foundation as solid as the pier on which they mounted their first Great Refractor.

III

JOSEPH WINLOCK,
THE THIRD DIRECTOR
1866–1875

I

The first era of the Observatory, twenty-five fruitful years, ended with the death of George Bond. During the last months of his life the work at the Observatory had suffered, both by his decline and by the continuing drains of war. Staff members were gone. Only Safford remained, and the benefactors on whom the Bonds had regularly relied for extra funds could not be called upon during the national crisis. The loss of two of the Observatory's oldest friends was particularly felt. Josiah Quincy, founder of the Observatory and faithful member of its Visiting Committee for many years, died July 1, 1864. Edward Everett's death followed six months later. Other friends remained, none more devoted than J. Ingersoll Bowditch, but the relations with an entirely new regime would inevitably lack the personal intimacy that had bound veteran members of the Committee to the first two directors.

On February 25, one week after George Bond's death, the Corporation named Safford acting director "to perform all the duties of the office until otherwise ordered." If this last phrase held a somewhat chilling note of impermanence, Safford took it calmly. In his report to the Visiting Committee he covered the portion of 1864 which George Bond had left unfinished and loyally pressed for the completion of Bond's manuscript on the nebula of Orion. "No successor," he wrote, "could give to his own work the peculiar merits which belonged to whatever Professor Bond did." During the next four months he concentrated on that task, but since his status was uncertain he began no new work.

On June 15 the president "communicated the Corporation's votes to elect Joseph Winlock[1] director, that the Overseers may consent thereto

136

Fig. 29. Joseph Winlock, third director of the Observatory. From a portrait by D. C. Fabronius, at the Observatory.

if they see fit, and the same were referred to Messrs. [James] Walker, [William] Mitchell, and [Artemus Bowers] Muzzey."[2] It was a somewhat surprising appointment. Winlock (Fig. 29) had had no connection with Harvard College, and he was better known as a mathematician than as an astronomer. Born in Kentucky in 1826 of Virginia stock, he attended Shelby College at Shelbyville, from which he was graduated in 1845. His exceptional record in mathematics led to an immediate appointment there as professor of that subject and of astronomy. His first association with Cambridge he owed to Professor Benjamin Peirce, whom he met in 1851 at the Cincinnati session of the American Association for the Advancement of Science. Peirce promptly invited Winlock to join the corps of computers in the Cambridge office of the *American Ephemeris and Nautical Almanac.* He remained there from 1852 to 1857, when he accepted the professorship of mathematics at the United States Naval Academy, but shortly after the outbreak of the Civil War he once more became superintendent of the *American Ephemeris,* and held that post until his appointment as director of the Harvard Observatory.

When after several months the Overseers had not seen fit to consent to the nomination, the committee to which the matter had been referred asked for and received additional time to consider. On the same day William Mitchell, the member most closely identified with the previous administrations, resigned, to be replaced at once by Stephen M. Weld. Mitchell's letter of resignation indicates no direct connection between the choice of Winlock and the committee's hesitation. He merely explained that he was leaving Massachusetts to join his daughter Maria, who had just been named professor of astronomy at Vassar.

Important as it was to end uncertainty, the situation presented a dilemma to those remaining members of the committee who were best acquainted with Safford and who had looked upon him as the natural heir to the directorship. He had come to the Observatory as a boy of eleven and had delighted the elder Bond by his avid interest in observation and his remarkable mathematical prowess. "Henry Safford came up to see the new planet" [Neptune], Bond noted in his diary for November 9, 1846, "which we observed very much to his satisfaction; he called it 'great sport.' " Not long afterward Bond sent some of Safford's accurate calculations on the elements of a new comet to Leverrier and Hind as the work of the "son of a farmer in our neighbourhood," a modest lad, altogether without vanity though he had received the most sensational publicity as a "lightning calculator." After graduation from Harvard, Safford had divided his time between the Observatory and the office of the *American Ephemeris,* where Simon Newcomb, who came there in 1857, found him "the most wonderful genius" on the staff, "a walking bibliography of astronomy, which one had only to consult in order to learn in a moment what great astronomers of recent times had written on almost any subject, where their work was published, and on what shelf of the Harvard Library the book could be found."[3]

For over a dozen years the Observatory had enjoyed not only Safford's exceptional skill in computation but his willingness to take on many responsibilities. He had contributed to four volumes of the *Annals,* he had readily agreed to complete George Bond's last work, and he had contributed over a period of years a number of papers to scientific journals that displayed both astronomical knowledge and literary competence. Support for him was not easy to overcome.

While the committee deliberated, Professor Benjamin Peirce, who had promoted Winlock's appointment and who did not doubt the ultimate decision, wrote the president, urging him to send Winlock to Europe for a few months to "perform important services to American science." While there he could organize a series of international observations on behalf of the Coast Survey, make arrangements for determining longitudes between Europe and America, and for the Observatory order a new meridian circle to replace the defective instrument the Bonds had long struggled with. As the United States Treasury would pay Winlock's passage, Harvard would not be put to much expense.

Another Winlock advocate, J. E. Hilgard, acting superintendent of the Coast Survey, took up the cause. On December 23, 1865, he wrote President Hill:

The unusual delay in the action of the Overseers in the nomination of Professor Winlock for the vacant position as Director of the Observatory leads to the supposition that doubts are entertained of his entire worthiness for the place. The consideration of the important relations heretofore sustained by the Harvard Observatory towards the Coast Survey, it being the cardinal point to which all longitude determinations have been referred, will I trust justify the expression of an opinion on my part for which there would perhaps otherwise be no warrant.

I beg leave therefore to assure you and through you, the Board of Overseers, that Professor Winlock enjoys the entire confidence of the Chief of the Coast Survey and his astronomical assistants, whom they will be pleased for the sake of science to see placed in the responsible position for which he had been nominated; and that they are ready to act in concert with him in all matters in which such co-operation will be conducive to the advancement of science and the public interest.

The Harvard College Observatory

Cooperation with the Coast Survey was nothing new for the Observatory—the Bonds had worked with its officers from the beginning—but a suggestion of financial support at this meagerly provided period and the complete confidence of Winlock's powerful friends argued strongly for a favorable decision. On January 25, 1866, nearly a year after Bond's death and slightly more than six months since the nomination, the committee finally reported in a somewhat unusual manner to the Overseers:

> Taking into view all the circumstances, they recommend that this Board concur with the Corporation in the election of Professor Winlock . . . But in doing so they cannot refrain from expressing their sense of the eminent fitness of Mr. Safford, the present assistant observer, to hold any place in the Observatory and their regret that many good friends of the College have been disappointed that he failed of a nomination.[4]

By a vote of 12 to 4 the Overseers accepted the report, and on February 8, 1866, in a gracious letter to the president, Winlock expressed his "profound sense of the honor which such an appointment confers, and of the great responsibility that its acceptance involves." He asked merely to be informed "at your earliest convenience . . . in what manner I am to enter upon the discharge of the duties of the office."

If Winlock's several periods of residence in Cambridge and his close association with a number of professors and scientists had not exactly made him a Harvard man, they at least had helped overcome the stigma of being a total outsider. Indeed, when the short-lived Cambridge Branch of the Astronomical Society[5] was founded in 1854, inspired chiefly by Professor Peirce in anticipation of a national astronomical body, Winlock became a charter member and recording secretary. His minutes, especially those reporting the more intricate mathematical and technical discussions of this elite group, reveal not only his own central role but his skill in recording them. He was also recognized as one of the original incorporators of the National Academy of Sciences. The reservations of some Overseers and of the committee therefore apparently stemmed less from distrust of Winlock's abilities and experience than from loyalty to the Bonds. These men knew what George Bond had suffered from the "Coast Survey clique" and its Cambridge associates who had opposed his election to the National Academy of Sciences. They also thought it unjust to ignore the claim of a Harvard graduate, a brilliant mathematical prodigy, and an experienced Observatory astronomer. On the other hand, sponsors of Winlock's candidacy had not forgotten that Safford with Bond's blessing had challenged the accuracy of Benjamin Gould's polestar positions before the American Academy and that the paper when published bore George Bond's firm endorsement. Now that the old regime was out of the way, the opportunity seemed to have arrived to capture at last the post that, at the time of George Bond's appointment, had eluded Peirce and Gould by placing in it a man of their own circle.

Yet if pressures on Winlock's behalf came from what one historian of science has called "the axis of power within the scientific community," based originally on "the strong personal and scientific ties established between Bache in Washington and Peirce and Gould in Cambridge,"[6] Winlock himself was not a party to any maneuvering or to any persistent hostility to the Bonds. On the contrary, after his election he promptly showed the most magnanimous appreciation of the Bonds' work by seeing to it that many observations still in manuscript were published, although during his entire administration he never had enough money to print his own annual reports. Winlock issued, for example, the Bonds' zone catalogues of stars and observations of solar spots, thereby completing virtually all their work, and he made repeated attempts to persuade the Corporation to buy George Bond's library, an act he thought desirable "from considerations independent of the value of the books or of their usefulness to us." He was anxious to obtain William Bond's portrait for the Observatory and was instrumental in the purchase of a piece of land from Bond's daughter, both because she needed the money and to protect the area of the Observatory. He used and praised Richard Bond's escapement clock and the Bond chronograph, employed William Bond & Sons for repair of instruments, and incorporated in his elaborate series of engravings several beautiful Bond plates.

In every respect Winlock proved to be both generous in spirit and worthy of the post.

II

Although the delay in confirming the appointment was not disastrous, it hardly lessened the problems confronting Winlock. His preliminary survey of the Observatory might have discouraged a less vigorous man. The building lacked such ordinary conveniences as gas, running water, and heat. The one spot suitable for study or computation in winter was the clock and chronograph room, thinly partitioned, and poorly heated by a single stove. Winlock, hunting for some way to pay for installing gas, sold a supply of quicksilver for $80, but fortunately a benefactor came to the rescue with the $202 needed. The instruments had suffered from lack of money for repairs, yet he still faced the kind of stringencies that had so frustrated his predecessor. He had only the same small sum of $200 for running expenses, could not call on the Quincy or Sturgis income for anything but publication, and the Sears Fund was still subject to restriction. Even more discouraging, the College treasurer had charged the Observatory with a deficit of $1452. Nevertheless, after ordering an accurate clock and some minor apparatus, Winlock determined on two important instruments for the work he meant to do and devoted to them a major section of his first report, covering the period of his first eight months of office. It revealed both his awareness of the latest astronomical devices and a shrewd sense of the kind of appeal Harvard donors were accustomed to respond to:

It is greatly to be desired that before the expiration of another year a new meridian circle of modern construction will be purchased for the Observatory, one capable of determining both right ascensions and declinations, which can fairly come into competition with the beautiful instruments of this kind that are already in other observatories of the country. It is contrary to the interests of this observatory and of astronomical science that any instrument not of a high class should be tolerated here.

The meridian circle would replace the old apparatus. The second instrument would be an innovation: a spectroscope for that "new branch of astronomical science which spectrum analysis has founded." In fact, he had taken the liberty of ordering one from Troughton & Simms. To make certain that it would embody the most recent features he had arranged for its construction to be supervised by "the great Professor [William] Huggins, whose distinguished services to this branch of science are so well known in this country," and who would "introduce into it all of the improvements that his own experience has suggested." Thus, the introduction of the spectroscope by Winlock and the beginning of celestial photography here by George Bond brought together the two techniques that became an essential feature of this Observatory's work and that indeed revolutionized astronomy itself.

Spectrum analysis may be defined briefly as the method of determining the physical condition and chemical composition of a body through the analysis of the various colors or wavelengths of the light radiated by the body. In 1665–66 Newton discovered that a prism resolves a beam of white light into a "rainbow" of colors, but it was nearly 200 years before the spectrum was understood. Here it is sufficient to refer to the year 1814, when Joseph von Fraunhofer, the Munich optical genius, applied a telescope with slit to the solar spectrum and found hundreds of dark lines crossing it. Fraunhofer mapped over three hundred lines and designated the seven more prominent ones by the letters A to G, which are still in use, but beyond certain conclusions about the relation of the lines to the absorptive power of the sun and other stars and the marked presence in several stars of the line he called D, he left to later investigators the interpretation of their meaning. In this country John Draper not only was the first to pursue the subject but has the distinction of having anticipated by more than a decade the theories of Robert Bunsen and Gustav Kirchhoff on the Fraunhofer lines. In 1843 Draper constructed an instrument for his investigations which he called a tithonometer and which Bunsen (with his collaborator Henry Roscoe) used for early experiments that enabled him to arrive at important laws on the chemical action of light.[7] In 1847 Draper published his article, "On the Production of Light by Heat,"[8] in which he stated that "to a particular colour there ever belongs a particular wave-length, and to a particular wave-length there ever belongs a particular colour." In 1857, two years before Bunsen and Kirchhoff made their decisive experiment in Heidelberg, Draper wrote, "If we are ever able to acquire certain knowledge respecting the physical state of the sun and other stars, it will be by an examination of the light they emit."[9]

On November 15, 1859, before any announcement to the world, Bunsen wrote to his former colleague and intimate friend, the English chemist Sir Henry Roscoe:

At the moment I am engaged in a research with Kirchhoff which gives us sleepless nights. Kirchhoff has made a most beautiful and unexpected discovery; he has found out the cause of the dark lines in the solar spectrum, and has been able both to strengthen these lines artificially in the solar spectrum and to cause their appearance in a continuous spectrum of a flame, their positions being identical with those of the Fraunhofer lines. Thus the way is pointed out by which the material composition of the

143

sun and fixed stars can be ascertained with the same degree of certainty as we can ascertain by means of our reagents the presence of SO_3 and Cl. By this method, too, the composition of terrestrial matter can be ascertained and the component parts distinguished with as great ease and delicacy as is the case with the matter contained in the sun.[10]

On his eight-foot map of the solar spectrum (published in the *Transactions* of the Berlin Academy in 1861–62), Kirchhoff charted the lines and identified positively a number of elements common to solar and terrestrial matter—among them calcium, iron, magnesium, sodium, chromium, and others in varying quantities. Almost immediately translations of his communications to the Berlin Academy, *On the Solar Spectrum and the Spectra of the Chemical Elements,* appeared in England, containing his "theory of the chemical constitution of the sun . . . accompanied with four plates of the fixed dark lines in the Solar Spectrum from A to G, and the bright lines of the Metals, showing the coincidences." Roscoe's lectures[11] on the subject in 1861 created interest in England, and within a few years in Europe the literature on spectrum analysis applied to both terrestrial and celestial matter grew to extensive proportions.

In England the first and most ardent investigator to study the stars, planets, comets, nebulae, and sun with the spectroscope was Sir William Huggins. Favored with a private observatory at Tulse Hill, he was free to follow without delay the exciting path Kirchhoff's discovery had opened up. Almost half a century after the first news of it reached him he could recall that electrifying moment,

to me like the coming upon a spring of water in a dry and thirsty land. Here at last presented itself the very order of work for which in an indefinite way I was looking—namely, to extend his [Kirchhoff's] novel methods of research upon the sun to the other heavenly bodies. A feeling as of inspiration seized me: I felt as if I had it now in my power to lift a veil which had never before been lifted; as if a key had been put into my hands which would unlock a door . . . regarded as closed for ever to man—the veil and door behind which lay the unknown mystery of the true nature of the heavenly bodies.[12]

By 1866, as Winlock had correctly observed in his report, Huggins's work had become known in this country. In a "Lecture on the Results of Spectrum Analysis Applied to the Heavenly Bodies," delivered to

Joseph Winlock, The Third Director

the British Association for the Advancement of Science, he summarized five years of the research he had begun in 1861, and the Smithsonian Institution promptly reprinted it in the *Annual Report* for that year. As virtually the only medium here providing scientific information of this kind, these *Reports* were widely read in American scientific circles. It was a genuine coup therefore for Winlock to enlist the interest of the leading figure in this "new and distinct branch of astronomical science."

To his chagrin, Winlock soon learned that it was one thing to plan for the "new astronomy," but another to obtain the apparatus for it. Although in this country the Clarks were rapidly gaining in reputation and skill, they were as yet unable to compete with the older German and English makers. For his first spectroscope Winlock accordingly turned to that familiar source, Troughton & Simms. The far more costly meridian circle would have to await complicated negotiations about design and enough donors to pay for it. Meantime, he set about reorganizing the Observatory. Since he wanted to apply every penny to the new equipment, he determined to economize on personnel by limiting himself for the time being to one assistant.

Safford had left shortly after Winlock's arrival to become director of the Dearborn Observatory in Chicago, where he had the benefit of the superior 18½-inch Clark telescope that George Bond had lamented losing in 1863. As a result of the disastrous fire of 1871 in that city and subsequent financial losses, Jonathan Young Scammon, the benefactor who had paid Safford's salary, could no longer do so, and Safford's excellent work came to an end. His accomplishments while at Chicago were most fully described by Philip Fox, then director of the Dearborn Observatory, in a paper delivered at the celebration of the semicentennial of the Chicago Astronomical Society in 1912.[13] In 1874 Safford joined the Geodetic Survey in Washington and in 1876 was appointed Field Memorial Professor of Astronomy at Williams College, where the demands of teaching and duties as librarian of the college restricted but did not stop his astronomical research, for he continued to write and publish in various scientific media. He died in 1901.

The choice of young men interested in astronomy or with adequate training in it was limited. Winlock, who was not alone in seeking assistants, found that they usually preferred other departments of science or professions offering greater promise of advancement. A month after

The Harvard College Observatory

he took office, however, he brought to the Observatory Samuel Pierpont Langley,[14] long attracted to the subject but as yet a self-taught amateur. The Corporation appointed him on March 31, 1866, to serve "until the commencement of the next academic year for a compensation of $50 a month." In offering Langley his first astronomical post Winlock started him on his brilliant scientific career as astrophysicist, pioneer in aeronautics, third secretary of the Smithsonian Institution, and founder in 1890 of the Smithsonian Astrophysical Observatory. In 1955 this institution was transferred from Washington to Cambridge, where it is closely associated with the work of the Harvard Observatory.

By the time of his second report, for the period from October 1866 to October 1867, Winlock could offer the Visiting Committee a more cheerful view of conditions and prospects than his first had promised. A fortunate legacy of $20,000 from the will of James Hayward for the unrestricted use of the Observatory provided installation of heat, piped-in water, and other physical improvements. J. Ingersoll Bowditch, with his usual generosity, supplied additions to the meteorological equipment, new books and other publications enriched the library, now partially catalogued, and the second series of zone observations, made originally by Safford and Coolidge under the Bonds' direction, was finally published in the *Annals* (volume 2, part 2). The engravings of the sunspot drawings of William Bond and all of the Bond aurora observations gathered from meteorological records were being prepared for printing. In association with the Coast Survey, Winlock had also made connections with the system of telegraph lines of the country by a loop from the line between Boston and Fitchburg for determining the longitude between the Observatory and the Coast Survey station in Washington. This service he planned to extend to other points.

The report said nothing about the strenuous efforts made all spring to press for prompt delivery of the clock ordered from Frodsham or the spectroscope from Troughton & Simms, but a long series of letters to these dilatory makers shows how impatient Winlock was to get on with the work of the Observatory. At the same time, he urgently appealed to the treasurer and members of the Visiting Committee for definite action on the meridian circle, his "first and most pressing want." With the addition of a few small instruments he promised that nothing afterward would be needed to equip the Observatory "on a scale that would insure it for many years to come a foremost rank in all that

146

pertains to excellence and efficiency of astronomical apparatus." Could a subscription be raised? Could not funds be borrowed in advance of it? To Weld, chairman of the Committee, he wrote:

> We have an equatorial telescope which is not surpassed by any in the world for its peculiar class of work. And the new department of Astronomy that spectrum analysis has founded makes it more important than ever that this splendid instrument should be left free for those subjects for which its great optical excellence and the admirable workmanship of its mounting especially fit it. A meridian circle of high class would be of incalculable advantage in many ways, not only for its own work, but for its aid in the efficient management of the other resources of the observatory. It is so clearly indispensable that I feel sure that one must and will be obtained if not now, at least in a few years. And if so how much better to have it now and save those years.

He knew of several instruments of the kind already in use: at the Naval Observatory, at Albany, at the University of Michigan, one on order for Chicago, and probably one for Cincinnati. "How strange it would be if Harvard College Observatory should be compelled to do without one!"

Meantime, in July, the new clock "of unusual excellence" had arrived, followed in August by the first spectroscope, which Winlock described as "admirably contrived and beautifully made according to directions by Huggins and Professor Miller."[15] He had promptly called in the Clarks to adapt it to the equatorial. To do so had been a labor of extreme difficulty and delicacy, and though every part had been expertly adjusted he candidly admitted that he had not yet succeeded in seeing all the lines Huggins had described.

In the autumn of 1867 the Corporation granted Winlock leave to visit Europe for three months to investigate meridian circles of the latest design. At the expense of the Coast Survey, he was also to visit the principal observatories of England, France, Germany, and Russia to arrange details for final determination of longitude between America and Europe by telegraph signals, or by occultations of stars. Before leaving, he had the satisfaction of seeing a draft of the following pledge, dated November 13, 1867:

> Whereas it is essential for the more perfect prosecution of the work at the Observatory in Cambridge to put up therein a Transit Circle, &

147

The Harvard College Observatory

a few other instruments as proposed by Professor Winlock, we the undersigned agree to endeavour to raise the money necessary for that purpose & to secure that object we severally pledge ourselves to raise or to contribute one thousand dollars each in 1868.

Armed with this assurance, Winlock sailed on the appointed date, leaving Professor Lovering in charge of the Observatory. Visits to the great European establishments brought the warm reception his predecessor had twice enjoyed, considerable information about equipment, and definite arrangements for the cooperation sought by the Coast Survey. With the Struves at Pulkovo he renewed old Harvard ties and established a lasting professional and personal relation of his own. One material result of that meeting was an order for the construction of a portable transit instrument to be built under Struve's direction by the most expert workman there. Before returning in March, Winlock made preliminary negotiations with Simms for the meridian circle, and in May, assured of the sucessful outcome of the subscription, he was told to proceed with the entire apparatus. Its cost when mounted was $13,385, and while somewhat less than that, $12,450, came from subscribers, no one, recalling earlier contributors, will wonder that the list of donors again included the names of such consistent benefactors as Bowditch, Weld, Amory, Sears, Thayer, Barnard, Whitney, Lawrence, besides others newly drawn to support of the institution.

Estimates originally received from Repsold of Hamburg and Simms in England had differed hardly at all in price, but the German maker had refused to promise the circle in less than a year and a quarter and would make no concession in cost unless the object glass came from Merz and Mahler. Simms, on the other hand, blandly claimed to need only six months and agreed to the Clarks' providing the principal glass. This in itself would be a saving to the Observatory and show Winlock's confidence in the growing expertness of the American firm. Eager to put the circle to work as soon as possible, Winlock considered intolerable the wait of the additional months, and in his optimistic, not to say innocent, belief that he could have the apparatus in the short time stipulated, he confirmed the order to the maker who had well served the Observatory in the past. Simms's letter of July 7, 1868, precisely outlined the design of the circle and the terms of the contract to the last detail. It was to cost £970 and to be delivered in six months. Winlock's third report (October 1867–October 1868) therefore re-

148

flected his satisfaction with the present state of affairs and his confidence in the immediate future. The Great Refractor was never in better condition or more conveniently set up for the observers. He pointed with justifiable satisfaction to the improvements he had made and to the results they had brought in observations on multiple stars, asteroids, comets, nebulae, measurements of satellites of Neptune, and so forth. Indeed, he went so far as to say that the conveniences of these mechanical innovations were "unequalled in any observatory in the world." The gift of the meridian circle would bring to the Observatory the most important addition to the apparatus since the mounting of the Great Refractor, for no effort had been spared to obtain the best instrument that modern skill could produce. He confidently expected to have it mounted and in operation before the end of the year. Other equipment ordered would augment the facilities, particularly the transit instrument from Russia, and a chronometer for geodetic determinations. His work for the Coast Survey had given him valuable experience in geodesy, of which he became professor in 1871 in the Lawrence Scientific School.

In spectroscopy Winlock could report progress—so far, observation and study of the spectra of 15 stars and 12 nebulae. He also had three new assistants: Charles S. Peirce,[16] Edward P. Austin,[17] and Arthur Searle,[18] who replaced his brother George when that valued observer, a convert to Catholicism, left to enter the priesthood. Appointed on March 3, 1868, as a temporary member of the staff at $50 a month, Arthur Searle shortly proved himself indispensable to Winlock and to the efficient operation of the Observatory. Unlike the Bonds and Winlock, Searle had not chosen astronomy as a profession, but had drifted into it by chance. Although as an undergraduate he studied astronomy and physics, he then had a greater interest in music (which continued all his life) and in languages than in science. He mastered Latin, Greek, German, and French, and taught himself Italian by reading Dante. After graduation from Harvard in 1856 he taught school for two years, then took a course in chemical analysis at the Lawrence Scientific School. In 1861, with a group of adventurous friends, he sailed to California with the intention of making his fortune as a sheep farmer. When this venture failed he taught for a time at the University of the Pacific at Santa Clara. Called back to Boston by illness in his family, he worked for several years in a brokerage firm.

Although Searle knew as little of astronomy as he had known of sheep farming or stockbroking, he learned quickly. In an autobiographical fragment written in his eightieth year, he recorded that when he began the work he had supposed "that I was entering upon one more out of my numerous temporary occupations. But, as it turned out, I had reached my final station." For seven years he served Winlock with extraordinary ability and devotion, was acting director in the interval between Winlock's death and Pickering's appointment, continued the history of the Observatory from 1855 to 1876 in Volume 8 of the *Annals,* became Phillips Professor of Astronomy in 1887, and from 1891 to 1912 taught astronomy at Radcliffe College as well. Among other topics he wrote on phases of the moon, orbits, and especially the zodiacal light, a subject on which he became a recognized authority. Altogether, more than merely a recorder of two decades of Observatory history, he can justly be said to have contributed materially to fifty-two years of it.

Winlock closed his account of the year 1867–68 on a note echoed in virtually every report of the directors who preceded and followed him—a plea for more endowment that would enable the Observatory to take fullest advantage of the superb equipment provided by "the intelligent liberality for which the citizens of this community are famous."

Since Winlock's annual reports are to this day buried in the obscurity of handwritten notebooks and not one of the projects representing his own research was published in the *Annals* during his lifetime, the usual accounts of his directorship pay tribute primarily to his acquisition of instruments and his mechanical ingenuity in modifying or perfecting them, then pass on quickly to the longer and more brilliant regime of his successor. No one can question the value of his contributions to the efficiency of the Observatory or to the influence of his inventiveness on the subsequent design of instruments made by the most renowned firms. His innovations included a galvanic attachment to a Frodsham chronometer for use with the chronograph in determination of longitude; a simple, inexpensive method of determining an observer's "absolute personal equation by mechanical means" (involving the use of an artificial star, automatic recording of its transit over a given line in the field, and independent recording by the observer); rearrangement and extension of the electrical switchboard orginally installed by George Bond; and the ingenious plan by which he made one instrument serve two purposes, such as his adaptation of the Russian transit to use as a zenith telescope without interfering with its usual operation. In 1870, among other modifications of the spectroscope, he invented an automatic device that enabled him to register any number of lines from a given spectrum and any number of spectra for ready comparison. He is likewise credited with being the first to combine an adjustable plane mirror with a fixed lens of great focal length for photographing the sun. "This idea," he explained in a letter to Simon Newcomb, February 10, 1872, came

to me directly after my experience with the eclipse of 1869. On my return from Kentucky I ordered a lens of 40 feet focus length uncompensated of Messrs Clark & Sons. The first pictures were made with it about January 1870 . . . The idea was original with me & I do not know now of anything of the kind having been done by any one else, & have not seen anything printed on the subject, but I have been told that it was mentioned in Nature by Lockyer last summer after I sent him some negatives. It may also have been described in Henry Morton's paper in the Journal of the Franklin Institute. My object originally in making the experiment was to ascertain if it might be useful in observations of the Transit of Venus. The advantages of this method and the mode of measurement to be employed were all clear to my mind in the latter part of 1869 and I caused a micrometer to be constructed at that time capable of measuring 4 inch photographs [the image of the sun on the plate was 4 inches in diameter] with a view to the use of the instrument on the transit of Venus. I shall give some account of it in my report of the Eclipses. I am very glad to find that you think well of it, and I shall feel obliged to you for any demonstration of its advantages.

For a part of the longitude work on behalf of the Coast Survey in the spring of 1872 Winlock devised apparatus for determining the time of transmission of telegraph signals between Cambridge and San Francisco. The modifications he found desirable for the mounting of the meridian circle increased its stability and counteracted changes in temperature, and his original method of protecting the equipment from dust and injury with glass cases was copied by manufacturers, who made them a standard addition to their products. By 1873, therefore, Winlock's reorganization of the Observatory had led to the stage of efficiency he had been seeking since he became director. In December of that year he read a paper before the Cambridge Scientific Club,[19] an organization founded in Cambridge in 1842, whose earliest members had included many of Harvard's most distinguished professors of science—Asa Gray, Benjamin Peirce, Joseph Lovering, to name a few. After reviewing the state of the Observatory as he had found it and as he had since improved it, he declared, "I have no hesitation in saying that there is no observatory in the world with a finer meridian and equatorial apparatus and none in more complete working order."

He omitted altogether the darker side of the picture, the anguish his two most important acquisitions had cost him. His long correspondence about the agonizing delays in their arrival, or, in the case of a particularly powerful spectroscope ordered from John Browning of London, his failure to get it at all, makes painful reading even today. It also helps explain the somewhat cavalier manner in which his administration is usually treated. For the time allotted to him was short, and, aside from his personal mortification, the disruption of his cherished plans for nearly three years doubtless affected the extent and perhaps the full value of his results. Only a proper understanding of his extreme difficulties in the matter can provide a just estimate of his accomplishments.

From the time of his first negotiations in 1867 until the summer of 1870, when the meridian circle finally reached Cambridge, his letters reveal growing frustration and despair. He had approved all patterns while still in London, had promptly sent detailed diagrams of his modifications of the mounting, had authorized the Clarks to order the glasses, and had revised the arrangements in the West Transit Room, where he expected the circle to be set up within the year. But the six-months' deadline came and went with not the faintest sign from the maker. In January 1869 Simms wrote that the instrument was completed, but five months later it was "nearly finished." Then followed a long succession of appeals from Winlock that went unanswered. In March he begged "for an occasional word" that "would tend to keep up my spirits even if I must wait." In September the arrival of the collimators and micrometer cheered him, but when he inquired about the rest, he received no reply.

My position is becoming extremely embarrassing. I was so wholly unprepared for such extraordinary delay that I scarcely know what explanation to offer for the disarrangement of the plans of the observatory and the disappointment of those who were liberal enough to advance the money.

Joseph Winlock, The Third Director

In October he lamented, "I cannot bring myself to contemplate the loss of another year," and in December, after watching the arrival of every steamer in hopes of some word, he wrote, "You would excuse my frequent letters on the subject if you knew how important to this observatory [it is] that it should have the circle at an early date or at least that we should know definitely what we are to expect." When 1870 rolled around his embarrassment became acute: "I am almost ashamed to meet the friends of the observatory. I have no doubt that this unfortunate delay has cost us many times the money value of the Circle." In February, at the insistence of the chairman of the Visiting Committee, he made another effort to get information. Every time he ventured out, he explained, he was assailed with questions: When did you hear about your circle? When do you expect it? What is the matter? "I am unable to understand," he complained in April,

why you should be willing to do me the injury that you are doing in failing to write to me. Aside from the injury that this Observatory has suffered . . . you are subjecting me to mortification and daily humiliation. When I report your last assurances . . . they are received with derisive laughter.

His attempt to secure spectroscopes of the desired quality and power provides another sad chapter in the chronicle of his frustrations. Since it seemed hopeless to expect even to hear from Simms, Winlock looked elsewhere. In November he ordered from Spencer-Browning in England a direct-vision spectroscope with cylindrical lens for study of star spectra. It arrived five months later but was not what he wanted. He most urgently needed spectroscopes for his forthcoming expedition to observe the eclipse of August 7, 1869. Browning sent three small hand instruments with reasonable promptness, but no direct-vision spectroscope or the colored glass, chemicals, and other material ordered. Nevertheless, Winlock persisted. On June 26, 1869, he wrote Browning to make a spectroscope similar to or if possible even superior to the one Troughton & Simms had constructed for Huggins to investigate the displacement of the lines in the spectra of Sirius and other stars. "Let it be as powerful and as complete in every respect as it can be made, and let me have it as soon as possible." He wanted only the best, would pay in advance if desired, and set no limit on size or price, but time was of the essence. It never came.

The Harvard College Observatory

The English makers were not the sole cause of his troubles. He had also to cope with his own government and the kind of tariff that had so outraged George Bond in 1847 when the 15-inch telescope was on the way. Almost immediately after placing orders in London, Winlock began to take steps to obtain exemption from duty. In June 1867 he wrote directly to the secretary of the Treasury, requesting free admission of a list of instruments he enclosed. Apparently nothing resulted. In May 1868 Winlock suggested that probably the surest way to avoid duty on the meridian circle would be to apply to Congress for a special act of exemption. The appeal, he argued, would be more likely to succeed if made in his own name than by "a rich corporation like Harvard." On June 15 he began his campaign by writing to the Honorable Samuel Hooper, Representative from Massachusetts:

I have purchased in London for this Observatory an Astronomical Instrument, a Meridian Circle, and I expect to receive it in the month of September next. This instrument is a gift . . . from some liberal gentlemen of Boston. The import duties will be about $1000 gold. The object of this communication is to ask your assistance in procuring such legislation as will enable us to import it duty free.

The act of Congress which compels us to pay duty is Sec. 25, Act 30 June 1864, repealing sec. 23 Act 2 March 1861. If this legislation was designed to protect American manufacturers it should not apply in this case, for there is no American maker who would undertake the construction . . . The parts that could be made here have been made by our own opticians, Messrs. Clark & Sons. If the law cannot be modified now so that it should apply to all similar cases, I hope it is not too late in this session to have a bill that shall exempt this instrument.

This letter he followed up on June 27 with an even stronger statement, worth quoting at length because of the light it casts both on the state of manufacturing here and on the attitude of the government toward scientific activity:

The meridian circle which was ordered for this Observatory will cost in London £1300 . . . The purpose for which it is made and the uses to which it is to be applied are precisely those which on account of their importance to navigation and to the public good otherwise have been thought sufficient to justify the establishment and maintenance of observatories at the public cost by every civilized nation. A similar instrument was purchased by our government for the observatory at Washington . . . made in Berlin . . . and mounted in 1865.

Joseph Winlock, The Third Director

Nearly the whole of the great service that the Royal Observatory of Greenwich England has rendered to navigation and to astronomical science has been done by means of such instruments but of less perfect construction.

No instrument maker in this country that I know for certain here who had established a reputation that would entitle him for a moment's consideration in a matter of this kind was prepared to undertake this work, for the reason that the number of such instruments must be so small, that the demand will not justify the cost of preparations necessary for the manufacture and the few who have shown any ability to make such an instrument if sufficient inducement were offered are already so much occupied in supplying the demand for a smaller and more remunerative class of instruments, that they could not afford to make them. I know of but one attempt that has been made in this country to make an instrument of this kind. That was very much smaller than this. It proved a failure and after an attempt to use it, it was condemned, taken down and sold for a very small sum. But I supposed that the purchase abroad of just such an instrument at the public expense for our own government would furnish a sufficient apology for going abroad for ours.

While this subject is under consideration, I beg leave to call your attention for a moment to the law which imposes this burden on colleges. I believe that it will be apparent to any committee that will examine the subject that the revenue from this source is insignificant, that the tax is felt to be a very serious one by scientific men, and that the protection that it may appear to give to American instrument makers is not required, for the reason that all philosophical apparatus for which there is any important demand, all that are required for purposes of education are much better and cheaper in this country than anywhere else, and such instruments would be bought here even if the duty were removed . . . All of our scientific men prefer to patronize our own workmen when they can . . . It is in the purchase of instruments for scientific investigation such as are new or rare that this tax is severely felt, for always when we go to buy, we have just a little less money than we need, so that even a small tax is quite a serious matter . . . This observatory has paid duty during the past year, and expects to pay more in future on instruments which on any of the considerations of public policy which usually control such matters ought to be admitted duty free. And I know that other departments of this College have suffered in a similar manner.

He added that the bill, drawn in the fancied interest of a few makers of instruments and not for the public good, if repealed, would benefit all learning. Besides, he noted, a bill had already been offered and unanimously passed to exempt certain articles imported by American sharpshooters—a form of favoritism not known in recent times!

155

The Harvard College Observatory

Not until June 15, 1870, did this particular struggle end favorably. The $1000 retrieved by the exemption helped make up the deficit between the amount of the original subscription and the actual cost of the circle, but as duties were still imposed on other apparatus, Winlock's problems with the customs officers continued. Even when justifiably aroused, he always maintained a temperate tone, but on one occasion he reached the limit of his patience. In unquestionably the severest letter of his entire correspondence, he wrote to one officer on July 25, 1871:

I rec'd your impudent letter returning my check. I had received no notice except the one enclosed to you. This had been mislaid, and I forgot it until your card came. Then I made haste to send you the amount that I believed correct. Of course I submit and will cheerfully pay whatever penalty may be due to my negligence. But I assure you it is not a pleasant duty to an American citizen to pay a tax like this, with the consciousness that what of it is not stolen outright by thievish collectors goes simply to foster the insolence of petty officials feeding at the public crib.

III

Although Winlock's intense preoccupation with equipment has tended to obscure other aspects of his administration, even a brief summary of his activities will quickly dispel any notion that he valued mechanical efficiency for its own sake. Indeed, he took care to emphasize that he viewed his additions to the apparatus not as arising from an ambitious desire to fill the Observatory with more instruments than he had observers to use them, but as labor-saving machinery. The new meridian, for example, could provide four times as many good observations as the old circle had given imperfect ones, and, in addition to the greater accuracy obtained, it could lessen considerably the work of reduction. His improvements in the electrical connections had also made it possible for the great equatorial, too, to be used with more ease by one observer than it had been originally by two, and by the replacement of the 4-inch equatorial with a new Clark instrument for photographing the sun a complete drawing of it became for the first time obtainable.

No better witness of his genuinely scientific interests can be found than Arthur Searle, who wrote:

With the increase of the income of the Observatory and of its expenditure upon the instruments, repairs, and publications [though of others' work], came a largely increased call upon its officers for attention to matters of business and administrative details of all kinds . . . The means of the Observatory, however, were inadequate to provide for increasing its equipment, for the fulfilment of its new duties, and at the same time for the engagement of a corps of observers and computers, to be occupied exclusively with scientific work. Under these circumstances, it could not have been surprising if this strictly scientific work had been to a great extent postponed until a time when the new organization of the Observatory should be completed. But it will appear from the following account, as well as from future volumes of the *Annals,* that scientific researches were actively pushed during the whole of Professor Winlock's term of office, although some of them were necessarily left incomplete, owing to the pressing demands of other work.[20]

In that "following account" Searle listed the major activities initiated or directed by Winlock: observations of binary stars (until 1872), in which more than 5000 measurements were made in five years and 250 new stars discovered; special lunar investigations by Nathaniel Shaler, begun in 1871 and reported on in 1872; (unsuccessful) searches for new planets; determinations of positions of comets and asteroids; magnetic and meteorological observations; longitude determinations in cooperation with the Coast Survey, including those made at various times between Cambridge and such points as Washington, Albany, Omaha, Salt Lake City, San Francisco, Allegheny, Ann Arbor, Hanover, as well as a new series for redetermining differences of longitude between this country and Europe. Other important projects included star catalogues in collaboration with the Astronomische Gesellschaft; expansion of time services to cities and railroads; intensive observations of the sun (especially of spots and prominences), resulting in a series of handsome engravings sold by subscription; photometric, photographic, and spectroscopic researches; and the organization and direction of two expeditions to observe total solar eclipses.

When the long-awaited meridian circle arrived in the summer of 1870 Winlock was fully prepared to put it to immediate use on a cooperative program involving most of the important observatories of the world, and in his view "the greatest astronomical undertaking of modern times." In 1865 at a meeting in Leipzig the Astronomische Gesellschaft had proposed a revision of Argelander's great star catalogue, the *Durchmusterung,* by redetermining and cataloguing the positions of all stars in the northern skies down to the ninth magnitude and within 10° of the pole. Little happened on the new proposal until 1867 when at a second session the Association revised its earlier plan by setting the limits between $-2°$ and $+80°$ down to stars of the ninth magnitude and outlined definite procedures for putting the program into practice. Representatives of leading European observatories at once volunteered their collaboration, but by the time Winlock was equipped to take part all but one region had been assigned to some thirteen observatories, among them those at Kazan, Dorpat, Christiania, Leipzig, Berlin, Bonn, and Cambridge (England).

Winlock offered to undertake a region south of the equator, a subject of great interest to him, especially since it would offer a fresh field. Only the northern zone 50°–55°, however, remained, and in November 1870 he placed the observations in charge of William Rogers,[21] Edward Austin, and Arthur Searle. After 1871 the chief responsibility became Rogers'. As already noted, he was no stranger to this Observatory. Having won the approval of George Bond during his apprenticeship in 1863, Rogers returned to Harvard in 1870 after a period of teaching, just in time to take over the principal meridian observations. The results of his arduous labors occupy seven volumes of the *Annals,* the first of which (Volume 10) appeared in 1877 with a preface by Pickering. In 1873 Winlock wrote on the progress of the work: "Although we were the last in the field I think that no zone is farther along now than ours." With characteristic efficiency he had made a precise plan for the project to conserve both the time and the energies of his small staff:

> Part of the programme on which we agreed to work was that no observer should observe more than two hours at a time on his zone; so that there was time to spare, and besides as the stars get thinned out something else can be done while waiting for them to come along.

Observations in themselves, as Winlock was the first to admit, were but a comparatively small and easy part of the work of an observatory, whereas the clerical work and computation were of such crucial importance that few establishments, especially those of limited means, could hope to keep up with the number of observations the excellent modern instruments permitted. Exploring the subject in 1873, he found a serious lag everywhere between observation and publication, even in such well-established institutions as Greenwich and Oxford, where at least three years elapsed between the two operations. The most recent publications, dated 1870, for example, described observations for these years: Cambridge (England) 1850, Pulkovo 1845–46, Cape of Good Hope 1862, Paris 1867, Brussels and Melbourne 1868, Konigsberg 1865, and in this country the last Washington volume was "filled up with old observations of the year 1845." Obviously, some new method would have to be devised. Winlock very early saw the need for a central computing bureau to provide more rapid and uniform astronomical information, but he derived no benefit from the idea. The problem, as always, was money for inexpensive clerical assistance.

As matters stood, staff members, whose time was worth more, had to do both observation and computation, a situation Winlock meant to remedy by finding some means to supplement his income. The most promising way had been shown by Samuel Pierpont Langley. In 1869, faced at Allegheny with an observatory in name only, Langley circulated a pamphlet offering to provide at set fees a service "for the regulating from this Observatory, the clocks of the Pennsylvania Central and other railroads associated with it." Within a short time his venture proved such a boon that other poorly provided institutions seized upon the idea. For Winlock, to whom Langley regularly communicated his plans and progress, New England was a wide-open field.

Furnishing time was in fact not entirely new to this Observatory. Under the Bonds it had first hired a telegraph line by which clocks in Boston could be regulated from the Cambridge signals, and between 1856 and 1862 it owned its own line, bought from the old Northern Telegraph Company. The service was entirely free, but in November 1871 Winlock proposed changing that philanthropic arrangement. He needed money for at least one assistant, he wrote Bowditch, and he knew of but one way to get it: "The Allegheny Observatory near Pittsburgh secured one thousand dollars from the Pennsylvania Central Road and $500 from dwellers of Pittsburgh, making an annual income of $1500." Langley expected even more by extending his arrangement to additional roads, and other observatories at Ann Arbor, Chicago, and Albany were receiving compensation for similar service. If Boston would put up a wire from the Observatory he would keep a clock in the city hall within a quarter of a second of the true time. Incidental expenses would be negligible.

In December Winlock informed Langley that he had as yet made no definite arrangements and in any case preferred limiting his services to Boston and allowing others to distribute the time to the railroads. He changed his mind, however, when he found that the Western Union would connect the Observatory with its office in State Street, furnish all batteries, maintain the line, and give the Observatory exclusive rights, for the sum of $250 a year. If that initial sum were provided by the Visiting Committee, he proposed setting up a clock immediately in the Merchants Exchange "or some other conspicuous spot so that the public can see and learn to appreciate the method of communicating time."

We who now glance casually at wrist watches or at public clocks, and can readily dial by telephone day or night to check the time, find it a strain on the imagination to appreciate the meaning of such an innovation in the 1870's. If it seems equally strange that a service of such obvious necessity should have required many weeks of intensive effort to secure subscribers, doubtless the novelty of the scheme and its expense prevented instant acceptance.

The influence of the devoted Bowditch paved the way to railroad presidents, whom Winlock personally solicited, along with hotel managers, jewelry firms, and the city fathers of Boston, who were reluctant at first to pay for something they had always received free. A committee of the Board of Aldermen finally recommended a small appropriation, though Winlock remarked ruefully that Boston should act at least as liberally as Chicago, which was paying the Dearborn Observatory there $1000 a year. He argued that the service would prove its value to the public and could be indefinitely extended, and the returns, instead of going into the pockets of private individuals or companies, would support a scientific institution that had long been useful to the community and the national government. By July 1872 fifteen subscribers had joined the program. By 1875 the increased line connections permitted dispensing with the lines rented from Western Union, and the average annual income, though not entirely clear profit, approached $2400.

Winlock's program, actually in effective operation only a short time after his first proposal, proved of enormous benefit. It unified the times of all local railroads entering Boston, provided jewelers with ready means for testing the accuracy of their watches, added considerably to the convenience of the public, and earned for his scientific work so desirable a supplement to his annual budget that in 1877, under his successor, the Observatory issued a circular offering even wider extension of the service: "The system has been so well received by the public of Boston, and by the leading Railroads entering therein," the announcement read, "that it is now proposed to afford, as far as the Observatory is able to do so, the same benefits to the other cities of New England as are enjoyed by Boston in this respect." The income from this source reached about $3000 a year until 1890, when altogether new, competitive arrangements for furnishing time made it no longer feasible or necessary for the Observatory to devote staff or funds to it, but for more than twenty years, at relatively minor cost to cities, railroads, and business firms, it had provided an essential public service not otherwise available.

IV

Winlock took an active part in two eclipse expeditions, the first in 1869 when the Coast Survey invited him to organize and lead a party to observe the total eclipse of August 7. His well-known interest in the "new astronomy," his mechanical expertness, and his administrative experience made him an ideal choice. Even though he was occupied with all the internal problems of the Observatory, he readily accepted, not only because of his wish to maintain the established tradition of collaboration with government and other astronomers, but because he was eager to explore the possibilities of the spectroscope.

Only two previous total eclipses had been visible in this country during the century. The first, in 1806, had been observed at Salem, Massachusetts, by Nathaniel Bowditch and had drawn young William Bond to astronomy. In 1834 Robert Treat Paine, an indefatigable follower of eclipses until the last year of his long life, had journeyed to South Carolina to observe the second. The eclipse of 1851, described with such vividness by George Bond at Lilla Edet, Sweden, was only partially viewed here by William Bond and his friends, but it had brought little more than some daguerreotypes by Whipple. The forthcoming event, which would be observable over an enormous area on this continent, would be the first for which the country was properly equipped with instruments and experienced astronomers to use them.

Since an eclipse expedition at any period is a major undertaking that requires vast preparation and in 1869 presented specially great problems of transportation, communication, proper equipment, and expense, astronomers began early to seek every means of pushing the enterprise.

The Harvard College Observatory

In January they started their campaign for a government subsidy. The American Academy of Arts and Sciences in Boston joined with the American Philosophical Society in Philadelphia to endorse the project, but their committee's request for an appropriation seemed to the astronomers most concerned too weak in facts and in patriotic sentiments to persuade a perhaps reluctant Congress. Winlock was therefore asked to expand the arguments and to emphasize the great importance to American science and prestige of a properly supported program. The stronger case made for the expedition reads in part:

In 1834 the United States held a low rank among the Nations in respect to the Exact Sciences . . . In 1869 all is changed. The zeal awakened by the Great Comet of 1843 produced the Cambridge, Cincinnati, the National, the Ann Arbor, and the Chicago observatories, to mention no others. The organization of the Coast Survey [and] the establishment . . . of the American Ephemeris and Nautical Almanac . . . also contributed to bring about the change in our scientific position, and the Astronomy of America may today stand proudly by the side of that of any nation in Europe. We must not allow her to forfeit this high position.

The English Government last year [1868] sent an Expedition to India to observe the Total Eclipse visible there last August, and France and other countries vied with England and France in zeal to observe it.

On account of the unfavorable state of the weather, accidents to instruments, or perhaps incomplete preparations for observing new phenomena, the results expected from that eclipse were somewhat disappointing.

At the same time it established some most important new facts concerning the physical Constitution of the Sun, and opened a new line of observation that has been followed up in Europe and this country since, so that American Astronomers can avail themselves with great advantage of the Eclipse of this year, which comes so opportunely to complete the Observations of the Phenomena . . . only begun last year.

The path of the eclipse of August, the statement pointed out, would traverse the newly acquired territory of Alaska, the territories of Montana and Dakota, the states of Iowa, Illinois, Indiana, Kentucky, a corner of Virginia and Tennessee, and all of North Carolina, an unbroken stretch of 2000 miles. The scientific world of Europe, where

162

the eclipse would not be visible, would be watching and expecting great results, and this country must do everything that modern knowledge, ingenuity, and enterprise could achieve if we were not to be disgraced. Astronomers and other men of science here would not stint either personal means or effort, but this government, which already had done much to bring the astronomy of the country to its present honorable position, should not allow them to "meet unaided so great an exigence as the present." Among the many phenomena to be observed, even during a brief total phase of only two or three minutes, the list included "movements of the heavenly bodies, their contour, color, spectroscopic changes, gaseous clouds about the sun, meteorological, magnetic, and optical aspects of the earth"—all of which should help establish "truths of the highest scientific value and of large technological uses."

Finally, the appeal took care to remind Congress that there was a precedent for such a venture—in 1860 the Coast Survey under government auspices had sent an expedition to Labrador to observe the total eclipse of that year. Now, with the introduction of the spectroscope, much more could be accomplished. Observers could far surpass the "imperfect results" of European efforts in 1868, and Americans would have "the honor of first applying this marvellous means of discovery to the phenomena of a Total Eclipse with complete success."

If the petition exaggerated American accomplishments in astronomy, erred in attributing to the comet of 1843 the origin of the Harvard and Cincinnati observatories, and underestimated some of the results of 1868, it nevertheless provided a review of previous total eclipses. The expedition of 1860 had coincided almost exactly with the beginning of spectrum analysis and had opened up the whole question of the nature and behavior of the solar prominences. On that occasion Father Secchi and Warren de la Rue had taken the first photographs of an eclipse, study of which settled the fact that these "pillars of flame" were solar, not lunar, phenomena. The eclipse of 1868 had been visible only within a short range—in India, the Malay Peninsula, and Siam—and, although the weather was not entirely favorable, the results were not so imperfect as the petition represented. The spectroscope had triumphed. Applied for the first time to the prominences, it showed, Charles A. Young declared, "the strange objects which had caused so much speculation, to be great clouds or flames of glowing gases, among which hydrogen is especially conspicuous."[22] (The day after the eclipse, the French astron-

The Harvard College Observatory

omer Janssen, experimenting with the dispersive power of the spectroscope, found that observation of the prominences need not depend on eclipses. Somewhat later, with a powerful new spectroscope, Lockyer independently made the same discovery.)

Congress responded favorably to the petition drawn up by Winlock, with the result that the country became one vast observatory. Every astronomer, whether sponsored or not, planned to go to some favorable site. "The observatories," says one account, "must have been left undirected; the mathematical chairs of the colleges must have been empty," and, although private homes offered hospitality to visitors, "judging from the crowded condition of hotels . . . Saratoga and Newport must have felt the different set of the travelling current."

The most elaborately outfitted expeditions went from Washington, one under the auspices of the Nautical Almanac Office (transferred in 1866 from Cambridge), the other from the Naval Observatory in charge of Commodore B. F. Sands. The first group chose three stations in Iowa, where observers at Burlington included, among others, Charles A. Young for spectroscopy and Benjamin Apthorp Gould, who was to attend to photography.

At Mount Pleasant, Iowa, accompanied by a group of professors that included E. C. Pickering, Professor Henry Morton of Philadelphia took over-all charge of photography for the Nautical Almanac contingent. At Des Moines in one Naval Observatory party Simon Newcomb observed the corona and searched for a possible intra-Mercurial planet,[23] while Harkness attended to spectroscopy, and others concentrated on photography. A second Naval Observatory group stationed itself at Bristol, Tennessee, where several observers were especially concerned with the corona. In fact, since the eclipse of the year before had produced almost nothing of value for the study of that phenomenon, it became a major preoccupation of many astronomers in 1869. None was more eager to investigate it than Winlock.

For his own party he selected his native state of Kentucky, to which he journeyed a few weeks ahead in order to arrange for use of his old college grounds at Shelbyville and of the excellent telescope, which had been mounted originally under his supervision when he was a professor at Shelby, and which he had later borrowed for use in Cambridge at Gould's observatory, known as Cloverdon. Privately owned by 1869, it was later sold to the University of Missouri and is said to be the first instrument ever used by Harlow Shapley, who in 1921 became the fifth director of the Harvard Observatory.[24]

Winlock obtained from the Harvard Corporation $500 to pay for assistance and saved all the expeditions a substantial sum by initiating the drive for free transportation of both observers and apparatus. His letter to Cornelius Vanderbilt, president of the New York Central Railroad, a man not notably distinguished for love of the public, supplied the model for similar requests to other railroad executives. In asking for passes, Winlock explained that his group of ten from the Harvard Observatory was

> not a party of pleasure or business but one trying to do with all the means that it can command the work that must be done for the credit of the country.
>
> Under other governments, such expeditions are fitted out and sent out at the public expense; but here we must call upon liberal public spirited men for help in enterprises where the slender purses of scientific men and the limited endowment of learned institutions, unaided, are inadquate.
>
> The value of the apparatus contributed for the party and the cost of adapting it for the present purpose will not be less than six thousand dollars and the services of all of the observers are given gratuitously, so that I hope the effort to save the cost of getting to the place of observation will not appear to you an improper one.

To what extent Mr. Vanderbilt was, in Bowditch's phrase, "melted" by Winlock's letter, does not appear, though officers of the Chicago, Burlington, and Quincy and the Burlington and Missouri readily agreed to provide passes, and Thomas Scott of the Pennsylvania even furnished a private car.

At Bardstown, Kentucky, Winlock stationed Charles S. Peirce and Nathaniel S. Shaler, equipped with a 5-foot equatorial and two spectroscopes to observe the spectrum of the prominences; at Falmouth, nearby, Arthur Searle observed especially for phases of the eclipse, and Langley, with instruments from his own Allegheny Observatory, observed at Oakland. Others in the group included George and Alvan Clark, Jr., and John Adams Whipple, the daguerreotypist who had pioneered in astronomical photography with George Bond. Winlock's notes on the expedition outline his primary purpose and some of his results:

> Before 1869 the prominences had been proved to be solar appendages containing incandescent hydrogen, but their other constituents were uncer-

tain; and everything was still uncertain with regard to the corona. Accordingly, I decided, in 1869, that to observe the spectrum of the prominences and to obtain a good photograph of the corona, were the principal objects to which my efforts should be directed. With these I endeavored to combine such other observations, both photographic and direct, as could be made with the means at my disposal; and a large number of such subsidiary observations were in fact obtained, without in any way interfering with the completeness of those which I regarded as of the first importance.[25]

Winlock's study of the spectrum of the prominences at Shelbyville established beyond a doubt that magnesium was a constituent of them, an observation later confirmed by other investigators.

Having planned to make photography an essential part of his program, during the spring preceding the expedition he had experimented almost daily with photographs of the sun. As a result, under his direct supervision and with the special arrangement of the apparatus he had devised, his photographers secured eighty-six photographs. One of these (Fig. 30), taken with a 40-second exposure, Winlock considered the best image of the corona yet produced anywhere—an opinion that Young later confirmed: "The corona," he wrote, "was for the first time satisfactorily photographed . . . by Professor Winlock's party . . . Mr. Whipple obtained a picture which has hardly been surpassed even by more recent observers."[26] And in another place Young was even more emphatic: "Photography has secured a most beautiful record of the inner corona in the remarkable picture obtained by Professor Winlock's party of the eclipse of 1869. In this respect, neither of the pictures of the eclipse of 1870 can compare with it."[27] This photograph, with others of eclipses, is reproduced in the *Annals,* Volume 8.

Young properly credited "Winlock's party" with the accomplishment of that admirable photograph. In 1872, however, Winlock was considerably irked when he found that the German writer H. Schellen had given entire credit to Whipple. In the English translation of Schellen's work,[28] describing the results of the eclipses of 1869 and 1870, Winlock read:

J. A. Whipple, of Boston . . . arranged his telescope for photographing the corona at Shelbyville in such a manner that the prepared plate was placed in the main focus of the object-glass, and he employed forty seconds as the time of exposure. In this way a picture was obtained in which the prominences appeared only as bright spots, while the inner ring of

Fig. 30. Solar corona at the eclipse of August 7, 1869; engraving made from a photograph taken at Shelbyville, Kentucky, by Winlock, Whipple, and assistants.

light, as well as the outline of the whole corona, and the peculiar curve of its rays, are clearly shown . . . J. A. Whipple, with Professor Winlock and several assistants, procured at Shelbyville eighty photographic pictures, six of which were taken during the totality, one of them exhibiting a complete and magnificent corona.

The Harvard College Observatory

Winlock refused to let these statements go unchallenged. He wrote Schellen:

I observe that in this as in your other publications you do not seem to be aware that the Eclipse Expeditions to Shelbyville and to Jerez de la Frontera were organized by me, and were under my immediate direction. It is hardly fair to give the whole credit of the photographs of the corona to Mr. Whipple and Mr. Williams who had nothing to do with the astronomical part of the problem, but whose duties were confined simply to the preparations of the sensitive plates and their exposure under my directions. Mr. Whipple himself did nothing but assist Mr. Clark in exposing the Plate during totality. The actual photographic work was done by Mr. Pendergast and Mr. Williams. The instrument used belonged to this observatory, it was fitted for the purpose by me & was adjusted by me, and all the arrangements for accomplishing the object in view, that is, to get pictures of the corona were made by me.

Winlock's second expedition took him, again at the request of the Coast Survey, to Jerez de la Frontera, Spain, to observe the so-called Mediterranean eclipse of December 22, 1870. For this venture Congress had readily appropriated $29,000. He had hoped to include in his party some of the leading astronomers who had observed the eclipse of 1869, among them Henry Morton of Philadelphia, Langley, Harkness and Newcomb of the Naval Observatory, and Charles Young. Langley accepted at once, but two refusals cast an interesting light on the prevailing rivalry between the Naval Observatory and the Coast Survey. Harkness flatly declined to participate in what he called a "Coast Survey expedition," despite Winlock's insistence that it was not so much a Coast Survey expedition as an eclipse expedition, and that any results obtained would redound to the credit of every institution represented. Newcomb also had therefore regretfully to decline. He had already received orders from the superintendent of the Naval Observatory to proceed to Europe, and any observations made there must be reported exclusively to that establishment.

Winlock sailed from New York on November 3, accompanied by two assistants, Henry Gannett and Charles S. Peirce. Of the several stations in Spain, Winlock's alone would concentrate on photography. Charles Peirce he sent to Italy, but at Jerez he was pleased to be joined by his two favorite colleagues, Young for spectroscopic observations and Langley, who had sailed with him, for telescopic study of the structure

168

Joseph Winlock, The Third Director

of the corona. Others in his party were Edward Pickering, Alvan Clark, Jr. (still on the trail of that mythical planet!), and G. W. Dean of the Coast Survey. The larger number of instruments sent this time from the Observatory included two telescopes especially prepared by Winlock for photography. For his photographer he had chosen O. H. Willard of Philadelphia, to whom he had earlier sent directions:

> I have decided that the pictures taken during total phase shall not be enlarged. They are to be taken in the focus of the large object glass. Several enlarged pictures will be made before or after the eclipse to fix the position of the spots. You will understand that all pictures of the eclipse, or of the instruments or of anything else that may be required for our report of the expedition will be the property of the Coast Survey under the control of the Superintendent.

Arrangements moved with remarkable smoothness everywhere. Local officials and inhabitants at Jerez were friendly and hospitable, and all but the weather seemed favorable. Luckily, a few seconds before totality the threatening clouds about the sun suddenly opened to allow, if not perfect observation, at least enough to save the expedition from complete loss. The Coast Survey published Winlock's formal reports in 1869 and 1870, but his hope of incorporating a special description of the Observatory's part in the two expeditions in a volume of the *Annals* proved vain, and all we have of his more personal account and estimate of results appears in excerpts from his notes quoted by Arthur Searle in Volume 8 (pp. 57–63). After reviewing the major observations of previous eclipses, and especially those of 1869, Winlock wrote:

> It seemed to me that my attention would be most usefully directed in 1870 to the study of the corona, and especially of its outer portions. The constant improvement in the spectroscope and in its use, which went on between the two eclipses, made it evident that the brighter portions of the envelopes of the sun would soon be studied satisfactorily at all times without the necessity of waiting for an eclipse of the sun. Besides, I felt confident that other observers . . . would pay sufficient attention to the chromosphere, prominences, and inner portions of the corona; while there was some reason to fear that that outer and fainter light, which perhaps may never be capable of being well observed except during an eclipse, would receive comparatively little attention.

At Jerez therefore he attempted to learn as much as possible from his own spectroscopic observations about the outer portions of the corona

The Harvard College Observatory

and to try for good photographs of all the light in the sun's neighborhood during the total phase. For this purpose he had prepared two special telescopes, but, as often happens on the most elaborately fitted-out eclipse expeditions, the weather disappointed his hopes. Still, he managed to achieve one good photograph taken with a telescope of 6-inch aperture, mounted as an equatorial, with "lens corrected for the chemical rays." Comparing that image with photographs taken the previous year, he found

that the same general figure was represented in each. It could no longer be reasonably doubted that the coronal light was not, as it had previously been depicted, equally distributed in a radiating form about the sun, but that it had, roughly speaking, a quadrangular shape, extending farthest from the sun at four points between the solar poles and equator, while its extent was particularly small about the poles. Without considering any other fact exhibited by the photographs, this alone makes it out of the question to suppose the corona, or that part of it at all events which the photographs represent, as any thing but an appendage of the sun. This is the first decisive result obtained upon this subject, since the mere estimates of observers in previous eclipses, that the corona was concentric with the sun rather than with the moon, afford at most only a presumption of the fact.

Thanks to his ingenious device for automatically registering the spectrum lines, Winlock successfully determined the positions of four bright lines in the light of the outer corona, "which was thus shown with much probability to be a solar appendage at least as far as 25′ from the sun." In a brief summary of his conclusions, he wrote:

We have then, as results actually attained by the resources of this Observatory applied to eclipse observations in 1869 and 1870, first, additional knowledge, since fully verified, of the structure of the prominences; and secondly, proof of a peculiar and previously unknown form of the corona as a permanent phenomenon; besides additional knowledge of the elements constituting the corona . . . [and] that the coronal light is of two distinct kinds.

Winlock's experience with the two eclipse expeditions encouraged him to carry on vigorously his efforts to master the art of solar photography. Each day, as long as he could find the means, he experimented with photographs of the sun. Although far from a prolific writer—only four papers[29] on spectroscopy bear his name in the bibliography of that subject—he had in mind the preparation of a treatise on solar physics based on the examination and measurement of his photographs. It was to be similar in purpose, at least, to one published in England by Warren de la Rue, Balfour Stewart, and Benjamin Loewy[30] analyzing photographs they had made at Kew, and acknowledged by them, so Winlock stated, to be inferior to his own. In order to continue photography, however, he would have to apply for aid. In December 1871 he turned to the Bache Fund. This fund, as described by its chief trustee, Joseph Henry,[31] was left in trust to the National Academy of Sciences, the income to be devoted to "the prosecution of researches in physical and natural science by assisting experimentalists and observers, and the publication of the results of their investigations."

Henry replied promptly that $800 had been granted for "continuing daily photographs of the Sun and investigations in solar physics connected with them," but he took care at the same time to caution Winlock that any results of the observations must be published in a way "to reflect credit upon the name of Alexander Dallas Bache."

With that initial subsidy, though it was not the whole sum that would be needed, Winlock constructed a powerful new spectroscope and embarked on an uninterrupted program of solar observations and photography. He soon decided, however, that instead of preparing a highly specialized essay that would appeal only to a few astronomers he would combine the Bache grant with the income from the Sturgis Fund and offer to subscribers a series of engravings that would unite "scientific value with the interest of the general public." He thereupon sent out a descriptive circular of his plan. He proposed to issue at least thirty plates of the principal planets, moon craters, sunspots, solar prominences, nebulae, and spectra of variable stars—representations, he wrote, "as nearly as possible of the principal objects in the Heavens as they are seen with the powerful instruments of the Observatory." Each set would cost $10; each plate would be delivered immediately when ready, and upon completion of the series full notes and accurate scientific explanations of the astronomical importance of each engraving would follow. All art work was to be put in the hands of Léopold Trouvelot, who, Winlock rightly said, "combined in rare degree the qualities of an excellent observer with the skill of an accomplished artist."

Winlock's persuasive circular attracted over 200 subscribers, a gratifying response, but the expense of reproducing the engravings amounted to the unexpected figure of $6500. Less than a third of this sum would come from subscriptions. The deficit Winlock confidently supposed he could make up with the Sturgis income and an additional grant from the Bache Fund. Unfortunately, his plan of pooling the two struck an unexpected snag in Washington. When he applied for another $1200, Henry asked how he meant to acknowledge credit to Bache. In Henry's view, any scientific results from a Bache subsidy must be separately published and exclusively attributed to it. By the time Winlock heard of this restriction it was too late to make a separate accounting of the expeditures from each source, and a long and unhappy correspondence ensued that left the situation in doubt. In fact, the problem remained an unpleasant legacy for Searle, as acting director, to resolve.[32]

On June 4, 1875, in the ninth year of his administration, Winlock suddenly fell ill, and, although his condition had not seemed alarming, he died on June 11 at the age of forty-nine. He had brought the Observatory to an admirable state of efficiency, not only by his mechanical ingenuity but by his excellent judgment of men. Though he had moved cautiously in adding personnel to the Observatory on his always limited budget, he can be charged with not a single error in his choice of assistants. Some for various reasons chose not to undergo a long apprenticeship in a science that offered too little prospect even of a bare living, but a list of associates that includes Langley, the two Searles, W. A. Rogers, O. C. Wendell, Henry Gannett, Nathaniel Shaler, Léopold Trouvelot, and Charles S. Peirce is itself a tribute to Winlock's discrimination and his standard of performance for the Observatory. In particular he recognized the genius of Peirce, turned over to him the new researches in photometry, praised his work, left him free to carry it on without interference, and helped lay the foundation of that branch of astronomy for which Harvard, as Chapter IV will show, very shortly became distinguished.

Arthur Searle (Fig. 31), who had been most closely associated with Winlock, wrote of him:

His thorough and extensive knowledge of the sciences of mathematics and astronomy commanded the respect of all who became acquainted with him; and in those who knew him best, this respect was attended by the strong personal attachment which his modest, retiring, and amiable char-

Fig. 31. Arthur Searle, acting director of the Observatory, 1875–1877. (Courtesy of Miss Margaret Harwood.)

acter, enlivened, as it was, by ready and unaffected humor, was certain to inspire. Few scientific men have been more generally regarded with kindly feelings by others engaged at the same time in similar pursuits, or more beloved by their familiar associates and friends.[33]

Winlock had not won the appointment easily, but in the end he fully vindicated the judgment of those who had insisted on confirming him,

The Harvard College Observatory

and it is unlikely that any surviving member of the committee who had opposed him would have disagreed with the Overseers' appreciation of his administration at their meeting on January 26, 1876:

During the 9 years of his Directorship he gave himself to the Observatory without stint or reserve. By his ability as an astronomer and his thorough and conscientious application of it to his works, he maintained and increased the reputation of the Observatory; and the value of his labor is fully recognized by those competent to judge them. He added largely to the working power of the Institution by his ingenuity in perfecting its instruments. He added largely to its material resources by the confidence with which he inspired his friends—and more directly by the considerable addition which he made to its income in the compensation rec'd for furnishing the exact time to the City of Boston and to the railroads issuing from it. The integrity, the modesty, and the singleness of purpose with which, careless of applause, he sought only the performance of duty and the achievements of well directed toil deserve the deepest respect.

At his unexpected death Winlock left incomplete a rather considerable body of work. Except for a few papers that he had sent to several journals and his series of engravings which, though useful as a pictorial record of the skies in a period of imperfect astronomical photography, fell short of his hopes, Winlock saw nothing of his own in print. Fortunately, under the supervision of his successor, the principal results of research that Winlock undertook fill or form a part of a number of volumes of the *Annals*. The most substantial project, observations and catalogues of the zones assigned to Harvard for the revision of the *Durchmusterung*, were finally completed more than two decades later by W. A. Rogers. Volume 10, results for the years 1871 and 1872, appeared in 1877, followed at various intervals by Volumes 12, 15, 16, 25, 35, and 36.

Photometric Researches, Volume 9 (1878), contained the long-delayed work of Charles S. Peirce, to whom Winlock had entrusted the entire project. Volume 13, *Micrometric Measurements*, though in part involving later investigations, chiefly comprised those directed by Winlock on observation of double stars, nebulae, occultations, satellites of Saturn, Uranus, and Neptune (fully prepared for publication but never printed and here given substantially as Pickering found the manuscript). Other Winlock research appears in Volume 19 (meteorological observations from 1840 to 1888, and some observations of the aurora,

174

Joseph Winlock, The Third Director

1866–1874). Even as late as 1908, Volume 60, *Miscellaneous Researches*, recorded Winlock's discoveries of more than a dozen nebulae. If these investigations, completed with the care that Winlock gave to all he undertook, had been published in his lifetime, they would doubtless have revealed a far more active and varied program of scientific activity than is usually credited to his administration. Despite the difficulties under which he began his directorship and the delays and insufficient support he encountered, he left the Observatory more efficiently equipped and, by his further experiments in photography, promotion of photometry, and early efforts in spectroscopy, better prepared for his successor's more spectacular accomplishments in sidereal astronomy.

IV

E. C. PICKERING:

THE EARLY YEARS

I

After Winlock's death more than a year elapsed before the appointment of a new director. During the interim Arthur Searle took charge of the Observatory and its small staff. Charles W. Eliot, who had become president of Harvard in 1869, had been trying to expand the role of science in the University and felt a strong interest in the Observatory as a research institution. He did not find it easy to select a man both qualified and willing to take the place that the two Bonds and Winlock had so ably filled. Friends of Charles S. Peirce, who had done some research under Winlock's direction, attempted to obtain the post for him but they were unsuccessful, perhaps because of an antagonism existing between him and Eliot, perhaps because Peirce himself was content with his work for the Coast Survey and did not seek the appointment. In writing to his friend William James (November 21, 1875), he remarked that the directorship of the Observatory was

of all situations I know of the one which has the most thankless, utterly mechanical drudgery, together with vexatious interference from two different sources, certain members of the committee and the president. I speak of *Directorship* of it, for my own connection with it was most delightful. Winlock was charming. I am not quite through with it yet because my book of photometric researches isn't out . . . But don't let me speak as if I did not feel the warmest gratitude to you and my friends who wanted to get me into the observatory. I don't know that I would have declined it even although it does not seem to me altogether desirable.[1]

Simon Newcomb, then professor of astronomy at the Naval Observatory in Washington, also found the position unattractive and declined

176

E. C. Pickering: The Early Years

Eliot's offer of the directorship. Newcomb later recorded in his memoirs that the Observatory at that time

> was poor in means, meagre in instrumental outfit, and wanting in working assistants; I think the latter did not number more than three or four, with perhaps a few other temporary employees. There seemed little prospect of doing much.[2]

He might correctly have said the same for nearly all the thirty-odd public and private observatories existing in the United States before 1880. None of them had enough financial support or a large enough staff to operate solely as a research institution. In European countries astronomy was usually supported by the state. In America, however, only the Naval Observatory was so fortunate; astronomy was generally regarded not as a science in its own right but chiefly as a tool to be used in solving practical problems—the determination of latitude and longitude, geographical boundaries, accurate time, and the peculiarities of weather and climate.

Only the Harvard Observatory and the Naval Observatory in Washington could reasonably be regarded as the equals of their counterparts in England, Germany, and Russia. A dozen or so colleges in the East and the Middle West owned a telescope but the instrument was usually employed only to teach the elements of astronomy to a few students, display striking celestial objects, and determine the correct local time. The Dudley Observatory at Albany, long disturbed by disagreements between its first director, B. A. Gould, and its board of trustees, was just resuming work after several years of inactivity.[3] A few observatories—the Litchfield at Clinton, New York; the Shattuck at Hanover, New Hampshire; the Allegheny at Pittsburgh, Pennsylvania; the Detroit at Ann Arbor, Michigan; the Dearborn at Chicago; and the Cincinnati Observatory—were producing some good work but they were all hampered by lack of money to pay for equipment, assistants, and the cost of publication. In the West, only private observatories existed, such as that of Professor George Davidson of San Francisco. James Lick, an eccentric Californian who had acquired a fortune from land investments and wished to memorialize his name with some striking monument (he had finally been dissuaded from his first idea: building a pyramid larger than that of Cheops), at his death in October 1876 left a bequest of

$700,000 to construct the most powerful telescope in the world, but as yet the project had not reached the planning stage. Some of the best American research was that carried out by nonprofessional astronomers such as Lewis Rutherfurd, John William Draper, and his son Henry Draper (see Chapter V), who used their own funds to support their private observatories. No national association of astronomers existed in the United States. No technical journal was devoted solely to the publication of astronomical papers. Both the *Sidereal Messenger* and Gould's *Astronomical Journal* had stopped publication during the Civil War. Although Silliman's *American Journal of Science and the Arts* accepted some papers dealing with astronomy, most Americans sent their observations to be published in the *Astronomische Nachrichten* in Germany or in English journals such as the *English Mechanic and World of Science, The Observatory* (founded in 1877), or the *Monthly Notices of the Royal Astronomical Society.*

At a meeting of the American Association for the Advancement of Science in 1876, C. A. Young of Dartmouth had summarized these facts and discussed the history and the possible future of astronomy in the United States. After commenting on the notable achievements of Nathaniel Bowditch, the two Bonds, Gould, Newcomb, Hall, Safford, and Winlock, he commented that "we may justly entertain a certain amount of honest pride in our record; there is room, however, for a still more abundant humility, if we enter into detailed comparisons between our own scientific achievements and those of other nations. American Astronomy has not yet passed its infancy." He also expressed his confidence "that if the record of the century past can be called honorable, that of the century to come is to be glorious."[4]

One of the institutions that was to make a contribution to this "glorious" future was the Harvard College Observatory.

II

In the autumn of 1876, President Eliot chose and the Corporation confirmed the appointment of Edward Charles Pickering, the 31-year-old Thayer Professor of Physics at the Massachusetts Institute of Technology, as the new director of the Observatory, to take office on February 1, 1877.

When Eliot had become president of Harvard in 1869 the College, like most colleges in the United States, offered primarily a classical education—languages, history, mathematics, and philosophy. No laboratory

E. C. Pickering: The Early Years

courses were given in the physical sciences and there was no graduate school. A young man who wanted training in physics or chemistry must go to the Lawrence Scientific School, a Harvard affiliate, or to the Massachusetts Institute of Technology (now familiarly known as MIT) or, if he could afford it, to a European university. Eliot had deliberately set out to change this situation by instituting courses in the sciences and by encouraging research. His unprecedented selection of a physicist rather than an observational astronomer to direct the Observatory obviously strengthened the role of science at Harvard. Although the choice evoked much disapproval,[5] Eliot had not acted capriciously. He had known Pickering both as an undergraduate at the Lawrence Scientific School and as a colleague on the MIT faculty, and had guided him at critical stages in his career. The decision also suggests that Eliot realized what was not widely recognized at the time, that the direction of astronomical research was undergoing a crucial change.

Edward Charles Pickering (Fig. 32) was born on Mt. Vernon Street, on Beacon Hill, Boston, on July 19, 1846, the elder son of Edward and Charlotte (Hammond) Pickering. His first American ancestor, John Pickering, had emigrated from Yorkshire and settled in Salem, Massachusetts, in 1636, bringing with him the family coat of arms and the motto *Nil desperandum*. The family had prospered and its men had made notable contributions to the growth of the colonies. Edward's great-grandfather Timothy Pickering was a close friend of Nathaniel Bowditch, the great navigator, and served in the cabinets of both Washington and John Quincy Adams. Later Pickerings included merchants, naturalists, lawyers, and physicians.

For five years Edward attended the Boston Latin School, where he acquired a great distaste for the classics. The instruction, he remembered in later years, included no science, and the work consisted chiefly in memorizing Andrew and Stoddard's *Latin Grammar*.[6] Entering the Lawrence Scientific School at Harvard to study civil engineering, he was fascinated with each science in turn—physiology, chemistry, mathematics—and graduated *summa cum laude* on his nineteenth birthday. At the age of twenty-one he was elected to the American Academy of Arts and Sciences, one of the youngest men ever to be chosen. After two years of teaching mathematics at the Lawrence Scientific School he became an instructor in physics at the recently founded Massachusetts Institute of Technology, where during the next ten years he revolutionized the teaching of the subject.

FIG. 32. Edward Charles Pickering, fourth director of the Observatory, during the early years of his administration.

E. C. Pickering: The Early Years

When the Institute opened its doors in 1865, the founder and president, William Barton Rogers, had hoped to try out a new system of teaching physics, by laboratory methods, but because of lack of money and other problems had given up the project. For several years the students continued to learn passively by listening to lectures and watching experiments performed by the instructor. Then, in the autumn of 1868, with the warm encouragement of President Rogers, Pickering initiated the new system by fitting a small room with tables, water taps, and gas fixtures. Since no such course had ever been given in an American university, he had no model to follow but he planned a series of experiments and wrote instructions to enable the student to perform them. (A year later a similar course was started at Manchester University in England; the Cavendish Laboratory at Cambridge was not founded until 1874.) The first experiments dealt with the principles of measurement, the construction and use of apparatus, the properties of gases, the mechanics of solids, the nature of light, and the qualities of prisms. Later he introduced courses in spectroscopy, photography, and practical astronomy.[7] Eventually he collected and published the instructions he had written, and thus produced the first laboratory manuals of physics.[8]

In addition to his teaching, Pickering carried out basic research in physics. In the winter of 1869–70 he gave a series of Lowell Lectures, on sound, and he constructed an apparatus for the electrical transmission of sound, which he demonstrated before the American Association for the Advancement of Science in 1870, six years before Alexander Graham Bell, in the same MIT laboratory, perfected the telephone. Pickering did not attempt to procure a patent on his device; like many other inventive physicists of the day, including Daguerre, Joseph Henry, and S. P. Langley, he held to the principle that a scientist should not place restrictions on the products of his work but should share them freely. He was particularly interested in problems of light and optics, and his publications for the decade he spent at MIT included papers on the solar corona, the spectrum of the aurora, the qualities of prisms, and comparisons of prismatic and diffraction spectra.

On hearing of Pickering's appointment as director of the Observatory, a colleague at MIT sent his congratulations and added:

The Physical Laboratory of the Institute has been so fully the creation of your own thought and has seemed so dependent on you for its success, if not for its very existence, that it was a surprize to hear of a separation. Everyone interested in the progress of Physics must thank you for all that has been done by you to train up a generation of experimenters. May I express the hope that the time is now coming when the College Observatories of our land are to be something more than advertisements to be displayed in the pages of the Annual Catalogues; from this sweeping condemnation a few Observatories as Harvard, Dartmouth, Allegheny, and one or two more are to be excepted because scientific work is done at them.[9]

On February 1, 1877, Pickering and his wife[10] moved from Boston to Observatory Hill in Cambridge. With the title of "Director of the Observatory and Phillips Professor of Astronomy and Professor of Geodesy," he received a salary of $3400 a year and the use of the residence, which had undergone some badly needed renovation. Built nearly forty years earlier for William Bond and modernized to some extent by Winlock, the house still lacked many conveniences. There was no adequate guest room, the furnace smoked, and the wooden walls of the upper stories were so thin that the rooms were hot in summer and cold in winter. During the autumn President Eliot had thoughtfully suggested that the Director and Mrs. Pickering should visit the residence to see what was desirable in the way of paint, paper, or repairs, and advised them to consult with Mr. Waitt, the Superintendent of Buildings. However, Eliot wrote (November 3, 1876), "if you wish some repainting done in the house, I think Mrs. Pickering had better select the colors rather than Mr. Waitt. His taste in color is somewhat uncertain." The faulty furnace had been repaired, fresh paint and wallpaper had been applied, and other alterations had been made, some of which the director paid for out of his own pocket.

Pickering's first tasks (which he had begun unofficially several months earlier) were to evaluate the assets and needs of the Observatory, and to plan the type of research to be carried out. The assets, as Newcomb had realized, were not impressive. Harvard no longer possessed the most powerful telescope in the United States, for both the Naval Observatory in Washington and the Dearborn Observatory in Chicago now had larger instruments. Still, the equipment did include two telescopes of respectable size and fine quality: the Great Refractor, which had been in use now for more than thirty years and, although its optical qualities remained unimpaired, needed adjustment and repair; and the 8-inch meridian circle that Winlock had acquired with such heart-breaking difficulty. Other instruments included the west equatorial, the old meridian circle, the Russian transit, and a Zöllner photometer. Equally important assets were the able and experienced assistants—Arthur Searle, William A. Rogers, Leonard Waldo, and Joseph McCormack—but they were few in number and all were grossly underpaid.

The financial resources, although greater than those of most other similar institutions in the United States, were still inadequate. The Observatory received no money from the University (subsidies from industry or government were of course unheard of) and was almost entirely dependent on the generosity of its friends, past and present. The total income was a little more than $14,000 a year, derived from the sale of time service and of publications, and the earnings from endowments that totaled roughly $175,000. (Pickering thriftily began to exploit a new source of income by selling the grass cut from the Observatory grounds. It realized only about $30 a year, on the average, but in this period, when a copyist or a computer could be hired for 25 cents an hour, even $30 was not negligible.) The meager income was scarcely enough to pay the salaries of the employees (including the director), to cover the expenses of the care and cleaning of the buildings and furnishings and the purchase and maintenance of instruments, and to pay the costs of fuel, water, books, and printing. Finding a new source of money was therefore imperative.

A decade had elapsed since the last public appeal for funds, which Winlock had used to buy the meridian circle. Following the example of his predecessors, Pickering launched a new campaign, through the Visiting Committee, to obtain money to relieve the immediate needs of the Observatory. Within a year the full amount he asked for had been subscribed, thanks largely to the enthusiastic efforts of Alexander Agassiz, William Amory, J. Ingersoll Bowditch, and other members of the Committee. Seventy-one patrons of astronomy (eleven of whom were women) agreed to contribute sums ranging from $50 to $200 annually, to make up a total of $5000 a year for five years.

With adequate support secured for the time being, Pickering faced one remaining problem: What kind of research should the Observatory undertake? The revision of Argelander's great catalogue of star positions, the *Bonner Durchmusterung,* was important, but Professor Rogers, using the Great Refractor, and Professor C. H. F. Peters, at the Litchfield

The Harvard College Observatory

Observatory of Hamilton College, had nearly completed remeasuring the positions of the stars in the zone assigned to this country. Further visual observations and drawings of the sun, nebulae, and planets, such as those made by the Bonds and by Winlock, seemed unnecessary. Pickering wanted to choose a field that would produce fundamental astronomical facts, and he did not want to duplicate work being done elsewhere. Before making a decision, he surveyed the astronomical scene in the United States.

Studies of the sun were in the capable hands of C. A. Young, then at Dartmouth College (and later at Princeton), and of S. P. Langley, at the Allegheny Observatory. Lewis Swift, at the Warner Observatory in Rochester, New York, was studying the nature and behavior of comets. The search for new asteroids was being carried out by Professor Peters at Litchfield and by Professor J. C. Watson at Ann Arbor. Micrometric measurements of the positions of double stars and of satellites of the planets were being made by S. W. Burnham at the Dearborn Observatory in Chicago, by Ormond Stone at the Cincinnati Observatory, and by Simon Newcomb, E. S. Holden, and Asaph Hall at the Naval Observatory. Experiments in the relatively new science of spectroscopy and the use of photographic methods were going on at the excellent private observatories of Lewis Rutherfurd and Henry Draper in New York. There remained one important aspect of astronomical knowledge relatively unexplored: measurement of the brightnesses of the stars. Except for C. S. Peirce's preliminary study, which was still unpublished, precise photometry was receiving little attention. Pickering therefore decided to make a quantitative determination of the brightnesses of all the stars visible from Cambridge.

III

The first known star catalogue, listing the positions and comparative brightnesses of several hundred stars, was that compiled by the Greek astronomer Hipparchus about 120 B.C. It may have served as the basis for the more extensive catalogue prepared more than two centuries later by Ptolemy, who in his *Almagest* increased the number of stars measured to more than a thousand and divided them into six groups according to their apparent brightnesses. Although other star catalogues made in the centuries that followed adopted similar groups, no two astronomers

184

used exactly the same scale of magnitude. Since no mechanical device existed that could precisely measure the amount of light transmitted by a celestial object, the brightness assigned to a given star was only an estimate that depended on the eyesight, skill, judgment, and experience of the individual observer.

Sir William Herschel had made his own estimates of stellar magnitudes, some of which were published in the *Philosophical Transactions* between 1796 and 1799. His son Sir John Herschel, working at the Cape of Good Hope early in the nineteenth century, had devised an "astrometer" to compare a star's light with that of the full moon, but his method yielded only approximate values. About the same time Argelander and his associates at the Bonn Observatory in Germany had attacked the problem of photometry by making simple comparisons by a "step method." The results were published as the *Uranometria Nova*, a catalogue giving approximate positions and brightnesses for all naked-eye stars visible from Europe. Later, in 1863, the same workers had produced the monumental *Bonner Durchmusterung*, which gave positions and magnitudes for more than 300,000 stars and remained the standard work. Since then, various ways had been proposed for constructing magnitude scales on a mathematical basis but none had received universal acceptance, partly because no instrument existed that could measure the difference in light emitted by two stars.

The first reasonably successful instrument designed to measure a star's brightness was the "astrophotometer" constructed by Friedrich Zöllner in Germany in the 1860's. It used as a standard the light of a kerosene lamp shining through a pinhole. In the field of view of the telescope the image of the light was brought alongside the image of the real star. A system of rotatable polarizing Nicol prisms was used to diminish the light of the standard to that of the star being measured. A scale attached to the prism measured the rotation, from which the reduction in light could be calculated. (The modern equivalent consists of two Polaroid filters, one rotatable with respect to the other, so that the amount of light transmitted can be controlled.)

An early project in photometry had begun at the Harvard Observatory in 1871 when Winlock acquired a Zöllner astrophotometer and assigned to Charles S. Peirce the task of planning its use. Although Peirce is now remembered chiefly as a logician and philosopher, in his own day he was perhaps better known as a scientist. After graduating from Harvard in 1859 he joined the Coast Survey, for which he determined the longitude of many cities on the Eastern seacoast and, in addition to his routine duties, devised a pendulum method that enabled him to detect some of the inequalities in the earth's gravitational field. A brilliant mathematician, an imaginative scientist, he was attracted by the many problems of astronomy. In 1871 the Coast Survey (then directed by his father, Benjamin Peirce) assigned him to duty at the Harvard Observatory, where he undertook the photometric project for Winlock.

For this study Charles Peirce tried to establish an absolute scale of brightness in which equal numerical differences in magnitude would exactly correspond with equal ratios of light. He also tried to formulate the mathematical laws by which the human eye perceives light. In the next four years, using a portable telescope and working partly in Cambridge and partly in Washington, he measured the magnitudes of 494 stars. He also attempted to determine the shape of the Galaxy, by roughly estimating the distribution of stars in the Milky Way.

The photometric measures had been completed and Peirce had gone to Europe in the summer of 1875, where he arranged for their publication in Germany, when Winlock died. The work was still unpublished when Pickering became director because, although the photometric material had been set in type, the rest of the proposed volume was not yet ready for the printer.

Peirce's somewhat informal connection with the Observatory now became a problem. Pickering respected his abilities but Peirce had a difficult personality and his future usefulness to the Observatory was not certain. According to the arrangement made with Winlock, Peirce had contracted to prepare catalogues of photometric data, of nebulae, and of double stars, and to see them through the press for the sum of $1200, which he had received. Now, however, he argued that the amount was too small for the labor involved and that before completing the work he had a moral, though admittedly not a legal, claim against the Observatory for further compensation. He wrote to Pickering (April 6, 1877) that "in regard to further work on the printing, I regret to say that I am obliged to occupy my spare time in earning money and therefore am unable to do anything about it, gratuitously." Since Pickering was eager to see the work published, he suggested that the material be sent to the Observatory for the necessary editing, but Peirce responded (April 25,

E. C. Pickering: The Early Years

1877) that this "proposition to take control of the work out of my hands is contrary to explicit understanding and to evident justice. I shall address Mr. Eliot on the subject."

In a letter to President Eliot (April 28, 1877) Pickering reviewed the relevant correspondence between Winlock and Peirce and concluded that Peirce "claims more and has done less than his own proposal specifies." Unsympathetic to what seemed a clear breach of agreement, Eliot advised Pickering not to wait for the unfinished work but to order the printing of the material then in press. These early *Photometric Researches* finally appeared in 1878 as Volume 9 of the Harvard *Annals*.

Peirce's connection with the Observatory now came to an end.[11] Although sound pioneering work, his photometric results were not definitive, partly because of the magnitude scale employed and partly because the lamp used as a standard could not produce a constant light.

IV

Having chosen to make a fresh attack on the problem of stellar photometry, Pickering invented a whole new class of instruments. Also, before beginning the actual measurements, he made two important decisions. First, he adopted the magnitude scale suggested in 1854 by Norman Pogson, at Oxford. On this scale a change of one magnitude represents a change of 2.512 in brightness ("Pogson's ratio").[12] (A first-magnitude star is 2.512 times brighter than a second-magnitude star and thus 2.512 to the fifth power or exactly 100 times brighter than a sixth-magnitude star.) Second, Pickering chose the Polestar, α Ursae Minoris (then supposed to be of constant brightness), as the standard light and arbitrarily assigned to it the magnitude 2.1.

Turning to the instrumentation, Pickering tried various modifications of the Zöllner photometer. Working in close cooperation with George B. Clark, of the firm of Alvan Clark and Sons who built the successive models, he devised a series of instruments, which he labeled alphabetically in the order of their construction, to be used with the 15-inch refractor. Each new instrument constituted an improvement over its predecessor or served some special purpose. Photometers L and M, for example, were designed to measure the light of double stars. With the help of his assistants—Winslow Upton, Leonard Waldo, Joseph McCormack, Arthur Searle, Oliver C. Wendell—Pickering tried each one, and

The Harvard College Observatory

with various photometers he achieved good measures of the magnitudes of planetary nebulae, the satellites of Jupiter, and the recently discovered moons of Mars. These measures were published in Volume 11 of the Harvard *Annals*. After two years' work these experiments culminated in photometer P, a revolutionary new instrument. This "meridian photometer" was a telescope with two lenses placed side by side. It was mounted so that it was directed always at the meridian (the great circle that passes from north to south through the zenith). By an arrangement of mirrors and prisms the image of any star, at any point on the meridian, could be brought alongside the image of the Polestar. Thus each star could be measured at its point of highest visibility, as it crossed the meridian.

After a few false starts, the formal observing program began on October 25, 1879, and ended on September 17, 1882. About 4000 stars of the sixth magnitude and brighter were selected for measurement. They included all those in Argelander's *Uranometria Nova* and a large number from other catalogues. Two men served as a team; during the first hour the observer adjusted the star images and reduced the light of the Polestar to that of the star being measured, while his companion read the scales and recorded the results in the laboratory journal. At the end of an hour the two men traded places for a second hour's work. At first they could measure only about 40 stars an hour but, with experience, they were often able to achieve a rate of more than one a minute. Although several of the assistants helped with the measurements from time to time, the chief observers were Pickering, Searle, and Wendell. The work required physical endurance. Adjusting the images and reading the scales involved a certain amount of bodily contortion. Also, since the telescope shelter was not heated, during the winter months when the temperatures dipped toward zero, and sometimes below, all the men suffered from the pain of the cold and consequent ailments of the throat and lungs. Although several women were then employed as computers and copyists, they were not allowed to make photometric observations because, as Pickering told President Eliot (June 1, 1885), "the fatigue and the exposure to the cold in winter are too great for a lady to undergo."

Pickering had chosen Polaris as the standard star because it maintained a relatively constant position in the sky, and was supposed not to vary in its light. To check its constancy, however, he measured its

E. C. Pickering: The Early Years

brightness at the beginning, the middle, and the end of each period of work. A scrupulous observer, he noted in the final publication that the magnitudes of Polaris seemed at first sight to present discrepancies that sometimes amounted to as much as half a magnitude, which suggested variability:

> The residuals for May and June, 1881, are rather large, but their number is small, and the result is most probably due to accidental errors. The entire series of estimates shows no peculiarities which require us to assume that either of the comparison stars is variable.[13]

(In 1911, some thirty years later, Ejnar Hertzsprung showed that Polaris is indeed a variable with a period slightly less than 4 days. The magnitude ranges between 2.48 and 2.62—a change probably too small to have been certainly detectable by Pickering's instruments and methods.)

Volume 14 of the Harvard *Annals*, published in 1884, presents the results of this first photometric study, which required three years to complete. The volume describes the instruments and methods used, and tabulates and discusses the data obtained, but, like most scientific reporting, it says nothing of the men and women who performed the work.

For a glimpse of Pickering and his staff as human beings, however, we can turn to the text of the *Observatory Pinafore*,[14] which Winslow Upton,[15] one of the assistants, wrote during four rainy days of an August vacation in Vermont. Using Sullivan's lyrics and a rather confused plot that revolved around the photometric work of the Observatory, Upton named his characters for the actual persons associated with the research. The one exception was "Josephine," whose real-life counterpart was Joseph McCormack, a young assistant. Opening a door to the Observatory during the early years of the Pickering administration, the parody discloses a vivid picture: the cold floors and shabby furnishings of the rooms where the computers worked; Pickering's courteous manner, and the affectionate respect given him by his staff; the assistants' longing for more adequate pay;[16] and their frequent exasperation with the troublesome equipment.

The story begins when a deputation of citizens arrives from Rhode Island, hoping to engage the services of "Josephine" for Mr. Frank Seagrave's[17] private observatory in Providence. Since she has only been determining star positions, and has not learned the techniques of photometry, Pickering agrees to let her go. "Josephine" is tempted by the offer, which would allow her to leave the Harvard Observatory and to exchange her "dark and dingy place, all cluttered up and smelling strong of oil" for "a new luxurious home with Brussels carpet and no dust and damp, And all the apartments lit by student lamp." However, she is reluctant to "lose her station" at the 15-inch telescope. She and Professor Rogers therefore devise a scheme: she will learn how to use the photometers; she will also help steal the prisms of the latest photometer (which had been giving trouble), take them to the Cambridge optician Alvan Clark for reworking, and reinstall them. Pickering will then be so impressed with her usefulness, and so pleased with the improvement in the instrument, that he will let her stay at the Observatory. Through the villainy of Winslow Upton (who portrays himself as a bumbling Dead-Eye Dick) their plot is discovered, but all ends well.

A chorus of computers opens the play, singing that

> We work from morn till night,
> Computing is our duty;
> We're faithful and polite,
> And our record book's a beauty.

Then Winslow Upton and Arthur Searle enter the room. (Upton, then a young man in his twenties, had been working at the Observatory for two years. Receiving very little pay, he lived at the Observatory and occupied a small room off the stairway leading to the Great Refractor.) When Upton complains that he can't sleep because the assistants who work in the daytime make so much noise, Searle advises him to find a better room.

Upton: I shall when my salary is large enough.
Searle: I guess you'll die of old age in that room if you wait for a large salary before giving it up.

There are jokes about Leonard Waldo's title of "Doctor"—with an LL.D. he was the only person at the Observatory boasting a degree beyond the M.A. There are also references to Pickering's very formal manners. When he enters the computing room, he addresses the ladies kindly: "My gallant crew, good morning." He then begins to sing, "I am the captain of this little crew," while the chorus responds, "And a right good captain too."

> Though moving by my right in society polite,
>   And among many men of note,
> I am never known to wear, though the ladies vainly
>   stare
>   A tall hat or swallow tail coat . . .
>
> Academic titles all, I have never failed to call
>   In addressing you by name.
> Though "Mister" I may occasionally say
>   I never speak the bare surname.
>
> What, never?
> No, never.

Later in the play, when he thinks his "precious prisms" are being stolen, he breaks out with, "Rogers, it's too bad!" and the shocked chorus comments, "He said Rogers with no handle!"

Upton, as "an assistant observer on Photometer P," refers to the various specialized instruments and his own mishaps in using them:

> In the cool night air with "S" and "P"
> I wearied my eyes on photometree.
> Bright stars with H, faint stars with I,
> Blue doubles reserved for a cloudy sky.
> So many close doubles were measured by me
> That now I am observing with Photometer P . . .
>
> I turned the dome with so grand a shock
> That I broke two windows and the Elliott clock;
> I burst the gas pipe rolling the chair,
> And created a blaze for the winter's scare.
> For my worthy zeal they requested me
> To try my strength on Photometer P.

An apparent reference to Mrs. Fleming (Chapter XI), who later became one of the most useful members of the Observatory, occurs when Pickering remarks that the force of computers is large enough to make possible "a good dance in spare hours," and that he "had thought of inducing our Scotch maid to give them instruction in the Highland polka, but she has unfortunately returned to her native land."

191

The Harvard College Observatory

Professor Rogers, "Josephine," and Upton, aided by the chorus of computers, humorously describe some of the discomforts of their work:

> An astronomer is a sorry soul,
> As free as a caged bird;
> His sympathetic ear should be always quick to hear
> The directorial word.
> He must open the dome and turn the wheel,
> And watch the stars with untiring zeal,
> He must toil at night though cold it be,
> And he never should expect a decent salaree.
>
> His eyes should shine with learned fire,
> His brow with thought be furrowed;
> His energetic speech should be ever prompt to teach
> The truths which he has borrowed.
> His knees should bend and his neck should curl,
> His back should twist and his face should scowl,
> One eye should squint and the other protrude,
> And this should be his customary attitude.

At the beginning of the second act, Pickering enters alone to study the record sheets of the meridian photometer, and sadly comments on the problems of accurate photometry:

> Pole Star, to thee I sing,
> Bright pivot of the heavens,
> Why are all our magnitudes
> Either at sixes or at sevens?

To correct the faulty prisms, Rogers and "Josephine" secretly remove them from the instrument and steal away to take them to the optician, accompanied by the chorus of computers who sing:

> Haste along, with footsteps steady,
> We shall soon be out the dark,
> And a horse car waits all ready
> To carry us to Alvan Clark.

But Pickering, alerted by Upton, is watching the guilty pair as they leave, and when "Josephine" drops first one, then the other prism, the catastrophe makes him emit his strongest oath: "Oh Polaris!" When Rogers and "Josephine" explain that they had merely hoped to achieve an improved photometer, Pickering is delighted, and all ends well.

192

E. C. Pickering: The Early Years

Fig. 33. Scene from *The Observatory Pinafore,* as performed on December 31, 1929. Left to right: "Prof. Rogers" (Percy M. Millman); "Josephine" (Cecilia H. Payne); "Lady computers" (Henrietta Swope, Mildred Shapley, Helen B. Sawyer, Sylvia Mussels, Adelaide Ames); "Prof. Searle" (Leon Campbell, Sr.).

Upton's parody was not performed during Pickering's lifetime, perhaps because Joseph McCormack, cast as "Josephine," died of typhoid on February 2, 1880, a few months after Upton wrote the libretto. A performance did take place during the Shapley administration (Fig. 33), when the Observatory staff put on a production on December 31, 1929, for the members of the American Astronomical Society, then meeting in Cambridge. Cecilia Payne (Gaposchkin) played the role of Josephine.

v

Photometry was not the only activity at the Observatory during those early years. Pickering and other members of the staff began attempts to photograph the sky, determined star positions, looked for stars having peculiar spectra, calculated the brightnesses and periods of variable stars, compared the several existing scales of stellar magnitude, and observed asteroids, satellites, and comets. By arrangement with Western Union the Observatory regularly transmitted the correct time to cities, business firms, and railroad offices throughout New England. Bostonians could obtain the correct time at noon by watching the roof of the Equitable Life Assurance Co. on Devonshire Street, where a four-foot copper ball, held in place by a magnet, was mounted on a scaffold. A direct telegraph line, connecting a clock in the Observatory with the magnet, dropped the ball at precisely twelve noon each day. Daily meteorological observations—of temperature, humidity, rainfall, barometric pressure, wind velocity—continued until 1881.

The Observatory assumed an important new function when it became the American center for receiving and distributing news of special celestial events—the detection of a previously unknown asteroid, planet, or satellite, the eruption of a nova, the advent of a comet. Until 1882 the communication of such news was somewhat haphazard. No international bureau existed, although an informal network of observatories operated in Europe. In 1873, using a system devised by Professor Peters of the Litchfield Observatory, the Smithsonian Institution in Washington began to serve as a central office for the United States, cabling announcements of discoveries to the Astronomer Royal in England, and receiving and distributing announcements from Europe. This system worked after a fashion, but it was often slow and inefficient, partly because no one had devised a universal code that could transmit the position of a celestial object quickly, economically, and accurately. Messages were sometimes hopelessly garbled in transmission, and astronomers lost valuable time in trying to determine the correct information. S. C. Chandler of the Observatory staff, commenting on this undesirable state of affairs, remarked:

> The requirements of mercantile pursuits have long since lessened the chances of a mistake in a message of business importance, and the buying of fifty hogs or the sale of one hundred barrels of flour, can be done with ten-fold more accuracy than the announcement of the position of a comet.[18]

He also determined to solve the problem.

Seth Carlo Chandler was a brilliant amateur astronomer who served as an assistant at the Observatory from 1881 to 1886. After graduating from the Boston English High School in 1861 he became a private assistant to B. A. Gould, who was then making determinations of longitude for the Coast Survey. In 1864 Chandler joined the Coast Survey,

E. C. Pickering: The Early Years

where he worked for several years before leaving to become an actuary for an insurance company, but his early interest in astronomy continued and in 1881 he became a research associate at the Observatory, working in his spare time.[19] As an authority on variable stars and an expert in calculating the orbits of comets, he had long been concerned with the difficulties of receiving and sending astronomical information. To solve the problem, Chandler worked in cooperation with another amateur astronomer, John Ritchie, Jr., president of the Boston Scientific Society and editor of its monthly publication, the *Science Observer*. Together they devised an ingenious code, based on the use of a dictionary, by which cometary positions and movements could be telegraphed or cabled. The *Science Observer* for several years had used this code successfully to transmit cometary positions and orbits (computed by Chandler and Searle at the Observatory) to astronomers in the United States and Europe, thus supplementing the Smithsonian's announcement of discoveries.

But the basic communications system remained unsatisfactory. The change, when it finally came, was inspired at least in part by the very human wish to establish priority in astronomical discovery. Because of inadequacies in the announcement service, determining just who was the first to observe a new comet was often difficult or impossible. One devoted comet seeker was Lewis Swift, director of the Warner Observatory in Rochester, New York. For some time he had been urging that Harvard take over the responsibility for cabling the discovery of comets, and in discussing the problem with Pickering he wrote (March 18, 1882) that the Smithsonian had once

failed to cable one (mine of 1877) & in consequence it required much time & trouble to establish my claim, for Borelley found it 3 days after, & cabled it here. Although I got the medal, about half the time it is called Borelley's comet.

Establishing priority became a matter of money as well as pride when in 1881 Mr. H. H. Warner, who had provided the funds to establish the observatory bearing his name, offered prizes of $200 in gold to each discoverer of a comet, in the United States, Canada, Great Britain, and Ireland. Competition was keen, and an efficient announcement service became even more important. The first year, four prizes were awarded for the discovery of comets—two by Swift, one by J. M. Schaeberle

195

of Ann Arbor, and one by E. E. Barnard, then at Nashville.[20] The prizes were offered again in 1882. Swift again found a new comet, as did C. S. Wells of the Dudley Observatory at Albany, but for some reason the Smithsonian had not cabled the news immediately. Disturbed by the delay, Swift urged Pickering to take on the responsibility in the future, and, since the officials at the Smithsonian had no objection, he agreed.

Nevertheless, no basic change of system took place, either in this country or abroad, until the advent of the Great Comet of 1882 (which also stimulated the development of stellar photography). This spectacular sun-grazing comet, which soon became so brilliant that it was visible in full sunlight at noon, was so poorly reported that few astronomers learned of its existence until too late to observe its form and movement in the important first few days after its appearance. The comet was freshly discovered at least five times over a period of two weeks. European observatories first received news of the visitor by telegram from Professor A. Cruls, director of the Imperial Observatory at Rio de Janeiro, who discovered the object on September 10. Later messages showed that Australian observers had seen it as early as September 3 (and had relayed the news to England by mail). B. A. Gould at Cordoba, Argentina, had first observed it on September 5, and W. H. Finlay at the Cape of Good Hope in South Africa had found it on September 7. But as late as September 17 not all observatories had received the news, and Professor A. A. Common in England, while making observations on the morning sun, again independently discovered the comet. The news reached the United States through the Smithsonian Institution, and one of the earliest observations here apparently was that made by Chandler and Wendell at the Observatory on September 17.

The delay in sending the news had obviously caused the loss of much valuable information. European astronomers quickly took action to prevent a similar occurrence in the future. In the United States, after discussions among Spencer Baird, secretary of the Smithsonian Institution, Pickering, and Ritchie of the *Science Observer,* the Smithsonian indicated its willingness to relinquish the announcement service. On January 4, 1882, Baird formally transferred its control to the Harvard Observatory.[21]

Some newspaper accounts of the change implied, incorrectly, that the Smithsonian had been forced to give up the service because of com-

E. C. Pickering: The Early Years

plaints from European astronomers. In a letter to Pickering (January 30, 1883) Secretary Baird expressed his dismay:

I shall be glad to have you do us the justice to make known that it was entirely the result of our spontaneous feeling that the service would be better performed at Harvard Observatory, in Cambridge, as a centre, than at the Smithsonian Institution in Washington; and it was not until after the transfer had been ordered, and every arrangement made for the same, that we had any communication whatever on the subject from the other side of the water.

In announcing the transfer (February 14, 1883), Pickering paid full tribute to the value of the Smithsonian's role:

An association of about fifty European observatories has recently been formed, with its headquarters at the Royal Observatory, Kiel, Germany, directed by Professor Kreuger, who has taken charge of the business of the association. Connections by cable have been established with South America, South Africa and Australia, and the Harvard College Observatory has been requested to co-operate with it, in the United States, by receiving and distributing in this country the telegraphic information sent from Kiel, and by forwarding to Kiel by telegraph any similar information of importance collected from American astronomers. By the courtesy of Professor Baird, Secretary of the Smithsonian Institution, the function hitherto performed by the Institution, of collecting and transmitting announcements of discovery, has been transferred to the Harvard College Observatory.

In accepting this transfer, it is right that a public acknowledgement should be made of the service rendered to science by the Smithsonian Institution in undertaking the labor from which it now retires. For several years, its action has relieved a want generally recognized, although not otherwise provided for; while, as soon as astronomers were prepared to assume the task, the Smithsonian Institution courteously offered to facilitate the change which has just been made.

The success hitherto attained by Messrs. Chandler and Ritchie—both of whom are now connected with the Observatory—in their project of improving the mode of transmitting astronomical telegrams, encourages the belief that the system now adopted will prove expeditious and satisfactory. Mr. Chandler will continue his computations of cometary orbits, which will be distributed by telegraph, as heretofore, when that course seems to be desirable.[22]

The responsibility thus accepted was carried out for more than eighty years. The Chandler–Ritchie code underwent some changes and was eventually superseded (by a code devised by W. P. Gerrish), but the Observatory continued to serve as the astronomical news center for the

The Harvard College Observatory

United States throughout the administrations of Pickering and his successor, Harlow Shapley. During the directorship of Donald H. Menzel, however, the telegraphic service was shifted once more. In 1955 the Smithsonian Astrophysical Observatory moved from Washington to Observatory Hill in Cambridge, under the directorship of Fred L. Whipple, and is closely though informally allied with the Harvard Observatory. To operate its network of satellite-tracking stations, the Smithsonian had established an independent communications system that covered the entire globe. The International Astronomical Union therefore voted, in 1964, to transfer to the Smithsonian the duties of both the European and the American bureaus. Thus Harvard returned the service to its originator, and the Smithsonian became the world center for the distribution of astronomical information.

VI

In the summer of 1883 Professor and Mrs. Pickering sailed to Europe. The results of the first photometric study were safely in press, a second was progressing smoothly, and for the first time Pickering felt free to leave the work for a while. In the six years since becoming director he had not had a real vacation; he had never been absent from the Observatory for more than ten days at a time, and he had not visited Europe since 1870, when he accompanied Winlock on the eclipse expedition to Spain. President Eliot urged Pickering to get a good rest and to stay as long as he liked. On this trip he visited the chief observatories on the Continent, and in England he called on the leading astronomers and attended a meeting of the Royal Astronomical Society (he had been made a Foreign Associate two years earlier). In a speech before the Society he outlined his plans for future work at the Observatory: to repeat the stellar photometry, using more accurate instruments and methods, and to extend the measures to fainter stars; to classify the stars from photographs of their spectra; and to prepare "a photographic map of the whole heavens."[23]

One of the most important events of the summer was a visit to Colonel John Herschel, a grandson of the first Sir William Herschel, at Collingwood. There Pickering renewed the close relation between the Harvard Observatory and the Herschel family, which had begun in 1815 when William Bond called at Slough and was entertained by Caroline Herschel, and had been renewed when George Bond visited Sir John

E. C. Pickering: The Early Years

Herschel in 1851. Pickering was allowed to examine Sir William Herschel's original manuscripts, many of them in the handwriting of his sister Caroline.[24] Among them he found two never-published catalogues of star magnitudes. Strongly convinced that unpublished work was wasted work, Pickering obtained permission to copy the manuscripts. Later, after his return to Cambridge, he undertook to reduce the unpublished measures (as well as those in the four Herschel catalogues that had appeared nearly a century earlier) to the scale used in the *Harvard Photometry*. Eventually the material was printed in Part II of Volume 14 of the Harvard *Annals*.

During these weeks in England Pickering also called on the Rev. Cyril Pritchard, the Savilian Professor of Astronomy at Oxford, who had begun a photometric study with a different type of instrument. Instead of a polarizing Nicol prism he used a tinted glass wedge to diminish the brightness of the image of the standard star to that of the one being measured—a principle that George Searle and Cleveland Abbe at the Harvard Observatory had tried and abandoned some twenty years earlier.[25] In first announcing the new investigation to Pickering, Pritchard had written (December 27, 1881) that he began the work after reading

what you had done in the same sort of direction so excellently well: and I was charmed to find that the mantle of Wm. Herschel was now taken up and worthily worn by you. Accept my congratulations on the good issue of your speculations . . . I am throwing the force of this Observatory into the photometric mags of stars visible nudis occulis from the Pole to Dec. —5°. My experience of these wedge-aperture readings is very encouraging and no hitch has yet occurred in 1200 readings so I shall persevere to the end (life and health given).

A year later he had made good progress but, since cloudy weather so often interfered with observations at Oxford, he had gone to Cairo for a few weeks. There he completed his measures under the clearer skies of Egypt, having set up his 4-inch telescope and wedge photometer on the roof of Shepheard's Hotel. Although the letters exchanged between the two men at this time were courteous, they reflected an increasingly wide difference of opinion. Pritchard was convinced that his wedge photometer gave far more accurate measures than did the polarizing photometer. Pickering was equally convinced that, because of the difficulties of calibration, the wedge method produced larger systematic errors than did his own. His visit to Pritchard and the Oxford observatory did nothing to change the views of either man.

When the *Harvard Photometry* first appeared, in 1884, it received wide acclaim. A review of Part I in *The Observatory*, a publication of the Royal Astronomical Society, gave it high praise:

> To say that the volume recently issued from the observatory of Harvard College is, in this branch of astronomy, epoch making, is to do no more than justice to a work which must henceforth be regarded as at once the foundation and treasury of scientific stellar photometry.[26]

A few of Pickering's friends in the Royal Astronomical Society suggested that he should receive the Society's gold medal but, when Pritchard and his Oxford colleagues attacked the accuracy of the work, the first enthusiasm subsided. Friends of the two men became partisans and a controversy developed, based less on scientific grounds than on the geographical location of the participants. An exception was Arthur Ranyard (a member of the Royal Astronomical Society), who wrote Pickering (May 11, 1885) of his "disappointment" that the medal had not been given to Pickering that year:

> Prof. Pritchard wants to make out that your probable errors are bigger than his. I conclude that his measurements ought to differ one from another more than they do—if the atmosphere of Oxford was not quite exceptionally uniform.

At the next meeting of the Royal Astronomical Society (June 12, 1885) the dissension widened when Pritchard's friends, including W. H. M. Christie, the Astronomer Royal, assailed the Harvard work, alleging that it contained large systematic errors. At the same time, some of Pritchard's claims for the wedge method—for example, that the presence of full moonlight had no effect on the measures of starlight—seemed unreasonable to many of the astronomers present. The discussion became heated, but Professor Common lightened the tension by remarking (with complete accuracy, although his prophecy sounded ridiculous to many of his audience),

> I may mention now that I do not believe either in Prof. Pickering's photometer or Prof. Pritchard's. The only photometer we shall ever have to be of any use whatever is photography. (Laughter.) There is no personal equation there.[27]

200

E. C. Pickering: The Early Years

Early in December 1885 Pritchard published his own photometry, the *Uranometria Nova Oxoniensis*, giving the positions and magnitudes of 2784 stars. His friends in the Royal Astronomical Society at once proposed him for their gold medal. Since Pickering's work could not well be ignored, however, Arthur Ranyard was appointed to champion his cause. This was a peculiarly uncomfortable duty for Ranyard because, though a warm friend of Pickering's, he believed the whole system of awarding medals for scientific achievement to be a bad one. He wrote Pickering (November 21, 1885) that the council was "fast degenerating into a mutual recommendation and medal awarding association." In spite of his personal disapproval of medal-giving, he made an effective advocate and in a letter to Pickering (December 12, 1885) was able to report success:

We had our council meeting yesterday, Prof. Pritchard having published his volume three days before and sent a copy round to each of us. I do not send you the copy sent to me as I have no doubt you will soon receive one direct. There was a strong muster of Pritchard's friends and they evidently felt very confident. Huggins opened in favour of Pritchard. De La Rue followed, then Christie, Abney, Rand, Capron, and Hind, who made a poor apology for the Oxford work. I thought it better not to criticise what was said about Pritchard's claims but to urge the impropriety of rewarding Pritchard's work and leaving you out—and then took up one after another the points which had been urged in favour of giving the medal to Pritchard. It was not very difficult to say a good deal more for you. In the discussion which followed De La Rue soon struck his flag and proposed terms of peace. In his most bewitching manner he said that he had been induced to alter his opinion by what Mr. Ranyard had said as to the illogical position in which the council would be placed if they rewarded Oxford and by inference condemned Harvard, and he therefore proposed that a gold medal should be given to both. I saw that the Philistines would prevent a medal going to Harvard if they could not have one for Oxford and I thought that half a loaf was better than no bread and therefore temporized and accepted the offer. The independents ultimately accepted the proposition and I think that there is not much doubt that the award will be confirmed at next Council meeting and that Pritchard will go down to posterity bracketed in the list of our Medals with yourself.

The award was duly confirmed. The Society conferred its gold medal for 1886 on Pritchard and on Pickering jointly for independent researches in stellar photometry. When Ranyard cabled the news, Pickering at first hesitated to accept, lest his action be construed as an endorsement of the wedge method. Before deciding, he wrote to Ranyard (January 7, 1886) to ask what were the exact terms of the citation, and he remarked with unusual harshness that "the decision of the Astron. Society that my work is of the same grade as that of Professor Pritchard I regard as the severest criticism that I have ever received." Reassured that he would still be free to express his opinions on scientific matters, Pickering did accept the medal—in spite of Ranyard's expressed hope that he would set a good example for other astronomers by declining!

Among the letters Pickering received at this time was one from E. S. Holden, director of the newly established Lick Observatory (February 23, 1886). It expressed a somewhat different and perhaps more prevalent view of the merits of medal-giving:

> Let me be among the first to congratulate you heartily on your reception of the Gold Medal of the R. A. S. an honor which all of your colleagues know that you have *earned*.
>
> It is a pleasant thing to see your honors coming early when they will be an encouragement to more future work & not late, as a reward for past work finished & ended.

The first *Harvard Photometry,* of course, had been a preliminary investigation. When it ended, Pickering at once began a second study, using an improved meridian photometer with 4-inch lenses, to measure stars fainter than the sixth magnitude. The results were published in Volume 24 of the Harvard *Annals.* (A few years later this work came under severe attack; see Chapter IX.) The same instrument was then sent to Peru to obtain the magnitudes of stars in the Southern Hemisphere (see Chapter VIII).

The photometric work continued for nearly a quarter of a century, and Pickering never tired of it. (On May 25, 1903, he made his one-millionth setting of the photometer!) The brightnesses of all the stars visible in the entire sky were measured and remeasured, by various methods, to obtain the greatest possible accuracy. This colossal project culminated in 1908 with the publication of Volumes 50 and 54 of the *Annals,* the *Revised Harvard Photometry,* which gave the magnitudes of more than 45,000 stars brighter than the seventh magnitude. These volumes remained the standard catalogue of photometric magnitudes until, just as Professor Common had foretold, photography largely supplanted the visual method. (Many years later, photoelectric techniques proved to be up to 100 times more accurate than the photographic techniques, in special instances.)

A twentieth-century evaluation of Pickering's first photometric work, made some fifty years after its publication, concludes:

> Thus Pickering laid the foundations of precise stellar photometry: he gave us the first definite magnitude system . . . and he gave us the first catalogue of standard photometric magnitudes to form the basis of all subsequent catalogues.[28]

VII

In his annual report to the Visiting Committee for the year 1883, under the usual headings, Pickering summarized the work done and that planned for the future. The Bond determinations of star positions based on the *Durchmusterung* were still undergoing revision. The search for variable stars continued. Pickering had been trying to find a good method of studying the colors and spectra of stars but a spectroscope bought from Hilger, of London, for that purpose had not proved satisfactory.

The most significant item in the report appeared in a single paragraph under the heading "Miscellaneous." Pickering stated that he had begun some experiments in photography, by which he hoped to determine the colors and magnitudes of the brighter stars. As another objective (which he had briefly outlined at the June meeting of the Royal Astronomical Society), he proposed to make a photographic map of the heavens. He had begun a collection of astronomical photographs, starting with those made by Bond and Winlock, which would become a permanent, continuing history of celestial events. This paragraph is the first official indication of Pickering's intention to change the pattern of research at the Observatory, a decision that during the next several decades resulted in three major contributions to astronomy: the Harvard plate collection, which accumulated from a regular patrol of the sky; the first complete photographic map of the sky; and the monumental Henry Draper Catalogue of stellar spectra, constituting Volumes 91–99 of the *Annals*.

Photography had made great technical advances during the preceding decade. Like the telescope and spectroscopy, it was a new tool that was about to revolutionize astronomical methods. The invention of the telescope in the first years of the seventeenth century brought the sun and the planets so much closer to the earth that, within the next three centuries, observers were able to discover a large number of hitherto unknown celestial objects: the nine moons and the rings of Saturn, the four large and eight lesser satellites of Jupiter, the planet Uranus and his five moons, the two Martian satellites, and many hundreds of asteroids. (The existence of the planet Neptune, though confirmed by the telescope, had been predicted on mathematical grounds.) But astronomers had still been limited chiefly to determining the positions and motions of celestial objects; their physical nature remained a matter of speculation.

Then early in the nineteenth century came a simultaneous advance on two fronts, the gradual understanding of the spectrum produced when a prism spreads light into its component colors, and the invention of the photographic process. Using spectrum analysis, the astronomer could probe deeper than mere appearance and learn something of the chemical composition, temperature, and structure of the sun and other stars. With photography, he was able to use long exposures to collect the light from faint objects, and so bring remote parts of the sky closer; and he could make an automatic record of celestial objects so that the value of his observations need not depend on the accuracy of his eye or his skill at drawing. (In the twentieth century, our knowledge of the universe has been further increased by equally revolutionary advances: the development of modern atomic theory, the discovery of radio waves, and the construction of radio telescopes that disclose cosmic radio sources, quasistellar objects, and pulsating stars. Most recent of all are the orbiting observatories that operate outside the obscuring envelope of the earth's atmosphere.)

Experiments in celestial photography had been going on for some 40 years (Chapter II), but the difficulties of the collodion process had limited its usefulness to the astronomer. The wet plate could not be prepared in advance but must be coated just before exposure and developed as quickly as possible afterward, before the surface had time to dry. Since the plate dried so quickly, really long exposures were not possible and only the brighter objects in the sky could be photographed. On eclipse expeditions, the astronomer had to transport a mountain of equipment in addition to his telescopic apparatus—a tent to use as a darkroom, developing trays, carboys of chemicals to make the sensitiz-

E. C. Pickering: The Early Years

ing bath and the fixing and developing solutions, plenty of distilled water to mix with the chemicals. The vagaries of weather also posed a problem: at low temperatures the solutions and chemicals would often freeze, and even the glass plates were likely to become brittle and crack.

In spite of these drawbacks a few astronomers had continued to use photography to record important objects and events. The Naval Observatory in Washington regularly included photography in its observations of solar eclipses, and used it in the expeditions to observe the transits of Venus in 1874 and 1882. Benjamin Gould at Cordoba, Argentina, had used the collodion process to photograph groups of southern stars. In their private observatories in New York, Lewis Rutherfurd had obtained pictures of the Pleiades distinctly showing some fifty stars and Henry Draper in 1872 had made the first successful photograph of the sun's spectrum. In England, Warren de la Rue had instituted routine photography of the sun as early as 1858.

The technique could not come into general use, however, until the perfection of a good dry-plate process. Dry plates became possible, eventually, from the use of gelatin emulsions. Among the many persons who contributed to their development was William Abney in England, who made especially important experiments to increase the speed and sensitivity of photographic plates. Many advances were made in the 1870's, and in 1878 a dependable dry plate finally became available, chiefly from the work of C. Bennett, who used silver bromide held in suspension by gelatin. Although the difficulties of celestial photography were thus largely overcome, still no major change occurred in observational methods until after the advent of Comet 1882 II, the spectacular sun-grazing object that also stimulated the creation of an international bureau for astronomical information. David Gill, the director of the Royal Observatory, Cape of Good Hope, wished to take a picture of the celestial visitor and, since the observatory did not possess any photographic equipment, he borrowed a hand camera from a newsman. The comet required a long exposure. To avoid the blurring produced by the rotation of the earth, Gill tied the camera to the telescope in such a way that "any motion given to the telescope was common also to the camera."[29] The best three of the resulting photographs not only gave sharp pictures of the comet but also showed beautifully clear images of a multitude of stars. When the pictures reached Europe they created a sensation. A stunning example of what Bond had envisioned some

The Harvard College Observatory

25 years earlier, they showed once and for all that photography could map the sky "with unimpeachable accuracy."[30]

Although Gill's pictures must have helped confirm Pickering's decision to resume photographic work at the Observatory, he has recorded that the idea was suggested by his younger brother, William Henry Pickering.[31] William Pickering was born on February 15, 1858, some 12 years after his brother Edward. He studied physics at MIT and graduated in 1879, having already sampled the delights of astronomy by going to Colorado to observe the solar eclipse of July 29, 1878. After graduation he was appointed instructor in physics at MIT, where he established a photographic laboratory and taught the art to more than 100 students each year.[32] A skilled craftsman, William carried out many experiments designed to achieve more rapid photography. In particular, he tried to devise ways to capture the image of an object in motion—a galloping horse, a racing locomotive, a splash of water, a lamp chimney at the instant it was shattered by a bullet. Of this last experiment he wrote:

> The difficulty in making the exposure at the right time was not so great as might at first be supposed. I stood quite near the chimney, and held the gun, while my assistant's finger was on the trigger of the shutter. I counted, and at the word "Three" we both pulled.[33]

Whether this event took place in the normally quiet halls of the Observatory he does not say.

With a grant from the Rumford Fund of the American Academy, Edward Pickering acquired a 7-inch photographic lens. He planned and organized the experiments to be made, while William undertook the technical work of exposing and developing the negatives. The early results were exciting. One of the first pictures showed 462 stars in a field for which Gould's *Uranometria Argentina* listed only 55, and other photographs clearly depicted the nebulae in Orion, in Andromeda, in Lyra, and in the Pleiades. Encouraged by these results, Edward decided to expand the work. In 1885, with a new grant of $2000 from the Bache Fund of the National Academy, he was able to buy an 8-inch portrait lens, a photographic doublet, which was corrected and mounted by Alvan G. Clark and Sons. "This unique instrument,"[34] as Pickering called it, could photograph stars as faint as the fourteenth magnitude.

Many astronomers were beginning to recognize the advantages of the new photographic techniques, though few had the necessary knowledge, equipment, or staff assistance to adopt them as routine. Enthusiasm for the new method was not yet universal, however. Some astronomers feared that, in the process of drying after development, the emulsion on the photographic plate would shrink or be otherwise distorted so that the relative positions of the stars shown would be wrong. In speaking before the American Association for the Advancement of Science in August 1880, Asaph Hall of the Naval Observatory had concluded that while photography was useful for descriptive astronomy, such as a solar eclipse, it could not serve when accurate measures were desired. Ormond Stone, director of the Leander McCormick Observatory at the University of Virginia, also questioned whether photography could wholly take the place of the older methods. He thought that some types of work could be done more accurately without the aid of a camera, and wrote that "however interesting such experiments may be, it is questionable whether the introduction of a photographic plate instead of the eye will not complicate rather than improve observations made with the meridian circle."[35]

As early as 1883 Pickering had outlined a plan to chart the stars by photography. Lacking the money to implement his scheme, for some time he could only continue experiments with his telescopic camera. In the autumn of 1886, a few months after receiving the gold medal of the Royal Astronomical Society, he was invited, with other leading astronomers, to attend an international conference to discuss the preparation of a photographic atlas of the sky.

Immediately after obtaining his pictures of Comet 1882 II and its star-studded background, David Gill had suggested that a number of observatories ought to cooperate to photograph the entire sky. At the same time, he had ordered the construction of a photographic refractor so that he could begin systematically photographing the stars of the Southern Hemisphere. (In 1900 he was able to publish the *Cape Photographic Durchmusterung,* a catalogue of the positions of more than 400,000 southern stars, which Professor J. C. Kapteyn of the Groningen Observatory in Holland had measured from the photographs.)

Gill's proposal for a cooperative venture soon received support from Paul and Prosper Henry, of the Paris Observatory. In 1872 they had begun to prepare a star map, based only on visual observations made through their telescope. When their charting approached the crowded center of the Milky Way, however, the task of distinguishing and measuring the positions of individual stars had become nearly impossible. They had then tried photography, with amazingly successful results. On a photograph of the Pleiades, made with a 3-hour exposure, they could count 1421 stars; Rutherfurd's photograph of the same group made a decade earlier (with a wet-plate process) had shown only 50. Adopting Gill's earlier suggestion, the Henry brothers also urged that observatories should combine their efforts to produce a sky chart. At the request of Admiral Mouchez, director of the Paris Observatory, the French Academy of Sciences had therefore called an international conference of astronomers to meet in the spring of 1887, to plan the Carte du Ciel. Pickering could not accept the invitation, since by this time he was fully occupied with the work of the Henry Draper Memorial (Chapter V) and in planning the first Boyden expedition (Chapter VI).

The International Astrophotographic Congress met on April 16, 1887, in Paris.[36] Fifty-eight delegates attended, representing sixteen nationalities. The three Americans present were Lt. A. G. Winterhalter (U. S. Naval Observatory); W. L. Elkin (Yale Observatory); and C. H. F. Peters (Litchfield Observatory). Among other members were W. H. M. Christie, Astronomer Royal and director of the Royal Observatory at Greenwich; David Gill, director of the Royal Observatory at the Cape of Good Hope; J. C. Kapteyn, of Groningen University, Holland; A. Kreuger, editor of the *Astronomische Nachrichten* and director of the Royal Observatory at Kiel; and Otto Struve, director of the Pulkovo Observatory in Russia.

According to the plan agreed on, each participating observatory would photograph the stars in an assigned zone of the sky, and all observatories would use identical instruments so that the scale and quality of the results would be consistent. A major question to be decided was what type of telescope to adopt. Refractor or reflector? Diameter of aperture? Short or long focal length? There the conference was not in agreement.

Although Pickering was not able to attend, he had sent a letter (December 10, 1886) to Admiral Mouchez, outlining his recommendations and offering to undertake the photometric discussion of the results. His recommendation embodied essentially the plan he had devised three years earlier. The best telescopic camera for the purpose, he said, was a refractor of short focal length. It should have a double object glass consisting of four lenses, like the portrait cameras used by commercial photographers. In other words, it would resemble the Bache 8-inch but would be larger, having an aperture of about 11½ inches and a focal length of about 67 inches. At first, Christie and other English astronomers supported this view. Eventually, however, for the sake of uniformity, they accepted the prevailing opinion based on the experience of Gill and the Henry brothers. After much discussion the Technical Committee of the Congress unanimously chose to recommend a photographic refractor, similar to that used by the Henry brothers, of 13-inch aperture and 11-foot focal length, to photograph stars as faint as the fourteenth magnitude.

Pickering was disappointed in the decisions of the Congress. The plan would produce accurate results, he agreed, but the larger scale of the longer-focus instrument would require a much larger number of plates; hence the charts would accumulate far more slowly than by the techniques he recommended. To illustrate the kind of maps his method would produce, Pickering sent to Gill some star photographs made with the Bache. Gill replied (June 22, 1887):

Your letter to the Congress received much more consideration than you appear to think. It was fully discussed at the preliminary meetings and by the members of the conference . . . The question of the adoption of combinations giving very large fields, composed of 4 lenses was fully talked out at the informal meetings at which the *real* business is done.

The scale of magnitude & accuracy to be attained were really the points which went against the 4 glass combination. Nothing less than 14th magnitude stars would satisfy a large number of members of the Congress, and it was felt that if we got beyond 15 minutes of exposure we ran some risk of failure. To combine these conditions w^d have involved the use of lenses of 12 or 13 inches aperture, and 4 glass lenses of this aperture w^d have cost too much—indeed I do not know if we could get them made at all. Besides when the greatest accuracy of *measurement* is desired I hardly think you can use so large a field to advantage because of the changes in the differential refraction during exposure—the correction of a 4 glass objective is also a species of compromise, and the law of distortion is probably troublesome to deal with. In fact the Congress was determined to do the work on a scale of accuracy and magnitude which appeared to exclude the use of double combinations.

As planned, the Carte du Ciel would consist of two parts, for which each observatory would photograph its section of the sky twice: once on a large scale to produce the charts, which would comprise 22,154 plates (the paper reproductions would have weighed about 2 tons);

and again on a small scale so that the coordinates and magnitude of each star could be measured and catalogued. The project was expected to extend over about five years. However, some time was required to build, mount, and test the new equipment. When the astrographic congress again met in Paris, four years later, no charts had yet been published. Pickering, meanwhile, had decided to prepare a sky map single-handed. David Gill wrote to him from Paris (April 8, 1891).

We have had, on the whole a very satisfactory meeting here—
Underneath the surface there is a pretty strong feeling that your chart will be a much better and more practical one than ours—as a chart, and that our true function is the Catalogue—indeed there was a motion very nearly carried that we should leave the chart aside and push on the Catalogue plates—And this would have been carried I believe but for the fact that the French astronomers want money from the Government for "*personel*" to carry on the work, and that whilst the *chart* appeals to the general public—and millions of stars appeal to the popular imagination, the solid work of a catalogue does not do so. Thus a compromise was reached which permitted the Frenchmen to say that the chart was to be commenced at once, although the catalogue plates were first to be pushed on.

The first photographs for the astrographic chart were taken in 1892. By 1909 (six years after Pickering had issued a complete, though small-scale, photographic map of the sky), only the Greenwich Observatory at Oxford had completed and published all its assigned share: 1149 plates and the related catalogue of star positions.[37] Of the other observatories, about half had published some work; the others had not even begun. The Carte du Ciel still remains unfinished.

Pickering's own plan to map the sky by photography had as yet made little progress. Then within a single year occurred two major events that for the first time gave him ample money for assistants and equipment: in 1886 the founding of the Henry Draper Memorial for the study of stellar spectra, and in 1887 the acquisition of the Boyden Fund for the establishment of a high-altitude station.

V

THE HENRY DRAPER MEMORIAL

I

Dr. Henry Draper, along with such men as George Bond, Lewis Rutherfurd, Warren de la Rue, and William Abney, was a notable pioneer in the art of celestial photography. Like his father, Dr. John William Draper, who made the first daguerreotype of the moon, Henry was a New York physician far less interested in the biological than in the physical sciences. His early death cut short a brilliant career but, fortunately for the development of stellar astronomy, his work was carried on for many years through the Henry Draper Memorial, established by his widow's generous gifts to the Harvard College Observatory.

Born in 1837, Henry Draper (Fig. 34) attended the University of the City of New York and, after finishing his studies in the Medical Department, went abroad in the summer of 1857, the same summer in which Bond's photograph of Mizar (ζ Ursae Majoris)—the first double star ever photographed—was exhibited at the Royal Astronomical Society. Draper attended the meeting of the British Association in Dublin, and was then invited with others to see the 6-foot reflector—one of the earliest of the giant telescopes—that the Earl of Rosse had constructed at Birr Castle in Parsonstown, Ireland.[1] Inspired to try building his own telescope, Draper returned to New York and began to construct a 15½-inch reflector designed especially for photography and, with his father, established a private observatory on the grounds of the family home at Hastings-on-Hudson, New York. But the time available for astronomy was limited. In 1859 Henry Draper was appointed to the staff of Bellevue Hospital, in 1860 he became professor of physiology, and in 1866 dean of the Medical Department. Although his professional

211

Fig. 34. Dr. Henry Draper.

advance was rapid, medicine was clearly not his major interest. Of his publications appearing during his lifetime—a textbook of chemistry, papers on telescope making, celestial photography, and methods of photographing the sun's spectrum—only one (his doctoral dissertation) dealt with a medical subject.[2]

In 1867 Draper married Mary Anna Palmer, the daughter of Courtland Palmer, who had made a fortune in the hardware business and in New York real estate. The young Drapers lived in Mr. Palmer's house at 271 Madison Avenue, then on the northernmost fringe of the city, where Henry established his laboratory on the third floor. Later he built a second laboratory over the stables, which was connected to the house by a covered passageway.

Mrs. Draper, rich, red-haired, and a renowned hostess, was from the beginning an enthusiastic assistant in her husband's work. The year before their marriage he had begun constructing a new telescope, a 28-inch reflector, for the family observatory at Hastings. The day after the wedding he took his bride on a shopping trip downtown to choose the glass that would become the 28-inch mirror, an "expedition" that they always referred to afterward as "our wedding trip."[3] The new instrument took several years to complete; Dr. and Mrs. Draper themselves did all the work of correcting the mirror, a painstaking process that involved more than forty regrindings and repolishings. Teaching duties usually kept them in town during the winter but at the close of the school year they moved to their country house at Dobbs Ferry, and on nearly every clear evening they made the two-mile carriage drive to the Draper Observatory at Hastings to photograph the sky. Before dry plates became available, Mrs. Draper had the task of coating each glass plate with collodion just before it was exposed.

Using the newly-finished 28-inch reflector (which later came to the Harvard Observatory) and a quartz prism, Draper made many attempts to record stellar spectra on a photographic plate. As early as 1864 Professor Huggins, in England, had tried to photograph the spectrum of the bright star Sirius but had abandoned the work after he obtained only the continuous background, with no distinguishable lines.[4] Thus Draper achieved a major astronomical triumph when in August 1872 he finally obtained a plate of the spectrum of Vega, the brightest star in the constellation Lyra, showing four strong lines of hydrogen in the wavelength region 4350 to 3440 Ångstrom units. In the next few years he obtained the spectra of more than 100 stars.

When the U. S. Government was preparing to send out scientific expeditions to observe the transit of Venus in 1874, Draper was asked to organize the photographic work of the several groups. Even though it meant a long interruption of his own research, he accepted the assignment and carried it out so well that Congress awarded him a special gold medal bearing the inscription "Decori decus addit avito" ("He adds luster to ancestral glory")—a tribute to both Henry and John William Draper.[5]

Just when Henry Draper's association with the Harvard Observatory began is not known. In 1864 he sent to George Bond a photograph of the moon (made of course by the collodion process) that Bond in his letter of thanks (November 15, 1864) called "magnificent." He added:

> Will you allow me to suggest to you the great importance of attaining such a degree of sensitiveness in the plates that they will furnish, at moderate exposures, images of telescopic stars? There is no method known of determining the distance and angle of position of double stars, which is so exact as that by photography; but our experiments at this observatory have been arrested at about the sixth—seventh—magnitude.[6]

Draper replied (November 21, 1864) that, inspired by Bond's own work, he was already planning to try to photograph the stars, and suggested that "the best place for celestial photography would be somewhere on the west coast of South America—near the equator, for instance—in the neighborhood of Quito, and at a considerable altitude above the sea."[7] This remark has a peculiar interest because, a quarter of a century later, Harvard did establish such a station at Arequipa, Peru, which was supported in part by gifts from Henry Draper's widow.

After Bond's death, Draper had continued his friendly relation with the Observatory; he received volumes of the *Annals* from Winlock, and sent Pickering a note of congratulation on his appointment as director. Preparing to observe the solar eclipse of July 29, 1878, and lacking a suitable stand for the 5-inch telescope he planned to take, Draper appealed to Pickering, who lent him one.

Harvard did not send an expedition. Pickering's reluctance to spend money on eclipse expeditions, then and later, probably stemmed from his disappointing experience with Winlock's party at the 1870 eclipse, in Spain, when clouds had interfered with their plans. Winlock, apparently resolved to look on the bright side of failure, wrote afterward that he

> had prepared two telescopes for these photographs, and if the weather had been more favorable, the results would undoubtedly have added a great deal to our knowledge of the corona. Even as it was, the day being very cloudy, one good photograph was obtained.[8]

The Henry Draper Memorial

One good photograph seemed a poor return for the investment of so much time and money, and Pickering had been impressed with the astronomer's complete dependence on the chance of cloudless weather. In view of the scarcity of money for routine work at the Observatory, he did not feel justified in making an expenditure that might produce no results. However, he freely lent equipment to other observers, such as Henry Draper.

Observations of the 1878 eclipse were expected to be of unusual importance and photography was to be widely used. Astronomers hoped to settle some unanswered questions about the nature of the sun's corona: Was it self-luminous or did it shine by reflected light? Was it composed of solids, liquids, or gases? How far beyond the sun did it extend? They also hoped to find out whether there really existed an intra-Mercurial planet, tentatively called Vulcan: its existence had been postulated by Leverrier and it had supposedly been observed in 1859 by a French physician, Dr. Lescarbault, in its transit across the sun's disk.

Attempting to solve these and other problems, dozens of expeditions with their many pounds of equipment converged on the path of visibility, which extended from Wyoming and Colorado across Texas. Colorado was a favorite spot and her mountains were studded with camping astronomers: Asaph Hall and a group from the Naval Observatory, Arthur Schuster and J. Norman Lockyer of England, E. S. Holden and a second group from the Naval Observatory, Lewis Boss of the Dudley Observatory, C. A. Young of Princeton, Lewis Swift of the Warner Observatory, Winslow Upton (on leave from Harvard), William Pickering from MIT, S. P. Langley from Allegheny, and Maria Mitchell from Vassar.

In Wyoming another host of astronomers took up stations along the route of the Union Pacific Railroad. An expedition from the Nautical Almanac Office directed by Simon Newcomb settled at Creston, and at Rawlins was the large party of Dr. and Mrs. Henry Draper, which included "Mr. Edison, the electrician."[9]

On this occasion, as on similar ones, Mrs. Draper accepted an important but unenviable role. It was her special duty

to count the seconds during the eclipse and lest the vision might unnerve her, she was put within a tent and therefore saw nothing at all of the wonderful phenomenon. Here she sat patiently and accurately calling out the seconds while the glorious and awe-inspiring spectacle was unfolded.[10]

215

The Harvard College Observatory

Most of the expeditions had good weather this time, but their observations did not provide a clear-cut answer to the major questions. At first, the existence of a planet between Mercury and the sun seemed to be confirmed. Both Watson at Rawlins and Swift at Denver reported seeing Vulcan, but the positions they gave did not agree with that predicted and later evaluation showed that both men, in the excitement of the moment, had mistaken ordinary stars for the hypothetical planet. As for the constitution of the corona, Draper's photographs showed a strong continuous spectrum, from which he correctly inferred (though some astronomers disagreed with him) that the corona was not self-luminous.

Draper wrote to Pickering from Dobbs Ferry (August 3, 1878):

My dear Sir
    I have today returned the equatorial stand to Messrs Clark with the request that they restore it to its place in your observatory at my expense. It worked beautifully giving complete satisfaction in every particular and I hardly know how to express my obligation to you for the loan. You have doubtless heard that I have been very fortunate in my results having photographed the spectrum of the corona and demonstrated its continuous or rather non-incandescent-gas character. The spectrum is really that of reflected solar light.

The following summer Dr. and Mrs. Draper spent in Europe. In England he visited the laboratories of A. A. Common and William Huggins, whose work in celestial photography and spectroscopy paralleled his own, and discussed the technical difficulties and the problems of interpretation. At the June 13 meeting of the Royal Astronomical Society he presented what he believed to be proof that oxygen existed in the sun: an apparent coincidence between the bright lines of the oxygen spectrum, made in his laboratory, and certain bright lines in the solar spectrum. (Spectra taken several years later, with greater dispersion, showed the coincidence to be illusory.) Members of the Society were skeptical of his interpretation, although they praised his industry and the excellence of the photographs. Professor Huggins commented:

Personally, I confess, I must ask to be allowed to suspend my judgment a little longer . . . It may be perhaps a misfortune on my part to be of a temperament which in such cases, especially in the department of physics

216

in which I have worked myself, makes me want to look into the subject personally. I must say that before expressing a decided opinion either for or against the existence of oxygen in the sun, I should like to have the opportunity of a little sunshine on the matter.[11]

Back home in New York, Draper continued his experiments, and was the first (September 30, 1880) to obtain a photograph of the great nebula in Orion, the object of so many of William Bond's patient observations and drawings. The picture was much admired by American astronomers but when Draper presented it to the Royal Astronomical Society, through his friend Arthur Ranyard, it was received without enthusiasm. Common, who had also been trying to photograph the nebula and had not then succeeded, remarked that it showed less than a good drawing. Huggins, another rival in the race to photograph difficult celestial objects, probably was the first to record the spectrum of the nebula. In March 1882 Draper obtained spectra showing four lines in the ultraviolet, but in a postcard to Pickering dated March 10, 1882, Huggins clearly implied that he had made such pictures a year earlier:

Upper Tulse Hill. S. W. Last Tuesday (7th) I obtained a photograph of the *spectrum of the Nebula in Orion*. Besides the 4 lines which I found year ago in the visible spectrum there is a line in the ultraviolet λ3730. Will send you further particulars.

For Draper, interest in astronomy had now completely supplanted interest in medicine. He hoped to continue his photographic experiments, he wrote Pickering (March 30, 1881),

as soon as we get up into the country. At present my observations labor under the difficulties of a journey to Hastings of 20 miles back and forward and that at the end of a day's work is tiresome.

In the summer of 1882, at the close of the academic year, he resigned his professorship. Still a young man, financially independent, he planned to devote the rest of his life to astronomical photography, particularly to the study of stellar spectra.

The last meeting between Pickering and Henry Draper took place at Mrs. Draper's dinner for members of the National Academy on November 15, 1882, a dinner that has been called "one of the most brilliant ever given in New York."[12] Draper had returned only three weeks earlier from a vacation, a hunting trip with friends in the Rocky Mountains during which he had ridden some 1500 miles on horseback. On the trip the party were caught in a blizzard and were forced to camp without shelter, in severe cold. Although the ordeal had left Draper exhausted and he was not well when he returned to New York, he worked intensively to prepare astronomical exhibits and physical apparatus to show his guests.

The National Academy had gained in stature since the days of its founding when George Bond and John William Draper had been excluded (Chapter II), and its members now included some of the most brilliant men of American science. Among those attending the dinner were Spencer Baird of the Smithsonian Institution, the physicist Wolcott Gibbs and the zoologist Alexander Agassiz of Harvard, Asaph Hall of the Naval Observatory, S. P. Langley of the Allegheny Observatory, the physicist Willard Gibbs of Yale, F. A. P. Barnard, president of Columbia College, Benjamin Silliman, Jr., editor of the *American Journal of Science,* C. A. Young of Princeton, George F. Barker of Pennsylvania, Cleveland Abbe of the U. S. Weather Bureau, C. H. F. Peters of the Litchfield Observatory, and Edward C. Pickering of the Harvard Observatory—a scintillating company. In the huge dining room of the Madison Avenue house the forty academicians and other friends of the Drapers sat at a table lighted with Edison incandescent lights, some of which were immersed in bowls of water. During the dinner, Draper suffered a severe chill and was not able to eat, but he continued to talk with his friends until the end of the evening. After the party he had another chill and a doctor was called. Five days later, on November 20, Henry Draper died of pneumonia. He was just 45.

II

A few weeks after Draper's death Pickering wrote to Mrs. Draper (Fig. 35), offering all possible help in finishing the work on stellar spectra. The ensuing exchange of letters has particular interest since it delineates the evolution of the Henry Draper Memorial, a project for which Pickering and his assistants eventually photographed and classified the spectra of more than 400,000 stars in both the northern and the southern skies. To publish all the correspondence here is unfortunately not possible but the first two letters will be given in full.[13]

The Henry Draper Memorial

Fig. 35. Mary Anna Palmer Draper.

January 13, 1883

My dear Mrs. Draper,

Mr. Clark tells me that you are preparing to complete the work in which Dr. Draper was engaged, and my interest in this matter must be my excuse for addressing you regarding it. I need not state my satisfaction that you are taking this step, since it must be obvious that in no other way could you erect so lasting a monument to his memory. I fully appreciate the difficulty of your task. There is no astronomer in this country whose work would be so hard to complete as Dr. Draper's. He had that extra-

The Harvard College Observatory

ordinary perseverance and skill which enabled him to secure results after trials and failures which would have discouraged anyone else.

The portion of his work in which I am especially interested is that relating to photographing stellar spectra, and I wish to renew to you the offer I made to him when we last met. I shall not soon forget, when the dinner to the National Academy was nearly over, amidst his many hospitable duties, that he found time for a few minutes talk with Dr. Peters and myself. I urged upon him the importance of an early publication of his stellar spectra (of which I had heard, but have not yet seen)—I called his attention to the similar work now in progress by Dr. Huggins, and that a delay might seriously diminish the value of the investigation. I finally stated that we had the necessary apparatus for measuring his spectra at Cambridge, and I offered to make the measurements myself and send him the results, if this would enduce their early publication. He told me of the change in his affairs, by which he could have more time for his original work, and that he hoped shortly to complete this part of his work. As great delay may ensue if you wait to have the necessary measuring apparatus constructed, it occurred to me that I might aid you in this matter. I should be greatly pleased if I might do something in memory of a friend whose talents I always admired, and regarding whom it will always be a source of regret to me that I should not have met him more frequently. I am aware that you must have many advisors in this matter, every astronomer seems to feel the loss a personal one. Dr. Peters was with me when we heard the sad news—at the Clark's—observing with their great lens. But he took no further interest in the work, in nothing but in the immediate effort to get a New York paper, hoping that the rumor might prove false. The rest of his stay with me was saddened by his continual reference to our dear friend.

Whatever may be your final arrangements regarding the great work you have undertaken, pray recollect that if I can in any way advise or aid you, I shall be doing but little to repay Dr. Draper for a friendship which I shall always value, but which can never be replaced.

Very truly yours,
E. C. Pickering

Mrs. Draper replied on paper edged with a half-inch black border:

271 Madison Avenue
New York, New York
January 17, 1883

My dear Prof. Pickering:

Thanks so very much for your kind and encouraging letter. The only interest I can now take in life will be in having Henry's work continued, yet I feel so very incompetent for the task that my courage sometimes

The Henry Draper Memorial

completely fails me—I understand Henry's plans and his manner of working, perhaps better than anyone else, but I could not get along without an assistant and my main difficulty is to find a person sufficiently acquainted with physics, chemistry, and astronomy to carry on the various researches—I will probably find it necessary to have two assistants one for the Observatory and one for the laboratory work, for it is not likely that I will find any one person with the varied scientific knowledge that was peculiar to Henry. I am willing to pay good salaries, so as to secure first class men, but I should want men willing to identify themselves with the work already commenced and develope it—I should like first to continue the stellar spectrum work, for that is comparatively simple. I am exceedingly obliged to you for your kind offer, to have the spectra measured and will certainly avail myself of it. I want to be careful not to publish anything that Henry did not consider ready. He wished to get the spectra of several of the winter stars, and this we intended doing this winter. It is so hard that he should be taken away just as he had arranged all his affairs to have time to do the work he really enjoyed, and in which he could have accomplished so much. I cannot be reconciled to it in any way.

Of the stellar spectra with which he was satisfied in definition etc. I have Vega, Capella, Alpha Aquilae [Altair], Arcturus. Of Antares, I have as good a spectrum as we could get for it is a very red star and being very far south it was difficult to get a satisfactory exposure on it. We have taken a great many others but those I mention are really fine. Whether it would be best to have those measured and described or wait until I can perfect others is a matter in which I will have to ask Henry's friends to advise with me—What do you think would be the wiser plan—I want to be very careful not to do anything that would in any way bring discredit on the work Henry has done. It will undoubtedly be quite a time before I can add anything to the collection for if I find someone to help me, it will take a long while to get any results for whoever comes to the Observatory even as an assistant will have a good deal to learn.

The whole undertaking seems so beset with difficulties, that I feel sometimes unequal to the task, and if it were not that Henry's friends gave me some encouragement I am afraid I should almost be tempted to give it up.

My real plan is, as soon as possible to get the work running under my own direction, then when I can buy the place at Hastings where the Observatory is, to do so,—then move the laboratory and all the apparatus there and eventually endow the whole as an institution for original research in astronomical physics to be called the Henry Draper Astronomical and Physical Observatory. As long as I could I should keep the direction of the institution myself. It seems to me the only suitable memorial I can erect to Henry, and the only way to perpetuate his name and his work.

I hope you have not lost all patience with my long letter, but your own letter was so very kind, that it tempted me to explain my plans to you in the hope that you would give me some advice in the matter. I am so unusually alone in the world, that without feeling that those friends who were interested in Henry's work would advise me, I could not do anything. I do not care to talk to people generally about my plans, but if you should see Mr. George Clark I would be glad if you explained my ideas to him, for in my letter to Mrs. Clark I merely referred to continuing the work without going into particulars—

Thanking you again for your kind interest and sympathy I remain,

Yours very truly,

Anna P. Draper

Pickering responded (January 20, 1883) with sympathy and encouragement.

As you ask for advice, I will put it in more detail than I should otherwise venture to do. I should certainly publish work already done, even if not entirely complete. There must necessarily be a great distinction between the past and future. It must be a long time before you can expect that your assistants will attain results equal to Dr. Draper's, however skillful they may be.

As regards the stellar spectra, the first thing would be a list of all the plates. The number of photographs of each star, except of course the failures. The number where the impressions are good, medium or bad should be specified, as a poor photograph may have value if we have no other. It would next be necessary that I should examine the plates on the measuring machine. Are you inclined to send me some as samples? . . . Dr. Huggins results indicate that the stars may be grouped into well defined classes. If this result is confirmed the work will be much simplified . . . Do not hesitate to write in detail, in no way can I aid astronomical science more effectively than by any assistance I can give you in this matter.

Early in February Mrs. Draper, with her friend Professor Barker,[14] visited the Observatory in Cambridge. She brought with her Draper's photograph of the nebula in Orion (Pickering hoped that a careful examination would show hitherto unknown stars, too faint to be seen in the telescope), and 21 photographs of stellar spectra, none longer than a quarter of an inch. Small as they were, Pickering undertook the work of examining them under a microscope, using a micrometer to measure the positions of the lines in relation to those of the sun's spectrum and to calculate their wavelengths. Preliminary measurements were encouraging. Pickering wrote Mrs. Draper (February 18, 1883) that "there is much more in the photographs than appears at first sight . . . A proper interpretation of the results proves more difficult than the measures." He suggested that she publish a list of all the spectra Draper had photographed, together with enlarged prints of the best and the wavelengths of the lines they showed.

By March, Pickering had completed his measures of the spectra. He drafted a paper describing the results and sent it to Mrs. Draper for her comments. After thanking him for the trouble he had taken, she ventured one remonstrance (April 3, 1883):

You suggested that I might wish to make some changes in the text of your manuscript, and as I want to have Dr. Draper's views of the spectra represented, I will take advantage of your permission in regard to the classification. Dr. Draper did not agree with Dr. Huggins in the opinion that Alpha Aquilae and Alpha Lyrae belonged to the same class. The bands in both are the same, but in α Aquilae between the bands there is an immense number of fine lines, which we could never get in the spectrum of α Lyrae. Prof. Young was so impressed by this peculiarity in the spectrum of α Aquilae that he asked Dr. Draper to give him one of the photographs of the spectrum of that star, which he did. In view of this I should not like to accept Mr. Huggins' classification as the standard when Dr. Draper did not agree with it.

You will not I hope be annoyed at my criticism, but I feel in publishing any of Dr. Draper's work that I want his opinions represented as nearly as possible, now that he is not here to explain them himself.

(Later work showed the correctness of Draper's view. The spectrum of α Aquilae contains many metallic lines not visible in that of α Lyrae, and α Aquilae is of a later spectral type.)

At the meeting of the American Academy on April 11, 1883, Pickering discussed his measures of the Draper spectra before a distinguished audience that included Otto Struve,[15] director of the Pulkovo Observatory near St. Petersburg; B. A. Gould, formerly director of the Dudley Observatory at Albany and now director of the National Observatory at Cordoba, Argentina; and Asaph Hall of the Naval Observatory, who in 1877 had discovered the two satellites of Mars. Mrs. Draper was not willing to publish the spectral work, however, until other material could be added—pictures of the Hastings Observatory, excerpts from Draper's research notebooks, and a detailed introduction by Professor C. A. Young.

That summer, Professor and Mrs. Pickering were in Europe (Chapter

The Harvard College Observatory

IV). After returning to Cambridge he resumed his correspondence with Mrs. Draper regarding the manuscript. Finally completed and published in February 1884, the account contained data on 78 of Draper's spectra and gave the measured wavelengths of 21.[16] Copies were mailed to all the leading astronomers, most of whom accepted the work as a sound and useful piece of research. An exception was Professor Huggins, who had devoted years to the measurement of stellar spectra (although his publications did not give the details of the method on which he based his calculations). Some of his wavelengths, particularly for the star α Lyrae, did not agree with those Pickering had determined from the Draper plates. In an extremely critical letter (March 12, 1884), Huggins wrote of Table VIII given in the Pickering-Draper publication that

the comparison of your measures of the lines of the typical spectrum of the white stars shows a fair agreement to a little beyond the end of the visible spectrum, (to line β) but beyond, your wave lengths are *very wild indeed*.

I have no doubt of the great accuracy of the positions of all these typical lines as they were remeasured and verified in many different ways. I am *sure* no one is out as much as *one* [Ångstrom] *unit,* I *believe* not .5 . . .

There can be *no doubt whatever* that your curve is quite *inaccurate* and increasingly so as the wavelengths get smaller. I should be glad if you could see your way to look into this, because it would be better that you should discover the error and publish the correction, than that the matter should be pointed out by others . . .

It would be a great drawback to the advancement of science if your measures were allowed to pass without criticism on this point, but as it is obviously of the nature of a slip, I should be so glad if you would *yourself* look into the matter and yourself publish the correction as soon as you find out the cause.

Pickering forwarded this letter to Mrs. Draper with a dry comment (March 31, 1884): "Fortunately, Dr. Huggins' arguments, that results must be wrong if they do not agree with his, will not generally be regarded as conclusive unless supported by facts."

On the same day he replied to Huggins, with commendable restraint, and ventured a mild criticism of his own:

Your letter of March 12 duly reached me, and I am much indebted to you for your kind promise to send copies of your photographs of stellar spectra for our museum. I fully agree with you that it is important to determine the cause of the difference in your results regarding the wavelengths of the lines in the stellar spectra and those I have deduced from Dr. Draper's photographs. Before the correctness of either can be regarded as established, they should be checked step by step. This may be done for my results, since the material for the reduction is given . . . so fully . . . that if wrong any reader can detect the error.

Your publication[17] does not enable the reader to verify your reductions, and he has no means of checking your results. I hope that you will supply this deficiency by a fuller publication and thus remove what seems to me the weakest part in your very valuable contribution to the subject . . . The determination of the true wavelength is doubtless important; whether the error is yours or mine is of little consequence. Any aid I can render you towards the solution of the first of these questions, I will gladly give.

But Huggins seemed unwilling to consider the possibility that his own work could contain errors. He returned to the battle in an undated letter, probably written in mid-April.

I received your letter . . . I cannot tell whether you have been led astray by your formula or by measuring wrong lines, probably the photographs were very badly defined, or there may have been some shift in the apparatus.

Some of our best men to whom I have shown your paper do not seem to think your method serious enough to make it worth while for me to take any notice of the paper, but considering the number of astronomers and bookmakers who have not any great practical knowledge of the subject, I hardly know whether I ought to let the matter pass. It is always a painful matter to me to find fault publicly. I have written to you plainly indeed, because I felt you would take it as a mark of friendly feeling.

Pickering checked his work but could find no source of error. He concluded that the different temperatures at which the photographs were made could account for the discrepancy between his results and those of Huggins. In any case, he believed argument to be a waste of time that ought to be spent in finding out what the facts were, and he did not reply to this letter. But Mrs. Draper expressed her indignation (April 30, 1884) at Huggins' attitude:

I felt very sorry that you should have been subjected to such an ungentlemanly attack, through your interest in Dr. Draper's work. If Dr. Huggins did not find the results of your measures agreed with his, there was no objection to his saying so if he had expressed himself in a more courteous manner. I was glad to see your excellent letter sent in reply—If anything could impress a person like Huggins, it would be the gentlemanly way in which you maintained your position in answering his attack.

A modern appraisal[18] of the controversial data has shown that neither set of measures was accurate, although Pickering's was slightly nearer the correct values. Huggins' wavelengths for the lines in question were consistently too low by about 3 angstroms, while Pickering's were too high by about 2 angstroms. Given the quality of the photographic plates of the day and the very small scale of these early spectra, more precise measures could scarcely have been achieved.

III

More than two years had now elapsed since the death of Henry Draper, and Mrs. Draper's plans to find an assistant and establish her own observatory were at a standstill. Although she had been a devoted and useful aide to her husband, she had no basic training in physics or astronomy and hence could accomplish very little alone. Pickering had suggested the names of several competent men, including H. M. Paul, then professor of astronomy at the University of Tokyo, W. C. Winlock of the Naval Observatory (the son of Joseph Winlock), and T. C. Mendenhall, then professor of physics at Ohio State University. None had seemed exactly suitable to Mrs. Draper. Finding the right man (that is, another Henry Draper) was proving extraordinarily difficult.

Pickering sympathized with her genuine wish to resume work in astronomy. At the same time, his own plans for the Observatory were hampered by lack of money. The five-year subscription begun in 1878 had come to an end, and within the year he had been forced to discharge several experienced assistants because he could not pay their salaries, small as they were. Friends of the Observatory had always been generous but the income from their endowments was not large enough to allow an expanded program. Like other directors of research institutions at this period, he could only try to raise funds by again appealing to the public for donations or by approaching some wealthy patron, who might be willing to help advance the cause of science in return for having his name attached to an instrument or a building.

Faced with the two problems of ensuring the continuation of Draper's spectral research and obtaining funds for the Observatory, Pickering

226

The Henry Draper Memorial

at last wrote to Mrs. Draper with a tentative suggestion for collaboration (May 17, 1885):

I am making plans for a somewhat extensive piece of work in stellar photography in which I hope that you may be interested . . . There is every prospect that I shall be able to do this with an appropriation from the Rumford, Bache, or some other fund. But the work is so nearly in the line of that which will I hope eventually be undertaken at the Henry Draper Observatory that I wish to ask if you would not like to enter upon some plan of cooperation by which Dr. Draper's name should be associated with the work. My plan is that the Clarks (who are much interested in the problem) should make the best lens they can for the purpose. That would probably be a double lens like that of a camera and of 8 inches aperture and 44 inches focus. With this an assistant under my direction would take photographs each covering a region of at least ten degrees square, which would furnish maps of the heavens from which the position, brightness variability and (indirectly) the color of the stars could be determined. We have also some experiments in progress which encourage the hope that the spectrum of all the brighter stars in these regions could be photographed . . . I think there will be no difficulty in carrying out this plan without your aid. On the other hand, if it commends itself to you, I am confident that we could make it conform to such conditions as you might impose . . .

A vast field of work is now opening in astronomical photography, and I should be very glad to aid in any plan by which Dr. Draper's name should continue to be connected with it.

In her reply (May 21, 1885) Mrs. Draper sounded puzzled but not discouraging:

Thanks for your kindness in remembering my desire to be interested in some work, with which Dr. Draper's name could be associated, and his memory kept alive. I will be glad to cooperate, if I can, in what you suggest, for its bearing on stellar spectrum photography appeals to me very strongly.

I do not quite understand from your letter in what way you thought I could be of service, whether by contributing towards the experiments being made at Cambridge, or by having similar work carried on here. If you will kindly let me know your ideas more fully, it will give me pleasure to assist if I can—

But Pickering was not yet ready to propose a definite plan. Mrs. Draper went to Europe for the summer and, although they exchanged a few letters during the autumn, the project remained vague. On Christmas Day, 1885, he wrote that he was sending her specimens of the

The Harvard College Observatory

Observatory's recent photographs, which he had already exhibited to the National Academy: the Pleiades, the nebula in Andromeda, and some stellar spectra. In her letter of New Year's Day, 1886, Mrs. Draper thanked him, rather wistfully, for the "exceedingly interesting photographs." On January 24, 1886, Pickering responded with an explicit offer of collaboration.

A remark in your note of Jan. 1 'that it seems impracticable to think of doing anything at the Hastings Observatory' has given me much concern. The various difficulties you have encountered in establishing the Henry Draper Physical Observatory have met with my hearty sympathy. In the hope that I may aid you I venture to suggest the following scheme, which at least has the advantage that it leaves you free to alter your plans at any time. Suppose that you should appropriate the interest only, instead of the principal, of a fund to astronomical physics, with the understanding that it will be continued only so long as satisfactory progress is maintained. It is very certain that no money would be wasted under these circumstances, and that every effort would be made to secure the greatest results. I can imagine no greater spur to all concerned, than the expectation that the continuance of a work would depend on its successful conduct. How would you like to begin at once to carry on some research in this way? In other words to do now, what would be done if your Observatory were already in operation. If this plan could be tried in Cambridge I would gladly give the necessary supervision.

Receiving a very encouraging reply, Pickering sent to Mrs. Draper (February 4, 1886) a detailed plan of cooperation:

That the Harvard Observatory should furnish the necessary supervision and the principal instruments consisting of the eight inch photographic telescope and prisms, and the fifteen inch equatorial so far as this could be used without interfering with its other work, also the rooms required for the computations, measurements, and photographic work. That the Draper Observatory should furnish the means for the current expenses of the work . . . I hope that your visit to Boston will take place soon that we may discuss these and other interesting topics.

On February 14, 1886, the date they later agreed on as the official beginning of the Henry Draper Memorial, Mrs. Draper wrote:

Your letter of February 4th has remained unanswered as I hoped in replying to it to be able to say when I should be in Boston,—but my plans are still uncertain.
I should be very glad to make such an arrangement as you suggest

The Henry Draper Memorial

if you think in publishing, the results could be in some way associated with Dr. Draper as a continuation of his work . . .

I would be willing if the plan could be carried out satisfactorily, to authorize the expenditure of two hundred dollars a month or somewhat more if necessary—

The project now took shape rapidly. Mrs. Draper presented to the President and Fellows of Harvard College the sum of $1000 to establish the Henry Draper Fund for the study of stellar spectra. The money was to be treated as income, not capital; Pickering could spend the entire amount and Mrs. Draper would supply more as needed. In return, Pickering and his assistants would photograph, measure, and classify the spectra of stars; the resulting catalogue would be published in the Harvard *Annals* as a memorial volume to Henry Draper.

A Circular dated March 20, 1886, announcing the establishment of the Henry Draper Memorial, was mailed to observatories, scientific societies and libraries, astronomers and other scientists in the United States and abroad, to the editors of *Nature* and *Science,* and to the New York and Boston newspapers.

At the Observatory, work began at once. New equipment was bought, new assistants were employed. In May, Mrs. Draper arrived in Cambridge to plan for the transfer of Draper's 11-inch photographic telescope from the Hastings Observatory to Cambridge, and for the construction of a new building to house it (Fig. 36). After arranging for the regular payment of funds to the Observatory, Mrs. Draper went to Europe for the summer, and on her return Pickering was able to report that he had already obtained stellar spectra over almost the entire sky visible from Cambridge; they included some 3000 identified stars.

Early the next year, Pickering sent to Mrs. Draper some recently obtained spectra. Although enlarged to unprecedented size—4 inches wide and 2 feet long—they were still sharp and clear. He remarked (January 18, 1887) that when the photographs were distributed they would probably "induce other astronomers to undertake the same work," and that he had devised a plan by which she might, if she liked, preempt the entire field. He suggested that they expand the work to include the spectra of faint stars, as well as a comparison between the spectra of stars and those of the chemical elements existing on the earth.

The current expenses of this investigation would be small if you were inclined to use the 28 inch reflector. If not, perhaps we could use our

229

The Harvard College Observatory

Fig. 36. The Observatory as it appeared in 1887; photograph taken from Madison Street: center foreground, 28-inch Draper reflector; right, 11-inch Draper photographic telescope; rear center, dome of the 15-inch refractor in Sears Tower.

15 inch refractor . . . If you propose to study terrestrial spectra, a delay may result in the ground being covered elsewhere . . . To carry out the entire plan in the way that I should recommend would involve an expenditure during the coming year of about five thousand dollars for current expenses and three thousand five hundred for permanent outfit. Of course this sum could be reduced indefinitely either by doing the work more slowly, or by taking up a part of it only . . .

I am daily expecting the transfer of the Boyden Fund [see Chapter VI] to this Observatory. If that is done my Brother will have a position here and we shall then be able to call on him more freely for advice and aid regarding the spectrum work . . . I hope that you will soon be able to appoint a time when you can come to Cambridge, for Mrs. Pickering and I want very much to have you under our own roof.

Mrs. Draper responded (January 23, 1887) with what amounted to a final abandonment of her dream of establishing a Henry Draper observatory, and an agreement to turn the work over to Harvard:

It scarcely seems possible that stellar spectra can be taken which will bear the enlarging of those that you have sent me, for even in the enlargements the definition is so good. It seems as if the prisms must be almost perfect—I wonder what Mr. Huggins will say when he sees them—I quite agree with you in feeling that I should like to appropriate the entire ground

230

that is possible, before any distribution is made of the photographs, and am entirely satisfied to devote nine or ten thousand a year to carrying on the work, and unless some very unexpected financial difficulty should occur, there is nothing to prevent my doing so—It is somewhat more difficult to make up my mind to remove the Observatory from Hastings, for if I take the 28 inch, I might as well put the 15 inch to some use, for although we have not used it lately, we were talking of having it mounted equatorially, for its definition is so good, as was shown by the lunar photograph—The struggle is to decide to remove the 28 inch, but it probably will have to be done sooner or later so I will think about it.

Pickering wrote to Mrs. Draper (January 25, 1887) of his great pleasure

that you are inclined to carry on the photography of stellar spectra on so liberal a scale . . . I quite understand your feeling regarding the removal of the 28 inch, and the transfer to Cambridge of the investigation of terrestrial spectra. I shall await with much anxiety your decision, but hope that you will make this additional sacrifice to science, if you are satisfied that the work had better be done here. Mrs. Pickering writes by this mail to tell you how very glad we shall be to welcome you here.

The next day Pickering sent news of the expanded Draper Memorial to President Eliot, who was about to leave for Europe, and a few weeks later received a note from Eliot's son Charles saying that his father had arrived at Gibraltar and

asks me to thank you for the news of the "Draper subvention" received just as the ship cast off from the Brooklyn pier.

The voyage was a good one and he is now presumably riding a donkey either in Spain or Africa.

It is doubtful that Pickering felt envious, even in the raw cold of a Boston March. At the Observatory, the road ahead now looked broad and clear, for the first time since he had become director, and the possibilities of research must have seemed almost limitless. The Boyden Fund had just been given to Harvard and, with Mrs. Draper's promise of so large an addition to the spendable income, Pickering could reasonably hope to realize his dream of obtaining spectra of all the visible stars and mapping the sky by photography.

231

The Harvard College Observatory

By the method commonly in use at this period, the image of a star was focused on the slit of a spectrograph and spread by a prism or a grating into a spectrum, which was then photographed. This method could record the spectrum of only one star at a time. Pickering decided, instead, to adapt for photography the method originally used by Fraunhofer and Secchi in their pioneer visual studies. A prism placed before the object glass of the telescope would produce spectra of all the stars in the field of view, which could then be photographed on a single plate. Photographic plates exposed for about 5 minutes would record the spectra of stars brighter than the sixth magnitude. Exposures of an hour or more would provide the spectra of fainter stars, down to about the eighth magnitude.

The first work planned for the Henry Draper Memorial involved the preparation of two extensive catalogues. The first, a continuation of the project begun earlier with the Bache instrument, would list the spectral types and magnitudes of all stars visible to the telescope in the region extending from the north celestial pole to 24 degrees below the equator.

The second catalogue would be based on photographs made with the 11-inch Draper telescope, which could produce plates showing much greater detail and a wider separation of the spectral lines than did the Bache. This catalogue would present a classification of some of the brightest stars in the same region, and would give a thorough analysis of their spectra, together with the wavelengths of individual lines measured from a comparison with those of the sun's spectrum.

The first rough systems of classifying stars by their spectra, devised in the 1860s, were based on visual studies alone and depended on gross differences in the position and grouping of lines. The early workers had recognized that stars of a given color tended to produce spectra of the same general kind. Lewis Rutherfurd, who was perhaps the first to formulate a classification, suggested separating spectra into the three types characteristic of white, yellow, and red stars. Father Secchi of Rome, and later Vogel of Potsdam, recognized four basic types. Class I contained the white stars, like Sirius, whose spectra showed the lines of hydrogen prominent against a continuous background. Class II, the yellow stars like Capella and the sun, produced spectra having many fine lines, among which those of iron and magnesium were conspicuous. Class III, which included the orange or red stars like Antares, showed strongly marked groups of lines, some of which were bright instead of dark. Class IV, the stars that were red but faint, produced an alternation of bright and dark lines.

No general agreement existed as to the physical significance of the four types of spectra, but the opinion soon gained ground that they must reflect some kind of evolutionary process. Secchi believed that the various colors of the stars represented differences in temperature. Vogel thought they might also indicate differences in age; this view, which is no longer accepted, has been summed up as follows:

The hot, white stars, rich in hydrogen, were considered to be in the heyday of their youth; the cooler yellow stars, like our sun, whose atmospheres abounded in metallic vapours, were of middle age, and lastly, the red stars, whose beautiful fluted spectra indicated a comparatively low temperature, were presumably in the evening of their life-time, the final stage before extinction.[19]

Some astronomers, such as E. M. Maunder, did not agree that the spectral types represented stages in stellar evolution but ascribed them only to basic differences in chemical composition. Huggins thought it possible that, under the different temperatures existing in individual stars, the gases might differ in their capacity to absorb light, or that the amounts of the chemical elements present might vary from star to star. Norman Lockyer considered that the spectra expressed differences in both chemical composition and temperature, and represented progressive stages in development. According to his views, evolution proceeded from cool, young, red, giant stars, to hot, middle-aged, white stars, to cool, old, small, and highly compressed red stars. One of the most original theoreticians of the period, Lockyer carried out laboratory studies of the spectra produced by chemical elements at high temperatures, and used the results to help identify unknown lines in the spectra of the sun and other stars. He found that when an element was vaporized at the comparatively low temperature of the flame in a Bunsen burner, at the higher temperature of the electric arc, or at the still higher temperature of the electric spark, its spectrum changed. The strengthened lines produced at the highest temperatures he called "enhanced," and theorized that either the higher temperature or the electric stresses used had broken down the element and dissociated it into simpler substances. Only years later did it become apparent that he had actually been observing the spectral lines of the ionized elements, atoms that had lost one or more of their normal complement of electrons.

This question of what physical differences accounted for the several types of spectra was one that Pickering did not concern himself with. His object, as it had been in compiling the photometric catalogues and as it was later in producing star maps, was to accumulate large amounts of data and to make them available to other astronomers. Such information, he believed, was a prerequisite to sound theory and to an eventual understanding of the structure and behavior of stars. Relegating speculation to the future, he organized the procuring of spectra so that it could be carried out as systematically as the work of an assembly line.

In the 1860s, when Rutherfurd, Secchi, and Vogel had begun their attempts at classification, using visual means alone, their material had necessarily been sparse, for many hours of fatiguing work at the telescope had been necessary to determine the spectrum of a single star. To map a single spectrum completely, Huggins had said, would have required several years. The first catalogue of stellar spectra, published by Vogel in 1883 and listing some 4000 stars, represented nearly 20 years of visual observation. Now, one photographic plate in only a few minutes could permanently record the spectra of as many as 200 stars, to be studied and classified at leisure.

Willard P. Gerrish, who had joined the staff the preceding year and had proved adept at devising improved instrumentation, was placed in charge of the photographic work. Each clear evening he or one of his assistants arrived at the Observatory at dusk and began to take pictures of the selected areas of the sky with both the Bache and the Draper telescopes, which were kept steadily at work until morning twilight dimmed the stars. Each night's accumulation of plates was carried in lightproof boxes to the darkroom of the photographic cottage, where they were developed, copied, and enlarged on glass. The enlarged copies were transferred to the computing room and mounted so that they were illuminated by daylight. The ladies of the computing corps then determined the identity of each star represented on a plate and calculated its celestial position for the year 1900.0. The assistant in charge, using the sun's spectrum as a standard, measured the position of the lines in each spectrum and from this study assigned the star to the appropriate spectral class. The work was slow. No calculating machines were available and all computations had to be carried out laboriously with pencil and paper, the chief aids being tables of logarithms. Nevertheless, so efficient was the routine that by the end of the first year the spectra of more than 8000 stars had been photographed, measured, classified, and catalogued. By the end of the second year the number had more than tripled.

During the first months of the Draper Memorial work the measuring was done chiefly by Miss Nettie A. Farrar, but at the end of the year she decided to leave to be married, as Pickering reported in a letter to Mrs. Draper (December 31, 1886). He added that, before leaving, Miss Farrar was "instructing Mrs. Fleming who has assisted me, and who will I think take her place satisfactorily." This opinion soon proved to have been an understatement. Williamina P. Fleming, although she had no particular training in mathematics, physics, or astronomy, displayed unusual aptitude. Encouraged by Pickering, she learned rapidly and within a few months had developed into one of his ablest and most enthusiastic assistants.

The spectral images produced by the Bache telescope were less than half an inch long. When enlarged with the aid of a cylindrical lens to about five times that length, they showed so much detail that Secchi's four general classes were not adequate to express the number of types that could be distinguished. Mrs. Fleming therefore subdivided Secchi's classes to produce 15 groups, differentiated largely by the intensity of the hydrogen lines. She used the letters of the alphabet from A to O (J was omitted because in German script it is indistinguishable from I) for these groups, plus the letter Q to indicate peculiar spectra that did not fit anywhere in the scheme.

After four years of work by the computing staff, the first of the projected volumes was published in 1890 as Volume 27 of the Harvard *Annals. The Draper Catalogue of Stellar Spectra*, chiefly the work of Mrs. Fleming, listed the spectral classes and magnitudes of more than 10,000 stars down to about the eighth magnitude. For some stars a single photograph had been enough for classification. For others, the decision was more difficult and required the examination of from two to a dozen or more separate photographs. Thus the ultimate classification of these 10,000 northern stars was based on the study of nearly three times that number of individual spectra.

While the first catalogue was in preparation, the second major project, an analysis of the spectra of the brightest northern stars, had begun. Under Pickering's supervision, the study was placed in charge of Miss Antonia C. Maury, Henry Draper's niece. More fortunate than most girls of the time, Miss Maury had attended one of the few existing women's colleges—Vassar—where she had studied chemistry and mathematics. After coming to the Observatory in 1888 she was attracted by the problems of spectral analysis and the challenge they offered.

Spectra for the first catalogue had been photographed wholesale. Although large numbers appeared on a single plate, each image was small and the dispersion was limited. For the second catalogue, however, Pickering's object was not quantity but quality: to obtain the best possible spectra of a few selected bright stars and subject them to intensive study. For this purpose the 11-inch Draper telescope was used, with from one to four prisms, to photograph the spectrum of a given star alone on a plate, with as wide as possible a separation between the lines. The resulting spectra were originally from 4 to 6 inches long. Enlarged and examined under the microscope, they showed an unprecedented complexity of structure. In 1872, when Henry Draper had obtained the first successful photograph of a star's spectrum (α Lyrae), the negative had recorded only four lines between the Fraunhofer lines G and H. So great had been the technical advances during the intervening years that photographs of α Lyrae now disclosed more than 100 lines in the same spectral region.

Since the plates showed so much more detail than those made with the Bache telescope, Miss Maury decided that the Fleming classification could not adequately describe the characteristics of the spectra. She therefore devised a more complex system of 22 groups, designated by Roman numerals. Classes I to VI included the brilliant white stars whose spectra exhibited strong "Orion" lines (later identified as due to helium) and those of oxygen, hydrogen, nitrogen, and silicon, with a few faint lines of metals. Classes VII to XI (Secchi's type I) were bluish-white stars whose spectra contained very few Orion lines and showed only faint metallic lines; the lines of hydrogen were prominent. Classes XII to XVI (Secchi's type II) were yellow stars of the solar type; the hydrogen lines were diminished in intensity; the lines of iron and other metals

236

were conspicuous. Classes XVII to XIX (Secchi's type III) were red or orange stars with strong calcium lines and banded spectra. Class XX contained variable stars showing fluted spectra. Class XXI (Secchi's type IV) were red stars whose spectra showed bright hydrogen lines and the bands of the carbon molecule; and class XXII (a "fifth type," which Pickering had proposed as a useful addition to the Secchi scheme) contained bright-line spectra like those of the planetary nebulae.

But even these 22 categories were not sufficient to define all the variety that existed. Some stars that apparently belonged to the same class, having the same color and lines that occupied identical positions, displayed marked dissimilarity in the width or the intensity of some lines. To take account of these differences, Miss Maury supplemented each class with three "divisions." In spectra of division a the lines were wide and well defined. In division b they were relatively wide but hazy, and of about the same intensity as in division a. In division c, the hydrogen and Orion lines were narrow and sharply defined, while the calcium lines were more intense than in the other divisions. Some spectra exhibited both the a and the c characteristics, and were designated ac. In her final publication, Miss Maury emphasized the importance of the c characteristic. She was convinced that it represented some fundamental difference in the properties of stars whose spectra otherwise seemed to be identical.

A photographic map of the solar spectrum, prepared on the same scale as that of the stellar spectra, served as a standard by which Miss Maury measured the wavelengths of the lines present. She identified them when possible, and assigned the star to the appropriate group and division.

The preparation of the catalogue went slowly, far more slowly than Pickering wished. As with the photometric work, he wanted to publish the material as quickly as possible, so that it would be useful to other astronomers. Refinements of measure and theoretical speculation he preferred to leave to the future. Therefore he sometimes showed impatience at what seemed to him a laggardly rate of progress, an impatience shared by Mrs. Draper. Miss Maury spent nearly nine years at this task. Periods of poor health, difficulties with eyesight, and her own determination not to overlook any bit of information contained in these richly complex spectra contributed to the delay.

The work was nearing completion when in March 1895 William

The Harvard College Observatory

Ramsay in England identified the spectrum of helium in the laboratory. This element, first detected by Norman Lockyer at the eclipse of 1868 from a bright yellow line in the spectrum of the solar prominences, had not previously been found on earth and had been supposed to occur only in the sun, a few other stars, and novae. Once identified as a terrestrial element and thus made subject to laboratory study, it became an important adjunct in determining the chemical constitution of the stars. The "Orion" lines, whose source had been a mystery, were now recognized to be those of helium. Although this new information was obviously important in classifying stellar spectra, to go back and rewrite the entire catalogue seemed undesirable. It was therefore published in its original form. However, supplementary notes identified some of the Orion lines with those of helium and briefly discussed the implication for stellar classification.

The Spectra of Bright Stars was finally published in 1897 as Part I of Volume 28 of the Harvard *Annals*. Based on Miss Maury's examination of some 4800 photographs, it gave detailed analyses of the spectra of 681 bright northern stars.

Part II of the same volume, a study of the spectra of 1122 bright southern stars, appeared three years later. The work of Miss Annie J. Cannon (Chapter XI), it was prepared from photographs taken at the Observatory's station at Arequipa, Peru (see Chapter VIII).

With Pickering's approval, to achieve simplicity Miss Cannon modified the system of classification used in Part I. Miss Maury had recognized the physical significance underlying the differences in stellar spectra, and her classes I–XXI represented a true sequence based on descending temperature. Valid though her scheme was, it was complex and difficult to use. Miss Cannon therefore decided to go back to the use of the lettered groups distinguished by Mrs. Fleming in the first Draper catalogue, but to rearrange them to accord with the temperature sequence represented by Miss Maury's classes. Miss Cannon's system, for example, reversed Mrs. Fleming's sequence A, B, to B, A, to reflect the fact that class B stars, although they displayed hydrogen lines that were fainter than those of class A, were at an earlier stage of evolution than stars of class A. Thus Mrs. Fleming's groups assumed the order O, B, A, F, G, K, and M. (To keep the correct sequence in mind, astronomy students traditionally rely on the mnemonic, "Oh, Be A Fine Girl, Kiss Me!") Subdivisions were indicated by Arabic numerals. Miss

The Henry Draper Memorial

Cannon did not adopt the collateral divisions set up by Miss Maury, although she listed some stars as "peculiar" and in detailed notes remarked that in certain spectra the lines were broad and hazy, very intense, or narrow with sharply defined outlines.

Miss Cannon's system proved to be so simple to use that in 1910, by vote of the International Solar Union, it was adopted as the official classification system at all observatories. With slight modifications, it is still in use and is known as the Harvard spectral classification.

Miss Maury continued to work at the Observatory at intervals, analyzing the spectra of certain variable stars that interested her, particularly β Lyrae. This eclipsing variable, discovered in 1784, is a very peculiar white star whose behavior has challenged the curiosity of many astronomers and given its name to a whole class of stars. It displays irregular fluctuations in light whose nature is still not completely understood. The spectra on the Draper plates indicated the presence of two components belonging to different spectral classes, and showed both absorption and emission lines. Later work, by Miss Maury and others, demonstrated that β Lyrae is a very close binary system of nonspherical components, with a supergiant primary revolving rapidly around a darker emission star of great mass. The spectrum of β Lyrae also displayed peculiarities thought to arise in a complex atmosphere that enveloped both components. The shifting pattern of spectral lines caught Miss Maury's fancy and she devoted much time to working out a physical explanation for these changes.[20] Certain irregularities in the pattern of its light fluctuation are still not wholly accounted for.

The basic classification system Miss Maury had devised seemed well on the way to oblivion when, early in the new century, it attracted the notice of Ejnar Hertzsprung, a brilliant young Danish astronomer. In 1905 Hertzsprung had attacked the problem of determining the absolute magnitudes, or real luminosities, of stars. This quantity could be calculated from a star's apparent magnitude if its distance were known, but at that time distances had been measured for only a few stars. However, stars with large proper motion were known to be relatively near, and those with small or imperceptible proper motion were obviously very distant. Since the proper motions of many stars had been measured, Hertzsprung was able to devise a formula from which he could calculate the distances and hence the absolute magnitudes of stars, with at least statistical accuracy. From these calculations he concluded that all white stars possessed high absolute magnitudes, but that red stars were of two classes, very bright (giants) and very faint (dwarfs). He could trace a regular sequence of stars in which change of color from white to red paralleled a decrease in absolute brightness, and he interpreted this to imply also a diminishing surface area and increasing density. But in addition to this main sequence he distinguished a collateral system in which some stars, whatever their color, were very bright, that is, all were giants. Concluding that two stars of identical redness, one a giant and one a dwarf, ought to show marked differences in their spectra, Hertzsprung found that Miss Maury's c division, spectra with narrow, sharply defined lines, provided the distinction he had hoped to find. None of the c stars had a measurable proper motion. Thus all of them must be giants.

This discovery fully justified Miss Maury's belief that the c characteristic indicated some intrinsic property of a star. In several letters to Pickering, Hertzsprung commented on the usefulness of the c characteristic, and in a letter of July 22, 1908, he expressed his disappointment that the newly issued Volume 50 of the Harvard *Annals* (the *Revised Harvard Photometry*), prepared by Mrs. Fleming and her assistants, had not retained the distinction:

In my opinion the separation of Antonia C. Maury of the c- and ac-stars is the most important advancement in stellar classification since the trials by Vogel and Secchi. But in the new catalogue the spectra of some of them as α Cygni and δ Cephei are not even mentioned as peculiar.

It is hardly exaggerated to say that the spectral classification now adopted [in Volume 50] is of similar value as a botany, which divide the flowers according to their size and color. To neglect the c-properties in classifying stellar spectra, I think, is nearly the same thing as if the zoologist, who has detected the deciding differences between a whale and a fish, would continue classifying them together.

As a physicist, Pickering had little faith in an attribute that could not be precisely measured and whose reality depended to some extent on the personal interpretation of the observer. In replying to Hertzsprung he said that, in photographs of spectra made with the objective prism, it was not easy to determine the actual breadth of the lines, since a slight change in focus, unsteadiness of the air, or other causes might all affect their appearance; and that several years ago he and Miss

The Henry Draper Memorial

Cannon had decided that the differences in width were too uncertain to justify a separate division based on them.

Hertzsprung replied respectfully but bluntly (August 17, 1908):

My best thanks for your lines of aug. 4.

The fact, that none of the stars called c by Antonia C. Maury has any certain trace of proper motion is, I think, sufficient to show that these stars are physically very different from those of divisions a and b.

I therefore am of opinion that we must lay the greatest stress on the c-peculiarities of stellar spectra notwithstanding the difficulty in determining these peculiarities (which not only consists in the sharpness but also in the differing intensities of the lines).

During the same period, without knowing of Hertzsprung's earlier work, Henry Norris Russell of Princeton had been moving toward similar conclusions regarding the red stars. Using photographs to determine parallax, he had measured the distances of 55 stars, from which he could calculate their absolute luminosities. To obtain their spectra and those of stars to be used for comparison he appealed to Pickering, who promptly supplied them from the photographic plates amassed during the previous 20 years for the Henry Draper Memorial. In writing (September 24, 1909) to thank Pickering, Russell tabulated some of the data that provided

very strong evidence that the *fainter* stars average redder than the brighter ones. I do not know of any previous direct evidence on this question.

I would not now risk reversing the proposition, and saying the red stars average intrinsically fainter. Some of them certainly do: but Antares and α Orionis are of enormous brightness, and the average may be pretty high.

Russell concluded from his analysis, just as Hertzsprung had done earlier, that two types of red stars existed. He wrote to Pickering (November 20, 1910):

Please accept my very hearty thanks for the spectroscopic data contained in your letter of the 18th. They remove the last apparent exceptions to the rule that all stars of small intrinsic brightness are of the redder spectral classes, and, I think, will help to establish the theory that these stars are in a very late stage of evolution.

In 1913 Russell offered a new theory of stellar evolution, which in some ways resembled Lockyer's earlier concept of a development based

241

on temperature. With more observational data than Hertzsprung had
had, Russell also represented the stars as points on a graph, which became known as the Hertzsprung-Russell (H-R) diagram, and was one
of the most important astronomical discoveries of the era. When he
plotted the absolute magnitudes of his stars against their spectral types,
an orderly pattern emerged showing two classes of stars. Those of one
class, which Hertzsprung had called "dwarfs," occupied places in a pro-
gression somewhere along a path on which, as the spectral type progressed
from O, to B, to A, and so on, both the luminosity and the temperature
decreased. Stars of the other class, which Hertzsprung had called "giants"
because of their very great luminosity, varied little in brightnesses even
though they belonged to different spectral types; they thus occupied
places outside the main sequence. To interpret the significance of this
pattern, Russell proposed the theory that stars began their existence
as highly diffuse, luminous, red giants. As they contracted in size
and became more dense they also became hotter, and hence changed
in color to form the blue and the white stars. When they had contracted
so completely that they could not undergo further compression, they
began to lose heat and thus became less luminous, evolved into the
yellow and the orange stars, and eventually became the red dwarfs of
low luminosity and very high density. (According to present-day interpretation of the H-R diagram, the giant, expanded red stars are not
young. Instead, they are nearing the end of their lives and are approaching extinction after becoming novae or supernovae.)

The giant red stars, and novae immediately after their explosion, all
produce spectra with the sharp, narrow lines that Miss Maury in her
catalogue had distinguished as the c characteristic. Full recognition of
its importance in the analysis of stellar spectra came in 1922 when
Commission 29 of the International Astronomical Union at its first General Assembly, in Rome, modified the official classification system (that
of Miss Cannon) for the first time in more than a decade. Among
the changes adopted was the prefixing of the letter c before a spectral
type to indicate the presence of narrow, sharply defined lines.

v

Although the wholesale photography and classification of stellar spectra were the primary objects of the Henry Draper Memorial, it yielded
many rewarding by-products. Pickering kept Mrs. Draper informed of

The Henry Draper Memorial

these unexpected finds, as well as of the progress of the regular work. One of the most startling was the first detection of a double star, not resolvable in the telescope, from its spectrum alone. Examining a spectral plate of ζ Ursae Majoris (Mizar) made in March 1887, Pickering found that the Fraunhofer K line appeared to be double, although in photographs taken during the succeeding months it always appeared single. Later photographs of the spectrum were studied at intervals and eventually Pickering wrote to Mrs. Draper (January 7, 1889) that Miss Maury had

made a discovery which I think is of no little importance. You may remember that last year we found the K line double in ζ Ursae Majoris and not in any other star. Now it seems nearly certain that it is sometimes double and sometimes single! It is hard to say what this means. Perhaps it will give the time of revolution of the star. Of course the first thing will be to determine the law of its variation.

Thereafter the K line sometimes appeared double, more often single. Pickering duly reported these events to Mrs. Draper, who wrote (May 27, 1889) that she was "delighted to learn by your letter that ζ Ursae Majoris has again given us the K line double,—the periodicity will be a simple matter but the laws that govern the phenomenon seem difficult . . . It is most interesting."

Eventually, after an examination of many spectral plates, Pickering concluded that ζ Ursae Majoris was in fact a binary system. Since the line seemed to become double every 52 days, the system apparently revolved with a period of 104 days, and the amount of separation between the lines indicated that the relative velocity of the two components was about 100 miles per second. After further study, Pickering concluded that the period was about 52 days,[21] and a decade later Vogel at Potsdam further refined the observations to obtain the much shorter period of 20.5 days. The discovery of this first "spectroscopic binary," which was announced at the meeting of the National Academy on November 13, 1889, was quickly followed by a second. As usual, Pickering sent the news to Mrs. Draper (December 8, 1889), shortly after one of her visits to see the work in progress at the Observatory:

Your visit brought us good luck in a succession of interesting discoveries. While you were here on Sunday night, we photographed β Aurigae. When

The Harvard College Observatory

Miss Maury examined the plate she found that although taken with a single prism only, the K line was clearly double. This was confirmed on Monday. It then became single remaining so until Friday, but last night it again became widely double. It apparently resembles ζ Ursae Majoris, but the period is about twelve days. [The actual period later proved to be 3.96 days.] Day before yesterday a still more remarkable case was found by Mrs. Fleming. When examining the Peruvian plates the spectrum of d Ophiuchis was found to be peculiar. After some study we found all the lines were double being spread about five times as much as the K line in ζ Ursae Maj. It is a striking example of the amount of material obtained from Peru that we found good spectra of this star on five dates. Apparently the period is about twelve days (although this is uncertain) and the motion about five hundred miles a second. Again, a photograph last night indicated a change in the bright lines of the spectrum of γ Cassiopaeiae . . . Now if all these results ensue in consequence of your recent visit here, is it not a sufficient argument in favor of your coming oftener?

The fourth Annual Report of the Henry Draper Memorial, which described the discovery of these first two spectroscopic binaries, was distributed as usual to all interested astronomers. After receiving a copy, Colonel John Herschel in a letter to Pickering (May 28, 1890) wrote:

I have just rec'd your last H. D. Mem. report. It is very like a pudding all plums—but I will ask you to convey to Miss Maury my congratulations on having connected her name with one of the most notable advances in physical astronomy ever made.

New facts about the stars began to flow from the Observatory in a steady stream. Brief papers in the *Astronomische Nachrichten* in Germany, the *Observatory* and *Monthly Notices* in England, and the *Sidereal Messenger* and its successor, *Astronomy and Astro-Physics,* in this country, announced the discoveries. In 1895 the Observatory also began issuing a series of Circulars, by which important discoveries made by the Observatory staff were promptly reported to the editors of astronomical journals in America and in Europe. Thus news of the detection of variable stars, stars with peculiar spectra, planetary nebulae, spectroscopic binaries, and novae could be published without delay. More than 200 Circulars were issued in the next 15 years.

An exciting find came in 1896 when an Arequipa photograph showed that the spectrum of the star ζ Puppis was "very remarkable and unlike

244

The Henry Draper Memorial

any other yet obtained." The continuous spectrum was traversed by three systems of lines, the third of which formed a rhythmical series closely resembling the Balmer lines of hydrogen. A few weeks later Miss Cannon found the same series of lines in the spectrum of 29 Canis Majoris. At first Pickering thought they must be due to some element not yet found in other stars or on the earth; later, he decided that since they so closely resembled a hydrogen series, they must represent hydrogen "under conditions of temperature or pressure as yet unknown."[22]

The real source of the "Pickering series" was not identified until after the development of modern physics and an understanding of the structure of the atom: In 1913, when the Danish physicist Niels Bohr was working out the details of his atomic theory, he determined that the lines of a helium atom that had lost an electron should appear in the spectrum close to the positions occupied by the Balmer series of hydrogen. This prediction was verified a few years later when Fowler, at London University, observed the lines of ionized helium in the laboratory. They were identical with those of the Pickering series.

VI

THE BOYDEN FUND

I

In 1887, shortly after the establishment of the Henry Draper Memorial, the Observatory received another large sum of money, the Boyden Fund, which enabled Pickering to implement his plans to construct a mountain station.

Thirty years earlier George Bond had suggested that "from some lofty mountain"[1] even very faint celestial objects could probably be photographed. Later he pointed out that, even if we could build more and more powerful telescopes, their greater optical power would

be rendered comparatively useless in ordinary climates and states of the atmosphere, by the incessant perturbations of the latter. The surface of the globe must be explored for the favored spots where a perfectly tranquil sky will afford the desired field for celestial exploration. If these were occupied, and faithfully improved, the fruits of the enterprise would be beyond all computation rich and interesting.[2]

Astronomers in general had long been convinced that a telescope placed on a mountain would give better "seeing" than at the low altitudes where observatories were usually built, but they had seldom been able to test this theory except during brief expeditions to observe eclipses or other celestial events. European astronomers had discussed the possibility of placing an observatory on Mt. Etna or Mt. Blanc, and S. P. Langley on his journeys to Pikes Peak, Colorado, in 1878, and to Mt. Whitney, California, in 1881, had been impressed with the tranquil air and excellent seeing. However, the effects of seasonal changes at

246

these and other mountain sites were unknown. Also, comparatively little was known of the exact geographical and meteorological conditions that produced the steady, motionless air in which the image of a star or other celestial object viewed through a telescope would not shimmer, dance, ripple, twinkle, or change shape.

No high-altitude observatory existed in the United States (the Lick Observatory on Mt. Hamilton, California, was still under construction) when in February 1887 the trustees of the Boyden Fund turned it over to Harvard University. With this bequest, designed explicitly to promote astronomical research at high altitudes, the Harvard Observatory built its first auxiliary station, on the California mountain then known as Wilson's Peak, and in the three-quarters of a century following established other stations in Peru, Chile, and South Africa.

Uriah Atherton Boyden was born in 1804 at Foxborough, Massachusetts. Although he had little schooling, he grew up in a family of practical inventors and, like his father and brother, displayed remarkable talent for solving mechanical problems. At the age of 29 he opened his own engineering office in Boston and became extremely successful. He supervised the construction of the Nashua and Lowell Railroad, designed the hydraulic works at Manchester, New Hampshire, and devised a turbine water wheel for the Appleton cotton mills at Lowell. Patents on his inventions—the hook-gauge, the Boyden water wheel, the Boyden turbine, and others—brought him a fortune. His abilities commanded wide respect, and when Congress established the National Academy of Sciences in 1863, Boyden was named one of the fifty charter members.

During his lifetime Boyden made several gifts to Harvard College, one to buy books on the physical sciences for the College library, another for prizes to be awarded to the two students who should have "acquired the greatest skill in mathematics at the Commencement of A.D. 1857."[3] In 1853 Harvard conferred on him the honorary degree of M.A.

Although somewhat deficient in his knowledge of physics, Boyden was interested in its problems, particularly the compressibility of water, the nature of heat, and the velocity of light. Astronomy also attracted him. During the Winlock years he regularly received the publications of the Observatory but he complained that the reports of the Visiting Committee were not informative and he criticized the quality of the

The Harvard College Observatory

research. In a letter to Winlock (August 27, 1867) he expressed his own idea of a worth-while project:

> Although I may not suggest any way of advancing Astronomy, I will mention a way in which perhaps astronomers may advance Optics; which way seems to deserve more attention than it has received. Perhaps observing stars and satellites which communicate light to Earth only at intervals may elicit proof of the comparative velocities with which rays of different colors travel; as if a star or satellite when it emerges from its eclipse does not seem to have its usual color; and at the commencement of its eclipse, its color seems complemental to its color at its emerging, it indicates that its rays of different colors occupy different durations in passing to the Earth.

In his later years Boyden became rather eccentric. When Pickering, as the new director, launched an appeal for a five-year subscription, Boyden was one of those approached for a contribution. J. Ingersoll Bowditch, a member of the Visiting Committee, later described the incident to Pickering and wrote (September 22, 1886) that Boyden had

> promised Mr Wm Amory & me to send us a check for five hundred dollars but declined to put his name to the subscription paper. When nearly all the money was raised I asked for his check—he declined saying in a very coarse manner that we were getting money under false pretenses—

Boyden never married. He died on October 17, 1879, and by his will left a sum of more than $230,000, almost his entire fortune, in trust to aid in the establishment of an astronomical observatory on a mountain peak "at such an elevation as to be free, so far as practicable, from the impediments to accurate observations which occur in the observatories now existing, owing to atmospheric influences."[4]

The will was contested, as is true of many wills leaving large sums of money for pure research, but was finally sustained by the courts in the autumn of 1881. The trustees were Boyden's old friend James Bicheno Francis, chief engineer for the Proprietors of Locks and Canals on Merrimack River at Lowell, Massachusetts; William Barton Rogers, the founder of the Massachusetts Institute of Technology and president of the National Academy of Sciences, 1878–1882; and William Goodwin Russell, a Boston lawyer and a member of the Board of Overseers of Harvard.

The Boyden Fund

The Observatory in 1881 was again in need of money, for the five-year subscription of 1878 would soon come to an end. (The establishment of the Henry Draper Memorial was of course still several years in the future.) Always alert to any possibility of adding to his resources, Pickering wrote to President Eliot (December 29, 1881):

Have you taken any action with regard to the Boyden Fund for astronomical research? While not directly available for our work it seems to me that this Fund might indirectly prove of great value to the observatory . . . I have devoted some time to the problem of mountain observatories and am prepared to present a scheme of work. A statement of it would however seem now to be premature. If you deem it advisable, I can communicate with the trustees regarding the matter, but it seems to me that in this case the aid of the Observatory should be asked rather than offered.

With the encouragement of President Eliot, Pickering opened the first attack in what turned out to be a five-year siege, by writing to the chief trustee, Mr. Francis (January 13, 1882):

It is with much satisfaction that I have learned that the will of Mr. Boyden has been settled and a large sum thus rendered available for an important scientific problem. The subject is one in which I have for some years been much interested both from the viewpoint of astronomy and of mountain exploration. Without wishing to intrude, I will say that if at any time my advice would be of service in this matter, I hope that you will let me know. If desired, I could put my views in writing, or meet you at such time and place as you might appoint.

But the disposition of the trust was not a question to be settled quickly or easily. Rogers was elderly and in poor health (he died a few months later while addressing the graduating class at the MIT Commencement), and Russell was not qualified to judge a matter involving scientific goals. Thus the chief burden of the trusteeship fell on Francis, who carried out his task conscientiously. For the moment he asked nothing of Pickering but the names of scientists who might provide useful facts. Armed with letters of introduction supplied by Pickering, Francis traveled throughout the country to ask advice from leading astronomers and physicists. He consulted, among others, Captain Richard S. Floyd, chairman of the Lick Trustees in San Francisco; E. S. Holden, who was supervising the construction of the Lick Observatory and later became

249

The Harvard College Observatory

its first director; Asaph Hall at the Naval Observatory; Simon New-comb, superintendent of the Nautical Almanac Office; Lewis Boss of the Dudley Observatory at Albany, New York; C. A. Young at Prince-ton; and the Clarks in Cambridge. Most of the country's astronomers agreed that the Boyden Fund should not be used to found a new institu-tion. Many favored a plan by which the National Academy of Sciences would control the use of the money and would allocate sums to indi-vidual observatories to support special research projects.

Still undecided after some two years of such consultations, Francis for the first time officially asked the advice of Pickering, who strongly disapproved of the plan to have the National Academy administer the fund. He argued that the treasurer, being a scientist and not a business-man, would not have the experience necessary to make safe investments, and that the officers of the Academy would not understand which of the proposed research projects were likely to yield important astronomical results. If the fund were turned over to Harvard, however, the Corpora-tion would invest it wisely and the income, under Pickering's direction, would be efficiently used for the purpose Mr. Boyden had intended.

At Francis' request Pickering wrote (February 9, 1885) a long and very detailed statement of his own proposal: to establish auxiliary Har-vard stations, including one in the Southern Hemisphere, so that stars in all parts of the sky could be photographed and measured with a single set of instruments.

In one section, labeled "Confidential," he commented freely on the policies of the National Academy.

The Academy plan commends itself at first sight as that most likely to lead to important results. This would be the case if the Academy in reality occupied the position it should in theory. Great dissatisfaction pre-vails with its management* (*Footnote:* *I wish here distinctly to except its past and present honorable Presidents.) even among its own mem-bers . . . The virtual control by Congress and the influence of Washington methods seriously diminishes the advantages of the Academy plan, which otherwise would doubtless be expected to render the fund more widely available than any other. The large amount of the Boyden Fund compared with the other funds held by the Academy would be a strong temptation to certain members to control it in order to exert influence in other ways. Congressional methods in their worst form would then govern its expendi-ture. The income would be likely to be distributed geographically after the manner of many Congressional appropriations. On the other hand it might be neglected, and the object for which it was established not be accomplished until the discovery of new methods had rendered it useless. The experience with other funds held by the Academy is not encouraging . . . Doubtless certain expensive expeditions would be undertaken with the Boyden Fund. Such expeditions are always popular among astronomers. It may be doubted whether the numerous expeditions to observe solar eclipses and Transits of Venus have given results proportionate to their cost. The income of the money thus expended would have maintained permanently stations at which a vastly greater amount of information could be collected.

Having done all he could, Pickering again had to wait for a decision, although, as he wrote President Eliot (February 17, 1885), "inaction in this matter is very distasteful to me." Another year passed before Francis and his cotrustee Russell finally determined to transfer the Boyden Fund to Harvard. Meanwhile, rumors of the impending bequest were reaching the friends of the Observatory, not all of whom reacted with pleasure. Mr. Bowditch expressed his opposition to accepting money from the estate of a man who had behaved so unpleasantly over the 1878 subscription. But Pickering had no such scruples. With a quarter of a million dollars at stake, with all it meant for the future of the Observatory, he could afford to overlook an earlier rudeness of the donor. Nevertheless, he did not wish to displease Bowditch, one of the staunchest supporters of the Observatory. In a conciliatory letter (September 23, 1886) Pickering wrote that he had

not forgotten the manner in which Mr. Boyden treated you and Mr. Amory . . . He maintained that I could not accomplish what I had promised to do; namely that the subscription while it lasted should double the work of this Observatory. Mr. Boyden's will gave me an opportunity to justify my cause. His trustees, William B. Rogers, James B. Francis and William G. Russell were certainly men of good judgment. I made a study of the problem of mountain observatories and proposed to them a plan for a mountain station as an adjunct to this. If I satisfied them that in no other way could equally important results be obtained, we surely might disregard what Mr. Boyden said of me in a moment of anger . . . How striking an instance of retributive justice, if through Mr. Boyden's means this Observatory should become the richest in the world! I hoped that this plan would satisfy both you and Mr. Amory that you had done wisely in placing confidence in me. I want to establish a station in Peru high up in the Andes where we could reach the southern stars. We could thus include all parts of the sky and this Observatory would be the first to make each portion of its work extend from one pole to the other. I shall be very much disappointed after all that you have done for this Observatory if you do not feel in this matter as I do. I have never forgotten, although you may have, how kindly you received me nearly twenty years ago, when I went to see you about a question of policy of the Institute of Technology. You withdrew your opposition in a way that made me feel it a personal kindness . . . I cannot but hope that you will add one more to the many favors you have done me, by giving your best wishes to this plan.

On February 14, 1887 (the first anniversary of the establishment of the Henry Draper Memorial), the trustees of the Boyden Fund formally transferred it to the President and Fellows of Harvard College.

Many astronomers wrote to express their pleasure in the good fortune of the Observatory, even though some had been disappointed in their own hopes of sharing in the Boyden bequest. Lewis Boss, of the Dudley Observatory, who had supported the National Academy plan and had hoped thereby to obtain support for his own research, sent a letter (March 15, 1887) of congratulations

upon the very large accession to your means. With an aggregate fund larger than all other astronomical endowments in this country combined, you will be able to advance the reputation of this country for astronomical achievements immeasurably. May the results make us all proud of the Harvard College Observatory!

The year 1887 marked a high point in the financial well-being of the Observatory under Pickering. When he had become director a decade earlier, he had taken over an institution with meager funds, small staff, and equipment in need of repair. Now, within a single year, it had acquired generous support. Mrs. Draper was contributing $10,000 a year to the Henry Draper Memorial, and the full income from the bequest of Robert Treat Paine[5] had become available for the first time. Furthermore the Harvard Corporation, always reluctant to part with capital, in an unusual concession had voted to allow the spending of $20,000 from the principal of the Boyden Fund, while the investment of the remainder would bring in about $11,000 a year.

Pickering now had the means to carry out his plan: to measure the magnitudes, map the positions, and record the spectra of the stars in both the Northern and the Southern Hemispheres. All that remained was to organize the work.

II

To provide an assistant to take charge of the Boyden Department, Pickering chose his younger brother William, who gave up his position as instructor in physics at MIT to join the Observatory staff. Solon I. Bailey (Chapter VIII), who had been working at the Observatory while studying for his M.A. at Harvard, also became an assistant. Other new helpers were employed, including four young women for the corps of computers supervised by Mrs. Fleming—Miss Annie E. Masters, Miss Jennie T. Rugg, Miss Nellie C. Storin, and Miss Louisa D. Wells.

Edward Pickering procured new equipment that included a 12-inch "pole-star recorder"; by photographing the trails of stars in the neighborhood of Polaris and of others near the horizon, it would provide a test of the steadiness of the atmosphere. From the Clarks he also ordered a 13-inch refractor, to be known as the Boyden telescope, whose crown lens was reversible so that the instrument could be used for either visual or photographic research.

The immediate problem was finding a promising site for the first experiments. A wise choice would require knowledge of the altitude, amount of rainfall, extremes of temperature, transparency of the sky, and degree of cloudy or sunny skies during the year at the places considered, but such information was difficult to obtain. About 1850 the Smithsonian Institution, under the direction of Joseph Henry, had established a system of receiving meteorological reports by telegraph from various parts of the country, but no national weather bureau yet existed. However, keeping records of the local weather had always been considered one of the normal duties of an observatory (as it had been at Harvard until 1885, when A. L. Rotch established a private meterological observatory at Blue Hill, south of Boston). Pickering therefore prepared the first Boyden Circular (March 1, 1887), quickly followed by a second and a third, which he mailed to astronomers all over the world. He briefly announced the acquisition of the Boyden Fund, stated that he planned to establish a mountain observatory, and asked for detailed meteorological information regarding desirable sites known to the astronomer receiving the Circular. He also wrote to a firm of photographers in Calcutta, to order pictures of the highest mountains in the Himalayas.

253

The Harvard College Observatory

The response to the Circulars was enthusiastic. Pickering received letters recommending mountain sites in the Andes and the Himalayas, and in South Africa, Australia, Japan, and Hawaii (where the Smithsonian Astrophysical Observatory now maintains a satellite-tracking station on Mt. Haleakala and the University of Hawaii has erected on Mauna Kea the highest major astronomical observatory in the world, at an altitude of slightly more than 13,000 feet). Professor Langley sent the meteorological records made during the Smithsonian expedition to Peru to observe the eclipse of September 7, 1858. But long before all the answers had reached Cambridge Pickering had arranged to make a preliminary expedition to Pikes Peak, Colorado.

Pikes Peak at that time was a federal reservation, where for more than ten years the U. S. Signal Corps had maintained a weather station, then the highest in the world, at more than 14,000 feet. As General A. W. Greeley, Chief Signal Officer, U. S. Army, wrote Pickering (March 14, 1887), this was "the only elevated station west of the Mississippi whose meteorological conditions are accurately known to this office." The observations had never been published, but Greeley made them available to the Harvard project. He also offered his full, official cooperation, the part-time assistance of one of the men at the station, and the use of the stone hut on the Peak. In return, Pickering agreed to publish the accumulated data for the years 1874–1888, at the expense of the Boyden Fund. They appeared in 1889 as Volume 22 of the Harvard *Annals*.

Professor Frank H. Loud of Colorado College, head of the newly established Colorado Meterological Association at Colorado Springs, also offered to cooperate with Pickering and wrote frankly (March 21, 1887) of the benefits that Colorado College would derive:

I should be especially glad if your choice should fall upon Bald Mountain,[6] and if some arrangement could be made by which Colorado College could be associated with Harvard in the work of maintaining observations at this point. We are at present a very young and feeble institution and there is not the least possibility of anything being done to utilize the advantages of our situation in an astronomical way. Were it however known that Harvard was coming to our very door, and were there a prospect that Colorado College would be able to secure a permanent observatory which would at once be made famous by the use of it under your direction by your skilled observers, it is not at all impossible that an enthusiasm might be aroused which would enable our friends to take a considerable share

254

of the burden of preparing the site, while in such matters as would relate to the present occupation and use of it, Colorado would of course be only too glad to leave the entire direction and control to Harvard, subject only to our right to be associated in the work in that distant day when there can be a science-making astronomical department here.[7]

Early in July members of the Observatory expedition reached Colorado Springs, where they made their headquarters. The 14 cases of equipment included the 12-inch polestar recorder, several smaller telescopes, photographic plates and supplies, self-recording meteorological instruments, and a sunshine recorder newly invented by Pickering. The party included Professor and Mrs. Pickering, William Pickering, Dana P. Bartlett (afterward professor of mathematics at MIT), Harry E. H. Clifford (afterward Gordon McKay Professor of electrical engineering at Harvard), and J. R. Edmands, a volunteer assistant at the Observatory who, like the two Pickerings, was a member of the Appalachian Mountain Club. Late in the month Mrs. Draper joined the group.

With the advice and assistance of Professor Loud, telescopes and other instruments were loaded on the backs of mules for transport up the mountain and on August 1 the men began the ascent. After climbing to Seven Lakes (11,000 feet), where they spent several days making observations, they moved on to the summit of Pikes Peak. There they mounted the 12-inch telescope, made observations to test the quality of the seeing, and took photographs in the intervals between a snow squall, a hail storm, and a violent electric storm. The Pickerings also visited Mt. Lincoln (14,300 feet)[8] and Mt. Bross (14,000 feet), where they set up self-recording meteorological instruments and surveyed other high points that could be reached by railroad.

By the end of the summer Edward and William Pickering had acquired a great deal of new information about the atmospheric conditions necessary for good seeing. The most important factors, they concluded, were the relative lack of clouds, transparency of the sky, steadiness of the air, and freedom from dew. Currents of hot, cold, or moisture-laden air, or even small inequalities of temperature in the air mass around the lens of the telescope or camera, could distort the image of the celestial object being observed. In particular, the height of the station was less important then the dryness of the air, for the violent and vertical winds that are common near mountain tops can readily cause clouds to condense out of moist air.

Before the party returned to Cambridge Pickering arranged for Loud and others to continue the meteorological observations during the winter. At the time, he planned a second visit to Colorado in the following summer but later gave up the project. Observing conditions at the several peaks proved somewhat less favorable than he had hoped. An equally strong reason, however, was that a group of Colorado citizens were applying political pressure in Washington to have Pikes Peak removed from federal jurisdiction and returned to the state. If successful (as they eventually were), they planned to build a road to the summit and create a profitable tourist attraction. The Peak would then be a very poor site for an observatory.

Instead of returning to Colorado, therefore, Pickering decided to investigate the mountains of southern California, and to search for a desirable station in the Southern Hemisphere.

III

The possibility of a mountain station in California had long attracted Pickering and in the spring of 1888 he was encouraged by a visit from the Rev. Doctor M. M. Bovard, president of the University of Southern California. Dr. Bovard wished to ask advice (which he received but in the end did not follow) on choosing a telescope for a proposed new observatory to be built in the neighborhood of Pasadena.

The Lick Observatory on Mt. Hamilton, constructed under the supervision of E. S. Holden, was newly finished and, equipped with a 36-inch refractor (made by the Clarks), then the largest in the world, in June 1888 would be formally turned over to the University of California at San Francisco. (It was said that James Lick, whose will provided for the erection of an observatory to contain "a powerful telescope superior to and more powerful than any telescope yet made,"[9] had hoped by this gift to prove or disprove the existence of animals on the moon.) Californians were experiencing great prosperity, land prices were booming, and rivalry among the cities was intense. E. F. Spence of Los Angeles had now offered to donate property worth about $50,000 toward the establishment of an observatory for the University of Southern California, to be erected on what was then known as Wilson's Peak, near Pasadena, if conditions there proved suitable. The citizens of Los Angeles hoped thus to outdistance those of San Francisco by constructing an observatory equipped with a still more powerful telescope, whose lens would be 4 inches greater in diameter than that of Lick's 36-inch instrument.

The proposed telescope was designed primarily for visual work. Pickering recommended, instead, a 24-inch photographic doublet, similar to the Bache—the same design he later used for the Bruce telescope (Chapter VII). The 24-inch instrument, being smaller than the Lick telescope, had no appeal for Bovard, but out of the ensuing correspondence grew a proposal for a cooperative venture: Harvard would establish a temporary station on Wilson's Peak to test the observing conditions and would share the resulting information with the Spence trustees of Los Angeles. In return Dr. Bovard, representing the trustees, would construct a road to the summit, provide buildings to house the Harvard telescopes and observers, and contribute a share of the expenses.

Given this opportunity, Pickering decided to combine the Wilson's Peak venture with an eclipse expedition. An eclipse of the sun was to take place on January 1, 1889, and northern California would lie in the path of totality. In general, Pickering believed that an expensive journey to observe an eclipse was an unjustifiable gamble of time and money against the doubtful probability of good weather. In this case, however, the Harvard party was going to California for other purposes. The additional expense involved in viewing the eclipse would be relatively small and the risk seemed justified.

Pickering himself stayed in Cambridge, but a group under the direction of William Pickering traveled to California in November 1888, and erected temporary shelters near the town of Willows, in the Sacramento Valley. The party included two young men recently hired as photographic assistants at the Observatory, Robert Black and Edward S. King; Mr. and Mrs. Solon I. Bailey; Winslow Upton, then professor of astronomy at Brown University; and A. L. Rotch, director of the Blue Hill Meteorological Observatory. The chief instruments were the 8-inch Bache doublet and the new 13-inch Boyden telescope, recently completed by the Clarks.

On the day of the eclipse the weather was favorable and the observers secured 47 excellent pictures of the eclipsed sun and the corona. Nevertheless, they encountered some disappointments not unfamiliar to any astronomer who has tried to carry out routine procedures and keep calm in spite of the unusual excitement and stress of an eclipse. In

257

The Harvard College Observatory

a letter to his brother Edward (January 3, 1889), William Pickering described one incident:

> You have doubtless seen by this time in the N. Y. Herald & World a full account of our observations and successes. I will now describe our failures, which were comparatively insignificant, but still annoying. In the first place when totality came on the spectators set up a yell. Although they were nearly 50 yards from our station & I shouted 'tip' as loud as I could, it completely drowned my voice where the counter was, we accordingly lost 19 seconds before he began to count. Owing to this accident we lost between 20 and 25 views, and only got 47 during totality itself . . . Our second failure was wholly my own fault. I forgot to uncap the reversing layer spectroscope. I had everything ready & set the clockwork going & all, but forgot the important thing. It was very annoying.

The eclipse over, Bailey started packing the Bache telescope for its trip to Peru (see Chapter VIII). Leaving King and Black to dismantle the station and transport the equipment to Pasadena, William Pickering made a brief visit to E. S. Holden at the Lick Observatory, and then went on to Los Angeles to prepare for the main object of the expedition, setting up a station to evaluate Wilson's Peak as an observing site. And here he ran into difficulties that were more than "annoying."

Nothing had been done to prepare for the Harvard observers. Dr. Bovard had built neither the promised road nor the shelters, which must now wait until spring. Nevertheless, on January 23 William Pickering and Alvan G. Clark, together with Dr. Bovard and other local residents, climbed to the summit where they spent the night in a deserted cabin. Observations made with a small portable telescope showed that the seeing was excellent. The men agreed to continue the experiment as planned, and settled that the summit would be reserved for the proposed Spence Observatory of the University of Southern California, while the Harvard station would be placed somewhat lower on the mountain.

William Pickering stayed on in Pasadena and Los Angeles for several weeks, trying to expedite the building of the road and the shelters. Before returning to Cambridge he obtained the help of N. C. Carter, of Sierra Madre, and by early spring a passable trail had been made to the site. All building materials, instruments, and supplies had to be carried up the path by burros or pack mules, but eventually a pier and shelter for the Boyden telescope were completed, and a small cabin was built for the observers, Edward King and Robert Black (Fig. 37).

FIG. 37. Harvard station on Wilson's Peak, California, about 1890.

The Harvard College Observatory

They obtained their first photograph on May 11, and thereafter kept the telescope going constantly every clear night, photographing stars, the moon, Mars and Jupiter, and the nebula in Orion. The seeing, King wrote to William Pickering, was almost perfect and the mountain was "an astronomer's paradise." With the ocean on the southwest and desert country to the northeast, the sharp temperature changes and high wind velocities so detrimental to good seeing were moderated. The station was also high enough above the valley to escape the fog and haze of lower altitudes, yet not high enough to encounter the full force of winter storms.

Living conditions that summer were comfortable but far from luxurious. The two men boarded with A. G. Strain, who had a cabin on the mountain and contracted to carry wood and water to the shelter. Black wrote to William Pickering (June 19, 1889) that the food "is not the most elaborate, but wholesome. Our greatest vexation is getting up and down the trail. One day each week is consumed in carrying the barometer down. Tourists are abundant, and especially in the early morning when one of us has to sleep. They pay little attention to requests or threats." Other lesser annoyances included dust, flies, and rattlesnakes. That autumn, with the assistance of Carter's teen-age son, the observers remodeled the shelter for winter occupancy and in November King[10] returned to Cambridge, leaving Black to run the station alone except for the assistance of young Carter.

Off the Peak, meanwhile, local rivalries and jealousies had created serious difficulties. The citizens of Pasadena, a town then little more than ten years old, were eager to outdo Los Angeles and contributed some money toward the expenses of the expedition. The Los Angeles group, Dr. Bovard and the Spence trustees, who had apparently become afraid of losing control of the proposed observatory either to a rival town or to Harvard, refused to contribute their promised share of the expenses. Dr. B. Homer Fairchild, a retired physician, wrote to Edward Pickering to set forth the claims of the town of Pomona. He warned Pickering that Dr. Bovard was unreliable, asserted that many California sites were superior to Wilson's Peak—Mt. San Antonio (Old Baldy), for instance, Mt. Disappointment, or Cucamonga Peak—and stated that if Harvard would choose one of these places the citizens of Pomona would donate the necessary land.

While these controversies went on, the expedition continued to incur expenses, most of which had to be met by the Boyden Fund. When Edward Pickering wrote to Dr. Bovard asking why he had failed to carry out the agreement, Bovard (July 11, 1889) bluntly expressed his doubts of Pickering, Harvard, and their good faith:

> If it could be made clear to us that your men were doing work for us we would pay for it. If you are not experimenting for the Scientific benefit to come from it or not seeking a place for an instrument for Harvard Observatory, but are making the chief point that [of] testing for a place for the Photographic instrument or the 40 inch for us then we could act intelligently on the question, But so many reports come to us representing that you are doing the work paying for it and the University of Southern California has no interest in it.

In reply (July 22, 1889), Pickering stated his position in detail:

> Your letter of July 11 is at hand, and I will endeavor to make the position of the Observatory with regard to its California expedition entirely clear. In the first place, as stated in my letter of May 13, the expense of sending the 13-inch telescope to California would not have been undertaken for the sake merely of the eclipse observations. Secondly, this expense would not have been undertaken in order to carry out other plans, in the absence of the expectation that it would be refunded by the University of Southern California. It was proposed, therefore, that the total expenses of the expedition should be divided between this Observatory and your University as described in my letter of June 9, 1888, and repeated in my letter of March 26, 1889.
>
> The advantage to be derived by this Observatory from the expedition was that of obtaining a series of photographic observations in a locality where it was hoped that the atmospheric conditions would be unusually favorable. The advantage to be derived by the University of Southern California was that the series of observations thus obtained would show whether the atmospheric conditions just mentioned were in fact as favorable to the work of large telescopes as had been anticipated. As I said in my letter of May 13, a full year's work with a rather large telescope was desirable to decide this question. It would be absurd to work for a year, or even for six months, merely to test the atmosphere, without taking the opportunity to collect other results of permanent scientific value. The chief inducement to this Observatory to undertake the expedition was, as just stated, the attainment of these results.
>
> The expedition was undertaken in full confidence that a large part of its expenses would be refunded in consideration of its utility in determining by its work the suitability of the location to the subsequent erection of a telescope of unusual power. These expenses, after deducting that part of them properly belonging to the eclipse observations, already amount,

as nearly as I can estimate, to four thousand dollars. It seems to me, therefore, that fifteen hundred dollars, as stated in my letter of May 13, is the smallest sum the payment of which, by the University of Southern California, could be regarded as a reasonable compliance with the plan for cooperation between that University and the Observatory of Harvard College, so far as the work had gone, at the time when I wrote.

Nearly a year elapsed before this letter received an answer, and the financial problem remained troublesome. Nevertheless, the enthusiastic reports from King and Black, plus the fine quality of the photographic plates that arrived in Cambridge, had convinced Pickering that the site was excellent. The photographs, when developed, far surpassed any taken in Cambridge. They clearly showed Saturn's rings, the equatorial cloud bands on Jupiter, and new detail on the surface of the moon and Mars. Photographs of globular clusters showed well-defined images of the stars, and were the first such photographs good enough to be used for a study of cluster variables. Particularly fine were the pictures of the great nebula in Orion, which appeared as a much more extensive and complex object than had been suspected.

Since atmospheric conditions on Wilson's Peak were clearly ideal for an observatory, Pickering consulted with President Eliot on ways and means to acquire the land. Two main obstacles existed: the title to the Peak was not clear, and the attorneys for the Harvard Corporation advised that the College could not legally buy land outside the state of Massachusetts. The Southern Pacific Railroad claimed the Peak as part of its land grant from the Federal Government. The Government argued that the land had been forfeited, and the case was in litigation. Even if the Government case should be upheld, there would remain three rival claimants for title to the land: A. G. Strain, with whom the observers boarded; Peter Steil, who also had a cabin on the Peak; and a Mr. Cowley of the Hollenbeck Hotel in Pasadena, who claimed a miner's rights.

In Cambridge, President Eliot and the Corporation devised a plan. Although Harvard could not buy land outside the state, the Corporation could accept anything given to them. Therefore, if a clear title could be obtained, Pickering could buy the land in his own name with money from the Boyden Fund, and would then immediately draw up a declaration of trust on the uses of the land, or deed it to Harvard. (Later legal advice agreed that the President and Fellows could, after all, properly buy the land.)

The Boyden Fund

In December 1889 Harvard's attorney, Francis V. Balch, after discussion with President Eliot and Pickering, wrote to Senator Henry Cabot Lodge in Washington to explain the problem. Balch proposed that a bill be introduced in Congress by which the Government would take the land on Wilson's Peak by right of eminent domain and give it back to the state of California, which could then sell the land to Harvard. (There was a precedent for this procedure; Lick Observatory had acquired its site on Mt. Hamilton in just this way.) Senator George F. Hoar of Massachusetts talked over the matter with the California senators and with Leland Stanford, who represented the Southern Pacific Railroad, all of whom seemed favorable to the project.

Balch then sent a draft of the proposed bill to Senator Lodge, and in his covering letter casually remarked (December 16, 1889), "I understand that the University of California at one time wanted and perhaps still wants a location on the Peak. We could arrange fraternally with them if the act left scope. It might run to Harvard College and to such other institutions of learning as it may see fit &c. &c."

A few days later (December 21, 1889) Senator Hoar wrote to Balch: "I have introduced your bill, changing it a little, saying that the State may grant to the President and Fellows of Harvard College and to the University of California and to such other institutions as it may see fit, etc."

Congress passed the bill in the spring of 1890 but, unfortunately for Pickering's hopes, its passage only created new complications. The "little" change had proved a fatal one. The bill should have read "the University of *Southern* California"; the northern institution already had Lick. With perhaps excusable provincialism, Balch had apparently assumed that the state had only one university. The incorrect wording not only confirmed Bovard's distrust of Harvard's motives but also aroused widespread opposition among other groups in southern California. It must have seemed that the Federal Government itself was conspiring with Harvard and the San Franciscans to rob the people of the Los Angeles area. Stanford changed his views and objected to selling the land to Harvard, the president of the Board of Trade of Pasadena telegraphed a protest to Washington, and Dr. Bovard wrote angrily to Pickering (May 19, 1890):

Your own scheme is made more and more manifest. That you have been decieving us from the beginning, is now plain. The latest is the Bill

The Harvard College Observatory

passed by the Senate, a copy of which is before me, bringing in the University of California and of course *Lick Observatory* as a joint owner in a site selected by us before you ever heard of Wilsons peak. Simply as it seems to me, without so much as "your leave sirs—" getting the United States Government to give to you our selected site—Of course we are left out of the count.

Pickering wrote many letters trying to explain the mistake but opposition continued. After visiting the Peak, Professor J. R. Edmands remarked in a letter (August 11, 1890) to Pickering,

I doubt whether there is a human being resident in southern California who has had occasion to know about Observatory affairs, who has not also become a partizan in behalf of some project, and against some person or persons . . . No one seems capable of rising above his own or somebody-else's petty jealousies or animosities.

The situation on Wilson's Peak had also deteriorated. The winter of 1889–90 was a severe one on the mountain. Snow, ice storms, and high winds frequently damaged the instruments. The site was barren and isolated, and the cabin had few amenities. Robert Black and the boy who assisted him had an extremely heavy work load. Alone, they must tend and read the meteorological instruments and keep them in repair, photograph the sky at night, make the regular trek down to the valley to ship the photographic plates to Cambridge, and trudge up the mountain again with mail and supplies. Early in January, while returning from a trip to the valley, Black was caught in a snowstorm and developed a severe cold that left him unable to work for several weeks. Young Carter, his only assistant, was dissatisfied with the long hours and small pay ($16 a month), and neither man ever got enough rest. At the same time, the number of photographic plates shipped to Cambridge was smaller than expected and they were not always of good quality. In answer to a critical letter from William Pickering, Black described some of the problems (March 26, 1890):

You spoke some time ago about the Pole Star plates being reversed. Our Pole star clock wont work. It got snowed and rained on too much all winter. In order to get good plates a good reliable clock is necessary. The anemometer will not work in a rain. We have taken that instrument down every storm, and fixed it over and tried it again. It will leak in spite of All we can do. We first "beeswaxed" every joint in the machine.

The Boyden Fund

That wouldn't keep out the rain. Then we tried oiled cloths around it, but they were not better than the beeswax. Lastly we paraffined every joint of it and every screw hole, and the last rain stopped it as usual. Our resources in that line are exhausted.

Also, dissension had developed between Black and Strain. When Black finally left Strain to board with Peter Steil, a rival claimnant to the land on the peak, Strain retaliated by reporting to Pickering that Black had broken his contract, and also was conspiring with others to rob Strain of his land. (Strain was then engaged in two lawsuits, one with Cowley for trying to get possession of the peak, the other with Steil for breaking down the gate he had placed across the trail near the observatory.)

Another major problem was tourists. Although the observatory was surrounded by a barbed wire fence with a padlocked gate that bore the notice, "No admittance," visitors got in anyway. Black wrote to William Pickering (May 3, 1890) that the tourists

are getting worse and worse. We have been able to get only ten hours sleep in the last two days on account of them. This morning a party of them came around the building when we were asleep—tried to pry open the door and windows—climbed the dome and hammered the tin, and climaxed the affair by moving the slides of the sunshine recorders down, and piling them on top of each other.

The weather in May was very bad, with high winds, rain, and fog—and consequent bad seeing. Worn out by the heavy schedule of work, young Carter went down to the valley for a few weeks' recuperation. Black was left to handle everything alone, even though he had not fully recovered from his illness of January. With little experience of the difficulties of field conditions, working in the established routine of the Observatory in Cambridge, William Pickering was angered by the relatively small number of plates coming from Wilson's Peak and wrote sharply to demand an explanation. Black, in turn resented the failure to understand his problems, and answered (June 27, 1890) that if his work was not satisfactory he would gladly resign. In the future, he would not even try to work longer than a 12-hour day. He added:

I have been on Wilson's Peak one year last April. During that time I have not been able to attend a concert, lecture, church or theatre. I have not been away farther than Los Angeles (and there *never* for pleasure) but once—that was last summer, one week in San Francisco. I am practically excluded from the world.

With so many problems confronting him, Edward Pickering sent J. R. Edmands to California to try to straighten out affairs. Edmands arrived at Wilson's Peak with a party of friends in midsummer in the midst of a heat wave. Later he described to Pickering the discomforts of their camp (July 29, 1890): a total lack of refrigeration, tepid drinking water, too little water for bathing or even for keeping the face and hands clean, no privacy for bathing, no individual towels. The tents had no tent-flaps so that the sun and the flies poured in. And, he added, "this a *camp with ladies.*"

After interviewing the several persons involved, Edmands concluded that Black had not broken any contract, and that Strain in fact had been overcharging for his services.

In the face of so many complications, Pickering finally decided to give up the station. At the end of August the telescopes were dismounted and taken down the mountain to be shipped to Cambridge. Black stayed in California. In his last letter to William Pickering (January 1891) he wrote that his health was broken and that he had "consumption." The rest of his history is not known.

Attempts to procure the land on Wilson's Peak continued for some time. In the spring of 1892 President Eliot, during a visit to California, ascended the mountain on muleback to inspect the site. But before the end of the year Pickering had abandoned his efforts to place an observatory on Wilson's Peak or elsewhere in California. By this time the station in Peru had been firmly established, and its expenses were claiming all the income available from the Boyden Fund.

Until a few years ago the base of the stone telescope pier still existed on Mt. Harvard (so christened by President Eliot during his visit there), west of and somewhat below the peak where the Mt. Wilson Observatory now stands. Fastened to the pier was a brass plaque, which Pickering had placed there in 1910, that read "Station of Harvard College Observatory, 1889–1890." Today the ground has been leveled off to make room for a TV installation. The plaque has been moved to another place, near Signal Point, and no trace remains of Harvard's first high-altitude station.

IV

The plans of the Spence trustees and the Los Angeles group to place an observatory on Wilson's Peak also failed. Dr. Bovard, against Pickering's advice, had ordered a 40-inch visual refractor from the firm of Alvan Clark and Sons, who commissioned Mantois of Paris to cast the two disks of optical glass. Before the annealing process had been completed, however, the United States had entered a depression and the land boom in California collapsed. The value of the Spence property soon shrank from some $50,000 to an amount far too small to pay even the $16,000 owed to Mantois. Building an observatory and equipping it had become financially impossible and the Spence project had to be abandoned.

The disks were shipped to Cambridge where they remained in the workshop of the Clarks, unfinished, and were offered for sale. In 1892 Charles T. Yerkes, a wealthy Chicago businessman, offered to build a new observatory for the University of Chicago and to buy the Mantois lenses from the Clarks. Thus the 40-inch refractor was finally constructed and mounted on the shore of Lake Geneva, Wisconsin, about 75 miles north of Chicago. It is still the largest of its type in the world.

The director of the new Yerkes observatory was George Ellery Hale, then a young man in his twenties. A native of Chicago, Hale had graduated from MIT in 1890. During his last two years at the Institute he had served as a voluntary assistant at the Harvard College Observatory, where, unlike the rest of the staff at this period, he concentrated on the study of the sun, and developed his first model of the spectroheliograph (invented independently about the same time by H. Deslandres of the Paris Observatory), a new type of instrument for photographing the sun in the light of a chosen spectral line. Using later and improved versions of the spectroheliograph, Hale at Yerkes carried out fundamental research on the structure of the sun. In 1904 he received the gold medal of the Royal Astronomical Society "for his method of photographing the solar surface and other astronomical work."[11]

Like the 40-inch lens, the site on Wilson's Peak came into use eventually, also through the efforts of George Ellery Hale. In 1903 the Carnegie Institution of Washington appointed him chairman of a committee to find a favorable location for a new observatory for the study of the sun. Hale visited a number of mountain sites in this country and abroad, and especially considered Mt. Etna, Mt. Whitney, and Pikes Peak. He also consulted with Pickering in Cambridge. In the end, he chose Wilson's Peak for the construction of the Mt. Wilson Solar Observatory (after 1918, when the 100-inch reflector was built, the word "Solar" was dropped), one of the finest in the world.

Thus, although Pickering failed to realize his dream of erecting a Harvard station on Wilson's Peak, he did have the satisfaction of seeing an observatory established there, directed by a man he had trained. And in 1921, after Pickering's death, the cycle was completed when a Mt. Wilson astronomer, Harlow Shapley, moved to Cambridge as the fifth director of the Harvard College Observatory.

VII

THE BRUCE

PHOTOGRAPHIC TELESCOPE

I

In drawing up his plans for an astronomical observatory on a mountain site, Pickering considered essential the acquisition of a powerful new telescope especially designed to photograph and chart the whole sky. Hoping to find a generous patron who, as in the case of the Boyden and Draper funds, might wish to associate his name with a major astronomical enterprise, Pickering issued on November 20, 1888, a strong appeal for $50,000. His circular, "A Large Photographic Telescope," pointed out the rapid change that photography had already brought about in astronomical research and described in detail the kind of telescope that experience with a smaller model, the 8-inch Bache, had shown to be best suited for the purpose:

The lens should be like that used by photographers . . . and should consist of two achromatic lenses. Its aperture should be twenty-four inches, and its focal length eleven feet, thus giving images of objects on a scale of one millimetre to a minute of arc. Its great advantages would be the large region covered by a single photograph, since five degrees square could be represented by it upon a plate twelve inches square. This is six or eight times the area covered by a telescopic objective.

Also, exposure times could be reduced and fainter stars recorded than were visible in the conventional telescope. Such a lens, he wrote, if mounted in a favorable location and kept constantly at work, would add more to our knowledge of the stars than could be obtained by a large number of telescopes of the usual kind. The resulting plates would provide unlimited possibilities for study, such as a search for

269

double stars, nebulae, asteroids, variables, and the distribution of stars over the sky. Moreover, the design of the lenses would be adapted to both visual and photographic use and with the application of a prism the telescope would actually become six instruments in one.

In his report for 1889 Pickering had the pleasure of announcing that his appeal had brought a response from a wholly unexpected quarter:

A gift of $50,000 was received last summer from Miss C. W. Bruce of New York, for the construction of a photographic telescope of novel form. If successful, it will materially affect the entire plan of work of this Observatory.

That matter-of-fact statement hardly suggests the stirring history of the benefaction, the remarkable personality of the donor, her relation with Pickering, and the notable results of her gift to Harvard and to astronomers all over the world.

Catherine Wolfe Bruce[1] was the daughter of George Bruce, a native of Scotland who came to this country in 1795 as a penniless boy of fourteen and who, after some years of operating a bookbindery and printing shop, established his own successful type foundry. By this occupation and by shrewd investments in New York real estate (like Mrs. Draper's father) he amassed the fortune that enabled his daughter to give nearly $200,000 to astronomers in the United States and Europe. Born in 1816, Miss Bruce attended private schools, where she developed her natural bent for the arts, foreign languages, and literature, though she is said to have shown an interest in astronomy at an early age. She traveled extensively and, as her delightful letters to Pickering show, possessed a subtle, witty, and richly stocked mind. In the spring of 1889, at the advanced age of 73, she met Pickering through some relatives of the astronomer-meteorologist Cleveland Abbe, and for the first time could give practical expression to her long-dormant interest in the only science that had ever attracted her.

On May 30, 1889, Pickering received the first intimation of definite action:

Mr. Clark told me this morning [he wrote Miss Bruce] of the generous gift you are planning to make to astronomy. I shall take pleasure in calling on you with reference to it at the time suggested, Monday morning, June 3, at half past ten. I am very glad that you propose ordering the 24-inch

The Bruce Photographic Telescope

photographic telescope, since I believe you could secure the most powerful instrument ever made, and the one with which the greatest amount of useful work could be done. Your telescope ought to be erected in the best possible climate and if it could be sent to the southern hemisphere after the northern stars are photographed its usefulness would be greatly increased.

With this letter Pickering sent his last annual report, pamphlets, and papers describing some of the work of the Observatory. To give her an idea of what might be expected from a 24-inch instrument, he also enclosed enlarged pictures of stars taken with the 8-inch Bache, the smaller version of the design he proposed for the new telescope. "If I can render you any assistance in making your plans, whatever form they may take," he concluded, "I shall be very glad." In noticeably refraining from any assumption that the proposed "generous gift to astronomy" meant *Harvard* astronomy, Pickering acted on the principle he had adopted earlier in the Boyden and Draper cases. At no point did he imply that he expected the telescope to be his, but he assured her that if Harvard should be her choice, he could promise wise planning, energetic administration, and entire deference to her tastes and wishes.

Even though Clark, after a second visit to Miss Bruce, reported that she favored Harvard, Pickering regarded nothing as settled. In fact, he wrote Mrs. Draper, the prospect of such aid "to our general plans for work" seemed rather dim. He would have to await his own forthcoming interview with Miss Bruce. On June 2, a day before the meeting planned, she informed him that she had decided on Harvard. He thanked her at once, promised unstinted effort, and enclosed an elaborate memorandum outlining the whole project for the 24-inch, to be known forever as the Bruce Telescope:

If we can find the best possible location, I hope that the Harvard Observatory will eventually maintain two permanent stations, one in the northern, the other in the southern hemisphere, as in Southern California and Chile. Nearly all our observations would then be made at these two stations. The Cambridge station would then be employed mainly for reduction and publication of results.

All preliminary testing would be done in Cambridge. Afterward, the instrument would be taken to the northern station for photographing

that portion of the sky, then moved to the southern site for photographing stars invisible here. Plates would be deposited in Cambridge and
stored in a small fireproof building for examination and measurement.
"You will see," he concluded, "that our plans for work are very extensive,
but we have two or three large funds[2] whose income is available, and
I do not believe that they could be better employed than in extending
the work of the Bruce Telescope if it proves as successful an instrument
as I anticipate."

Pickering's memorandum obviously envisaged a grand combination
of Boyden funds for the stations, the Draper Memorial income for stellar
spectra observations and computations, and the powerful new telescope
for a vastly expanded program of photography. He wrote Mrs. Draper
that he hoped the telescope would accomplish what he had been aiming
at for more than a year—"that the Henry Draper Memorial shall have
the use of an instrument of this form without expense to you." After
a visit to Miss Bruce on June 13 he reported, "All looks well in that
direction."

President Eliot, who meanwhile had cautiously investigated Miss
Bruce's financial situation, wrote Pickering on June 15, "She and her
brothers have the reputation of being very rich. I am advised by a
lawyer of experience in New York we need not hesitate to accept any
gift she may make to the observatory." Upon official acceptance by
the Corporation, Pickering thanked Miss Bruce (June 19, 1889) on
behalf of the Observatory and himself: "You have given us the means
of carrying out a plan which cannot fail to have an important influence
on astronomy. The Bruce Telescope should accomplish more than any
telescope hitherto constructed."

The president announced the gift at the Commencement dinner on
June 25, and the next day Pickering sent out a circular, "The Bruce
Photographic Telescope," giving in detail the differences between the proposed photographic telescope and the large instruments of more conventional design. The whole sky, he estimated, could be photographed in a
relatively brief time on about 2000 plates. "This generous gift," he
concluded, "offers an opportunity for useful work such as seldom occurs.
It is expected that the Bruce Photographic Telescope will exert an
important influence upon astronomical science by the large amount
of material that it will furnish." By August 5, he told Miss Bruce,
all mechanical details were settled. The lens disks, 3 inches thick

The Bruce Photographic Telescope

(1 inch thicker than those of the telescope at the Lick Observatory),
were ordered from Mantois of Paris, to be delivered within the year
to Clark for grinding and polishing. "As you may have noticed," he
added, "your liberality has been widely noticed in the newspapers."

The extensive publicity given to Miss Bruce's unique contribution
to astronomy brought unreserved congratulations to Pickering from fellow astronomers all over the country. From England, however, came
a slashing attack by the editors (H. H. Turner and A. A. Common)
of *The Observatory*, the Royal Astronomical Society's monthly review
of astronomy. In the August issue their article, "Another Photographic
Chart of the Heavens,"[3] harshly questioned the whole enterprise and
Pickering's personal tactics in undertaking it. Of Miss Bruce's liberality,
they wrote,

there can be but one opinion; the gift is not only lavish, but, according
to accepted canons, wisely bestowed; for Professor Pickering has shown
his ability to deal with previous gifts of this kind, and his great thirst
for administrative opportunity seems to be insatiable. The hearty thanks
of the astronomical world are due to Miss Bruce for her ready response
to an appeal from one of its most accredited representatives. We sincerely
hope that her generosity will be productive according to Prof. Pickering's
best hopes.

Of the manner in which the appeal was made, however, "not to
mince matters, we cannot speak with anything like the same approval,"
and "if troubles arise in the future from what we consider an inconsider-
ate and ill-advised action on the part of Prof. Pickering, he has only
himself to blame." He had gravely offended, in the view of these critics,
by completely disregarding the International Astrophotographic Conference, which had been held in Paris in 1887, for the purpose of organizing a cooperative program to chart the whole sky. Yet not a single
reference to that project appeared in Pickering's two circulars. Instead,
he had proposed his own similar scheme, but with a single telescope.

Had it merely been that he was not satisfied with the adoption by the
Conference of the particular plan he fancied . . . and wished to give this
plan a trial himself, no one could have had any serious objection, but
to ignore the adopted scheme altogether is quite a different matter. If
the public to whom Pickering appealed had no further information than
his circulars, they might fairly conclude from them that he was initiating

273

a scheme of his own. Should any evil-disposed person or persons charge Prof. Pickering, in thought, word, or deed, with deliberately excluding such mention of the Conference from his appeal as likely to be prejudicial to his chances of success, we cannot acquit him of having materially contributed to this result.

He therefore owed an explanation to the public and to the International Conference, which he had treated with scant courtesy by setting up a rival plan.

As for the instrument itself—"it is now only telling an old story to say that the construction of a photographic doublet of 24 inches aperture is an easier matter in print than in practice." Does Pickering expect to get disks of the required thickness? Can he avoid considerable distortion of star images from change of refraction? Some of these practical mechanical difficulties his well-known skill may surmount, but

if Prof. Pickering is right in thinking that he can chart the whole sky in two or three years with a single telescope, then the combination of seventeen observatories to do the same work is all wrong; that at least this is his own opinion, he has not hesitated to convey by not the most delicate of implications. He has thus thrown out a distinct challenge, and matched himself against the rest of the world in the same undertaking. More than this, he has taken upon himself the whole *onus probandi;* he has tacitly condemned the work of the International Conference before it is commenced, and claimed success for himself; whereas it is comparatively certain that the former work will be carried out, based as it is on definite work with a known instrument; while his own plan depends on that assumption, which has so often proved delusive, that a telescope may be multiplied by three. If Prof. Pickering fails in his rival scheme his position will be almost pitiable; and we cannot but feel that chances are against him.

These strictures would not have been necessary, the editors conclude, if in his circulars he had offered to supplement rather than oppose the work of the Conference, for example by photographing special portions of the sky or even suggesting improvements in the plan. Such a move would have given "a different meaning to the rest of the astronomical world" and the "trifling matter of a few words" would have found "that world filled as before with sympathetic and admiring well-wishers; as it is, he has repeopled it with critics whose attention is distracted from the research by the manner in which it has been endowed."

The Bruce Photographic Telescope

Turner, perhaps somewhat troubled by the effect of this devastating blast, warned Pickering on August 20, "In the August number of *The Observatory* we have criticized a circular of yours. We much regret, that through a mistake, a copy of the magazine was not promptly sent you, and have sent off yesterday such a copy." In a letter to William Pickering, however, thanking him for a "beautiful glass positive of Jupiter," he commented rather defensively:

> You will have probably by this time seen the August No. of the "Observatory" and our attack on Prof. E. C. Pickering. Of course he will hate us for it, but it may just warn him in time of how his new project looks to European eyes. Of course he may not care 2ᵈ about that; but one would wish to avoid unpleasantness if possible. I don't think the time has yet arrived when we can afford to have rivalry instead of cooperation.

Pickering informed Miss Bruce at once of the criticism the circulars had aroused in England and enclosed a copy of the editorial. He first explained that he had omitted any reference to the Paris conference to avoid invidious comparisons and because the wide publicity already given it did not seem to require further notice from him, then continued:

> As you are doubtless aware seventeen observatories have undertaken to photograph the entire sky with telescopes of the usual form thirteen inches in diameter. Several years will be required for this work. They propose to photograph stars to the 14th magnitude. Our 8-inch telescope photographs fifteenth magnitude stars. The Bruce telescope can hardly fail to take as faint stars or to surpass the Paris photographs. If it succeeds as I anticipate it should take five to ten times as many stars.

He went on to review the record of his own efforts to promote mapping of the sky—his proposal of 1883, recommending the use of a photographic doublet, published in *The Observatory*,[4] his more specific plan of 1886, "An Investigation in Stellar Photography," in the *Memoirs* of the American Academy of Arts and Sciences;[5] his strong appeal to the Conference in 1887 to employ a new kind of photographic telescope, printed in the *Bulletin* of that organization. "After these warnings," he added, "I do not feel that I am to blame if with one telescope we do more and better work than with seventeen of the usual form." He told her that they might expect further censure at the next meeting of the Astrophotographic Conference, and while he greatly regretted

on her account that ill feeling should have been aroused, he was consoled
by the thought that "the same objection applies to all improvements
and I do not see how we could have done otherwise if we wish to
advance science. It is the first time I have ever felt obliged to apologize
on account of the expected *excellence* of a piece of work."

His letter to the editors was no apology, however. Instead, he firmly
but without rancor upheld the correctness of his action. No reply would
have been necessary at all, he began, "but for an insinuation that reference to the plan of the Paris Astrophotographic Conference was deliberately excluded, as likely to be prejudicial to the success of the Circular."
As in his letter to Miss Bruce, he explained his reasons for omitting
mention of the Conference and recalled his own explicit proposals of
several years before. "Since, as regards scale and other important features," his original plan "was substantially the same as that now proposed
for the employment of the Bruce telescope, it does not seem that any
reference was necessary to the subsequent adoption in 1887 of plans
by the Paris Conference, for attaining a similar result with a telescope
of a different form." A map of the sky was only a portion of the work
planned for the Bruce telescope, but even if this were not the case,
he could see no objection to duplication by instruments of a different
form. He concluded his letter with the generous observation, "It is difficult to believe that the editors of the 'Observatory' intend their implication with regard to the work of the Bruce telescope in anything but
a friendly spirit, or that they wish to discourage an experiment whose
success can be determined only on trial."[6]

From the Cape of Good Hope David Gill wrote Pickering on November 4, 1889: "I am disgusted with the miserable, carping, envious attack
made upon you in the 'Observatory' article on the Bruce gift—and
still more so with its outrageous discourtesy and unscientific spirit. Your
quite dignified article in reply I much admire." Miss Bruce, to whom
Pickering had sent a copy of his letter, remained unperturbed, no further
reproof came from the autumn Astrophotographic Conference, and in
the end Pickering's course was fully vindicated. The success of the Bruce
telescope was such that in 1897, as Pickering wrote Miss Bruce, "Professor Turner of Oxford wants the Congress to leave the southern sky
to us. It was he who wrote, Does Professor Pickering expect with this
instrument to do as much as the seventeen other observatories taking
part in the Congress; now he has answered his own question."

The Bruce Photographic Telescope

Pickering's correspondence with Miss Bruce (or, when she was ill, with her sister Matilda) continued briskly for a decade. Although she lacked Mrs. Draper's understanding of astronomical matters, Pickering tactfully kept her fully informed of the problems and progress of the telescope, the work planned for it, and the high hopes he held of it. "I am more than ever convinced," he wrote her in May 1890, "that the great work of my life is to be with this instrument."

Miss Bruce rather relished the humor of finding herself drawn so unexpectedly into the astronomical world. "I am not competent to pronounce on the work," she admitted, though "of course whatever can be done to promote the work of that Pickering telescope will be clear gain. You know you have entire and undisputed control of that chunky Photo-Sterescope-Telescope." And as if to maintain the bliss of ignorance, she sent Pickering a clipping about a charge of insanity brought against Mrs. Elizabeth Thompson, the philanthropist whose fund for the advancement of science he had for some time helped administer.[7] The poor lady, Miss Bruce wryly commented, "perhaps had tried to understand what it was that she had given her money for, and as to the list that I saw, certainly some of the subjects were calculated to disturb a feeble woman's brain. I accept her case as a warning. Please tell Mrs. Pickering that I withdraw entirely from the work on Kepler's problem." In another letter, after reading a review of Professor Young's *Elements of Astronomy* that suggested it was adapted to the "humblest capacity," she wrote, "Well, there is in every lowest depth a lower deep, and I fear to fall into it. I think Young calls the vast spaces between the stars a vacuum and Fiske . . . in *The Idea of God* speaks of it as the luminiferous ether. I shall hold on to Young."

If she laughed at herself she could also look humorously on the scientific pretensions of other ladies. She had read in a book of travels in Russia that

when some very interesting astronomical phenomenon is about to occur parties are made up at Petersburg to go out to the Observatory, and they are careful to have someone attached to the court—and so the gay ladies victimize the astronomers, take possession of all the telescopes and insist on explanations . . . Now, how much interest in science, and how much for the fun of the thing and a live Lord thrown into the bargain, I know not.

The Harvard College Observatory

As an inevitable result of her lavish gift to Harvard Miss Bruce began to receive many requests from other institutions, one of which, the Lick Observatory, asked for a very large sum. Should she consider it? she asked Pickering. He replied that if she wanted to concentrate on a single establishment she could hardly do better, for the astronomers there were among the most respected in the country. His own view of the most effective way to help astronomy was somewhat different. He preferred aiding as many astronomers as possible, for he knew the severe handicaps they labored under—lack of assistants and hardly a penny for incidental expenses. The case of Professor Charles Young was typical. Having left Dartmouth for Princeton, he was too heavily burdened with a full schedule of teaching to carry on any independent research, though he had one of the best telescopes in the United States. With $500 he could pay an assistant a year's salary and restore to usefulness $100,000 worth of idle equipment. There were similar situations elsewhere.

If Miss Bruce would experiment with a modest sum for such needs, say $5000 for one year, Pickering volunteered to attend to all correspondence and evaluate the applications, though she must approve the final decisions. "This plan has always attracted me from its breadth," he wrote, "since it would help astronomy in all parts of the world. Apart from the value of the results the moral effect of showing that astronomers of all nations are working for one end is important, and if successful other donors might be induced to follow your example." He enclosed a tentative list of men he considered worth supporting, in addition to Young: Lewis Boss (Albany), J. K. Rees (Columbia), Henry Rowland (Johns Hopkins), Truman Safford (Williams), Edward Holden (Lick), W. W. Payne (Carleton—for the *Sidereal Messenger*), John C. Adams (Cambridge, England), David Gill (Cape of Good Hope), and two or three others on the Continent.

Confident that she could count on Pickering's judgment, Miss Bruce agreed to the plan, and on July 15, 1890, having overcome her original insistence on anonymity, he issued the announcement, "Aid to Astronomical Research," which basically embodied features he had proposed in his pamphlet of 1886, "A Plan for the Extension of Astronomical Research." Now for the first time he could put his long-cherished dream to the test. The terms of the offer were liberal:

Miss C. W. Bruce offers the sum of six thousand dollars ($6000) during the present year in aiding astronomical research. No restriction will be made likely to limit the usefulness of this gift. In the hope of making it of the greatest benefit to science, the entire sum will be divided, and in general the amount devoted to a single object will not exceed five hundred dollars ($500). Precedence will be given to institutions and individuals whose work is already known through their publications, also to those cases which cannot otherwise be provided for or where additional sums can be secured if a part of the cost is furnished. Applications are invited from astronomers of all countries, and should be made to the undersigned before October 1, 1890, giving complete information regarding the desired objects. Applications not acted on favorably will be regarded as confidential. The unrestricted character of this gift should insure many important results to science, if judiciously expended. In that case it is hoped that others will be encouraged to follow this example, and that eventually it may lead to securing the needed means for any astronomer who could so use it as to make a real advance in astronomical science. Any suggestions regarding the best way of fulfilling the objects of this circular will be gratefully received.

Along with the circular Pickering had sent to the leading astronomers in the United States and abroad personal letters asking if $500 would be useful to their work. Within five days replies began to arrive. By early August he had examined 55 applications, the number steadily increased, and on November 11, 1890, with Miss Bruce's approval of all his recommendations, he issued Circular II. Of the 86 requests received within the time specified (many arrived too late to be considered), he named 15 recipients of aid from the Bruce donation.[8] Pickering had once remarked that science should conquer boundaries and the "comparative selfishness of patriotism." In choosing only five Americans—Simon Newcomb, Edward Holden, Henry Rowland, Lewis Swift, and W. W. Payne—he showed that it was possible to do so. Among the ten Europeans included were Norman Pogson (Madras), David Gill (Cape of Good Hope), A. Safarik (Prague), Ludwig Struve (Dorpat), J. Astrand (Norway), John Adams (Cambridge), and, interestingly enough, one of Pickering's severe *Observatory* critics, H. H. Turner (Oxford). "Science," Pickering had eloquently argued in his pamphlet of 1886, "is an ennobling pursuit only when it is unselfish. The attempt to aid all astronomers and all observatories is a far broader and higher aim than local success." This principle he reaffirmed in his announcement of the awards:

The same sky overarches us all. It is to be hoped that the above-named . . . and other foreign institutions will obtain more important aid from neighbors when these become aware how highly the work of their scientists is appreciated in this country. The replies . . . have placed me in possession of important information regarding the present needs of astronomers . . . The income derived from a gift of one hundred thousand dollars would provide every year for several cases like those named . . . A few thousand dollars would provide immediately for the most important of the cases now requiring aid. The results of such a gift would be very far reaching and would be attained without delay.

He invited further correspondence with those who might wish to aid any department of astronomy, by outright gift or bequest, whether of large or small sums.

Miss Bruce's satisfaction with the results of the experiment led her to contribute to astronomy on an even larger scale. Between 1889, the year of her first gift to Harvard, and the date of her death, March 13, 1900, her gifts to 46 astronomers or institutions amounted to $174,275. The complete list[9] of those aided reveals both the trends of astronomical research at the time and Pickering's judgment of their value. She relied on him implicitly, and only when too much moved by some urgent appeal did she override his exacting scientific standard. On one occasion, when she asked his advice about giving $10,000 to Max Wolf (at Heidelberg) for a telescope, Pickering urged a rigid inquiry into its proposed design and purpose:

You know that I am always anxious, first, that as much money as possible should be donated to astronomy, and secondly, or at least of equal importance that not one cent of it should be wasted. The future of astronomy is to depend on donors like yourself, and I hope that you may be an example of one whose every gift has been judicious.

Wolf, with Pickering's approval, eventually received the gift.

As for the Harvard Observatory, he could report that it was in a healthy condition, best described perhaps as a "kind of wealthy pauperism," for a fourfold increase in endowment had brought with it a fivefold increase in expenses. It was a constant struggle to maintain an even balance, but "after reading the letters you and I received, and learning the needs of our fellow astronomers, no astronomer could feel rich. When I do," he added "you must infer that I am getting so old and feeble as not to see the urgent astronomical needs beyond the reach of this Observatory's income."

The Bruce Photographic Telescope

While administering her largesse to other astronomers and directing the expanded work of the Observatory, he kept up a steady stream of letters to Miss Bruce about the progress of the telescope. Like Winlock, he had to suffer the frustrating delay between ordering an instrument and receiving it. Letters to Mantois for news of the disks met the fate of Winlock's anguished appeals to Simms. "That miserable laggard Mantois," Miss Bruce wrote Pickering in February 1891. "If you think I could hurry him at all I would willingly try. My French is probably at least as good as his own and without really scalping him I will just shake the Tomahawk before him and find out what is the matter." A friend in Paris sent to needle the tardy maker reported promises of early action, but meanwhile both Miss Bruce and Pickering endured periods of profound discouragement. "Every day," she wrote, "I look for news of the arrival of the Mantois discs—and meantime have had a great fright about them." She had read that the Clarks had taken six years to shape and polish the 36-inch Lick lens, while "ours has only 24 in. . . . If you can suggest any comforting thought please let me have it." Yet when Pickering expressed his own annoyance at the long delay, she tried to encourage him: "I shall be only less glad than you when the discs arrive. Let your patience hold out a little longer—another two years or so—and what are two years in the calculations of an astronomer?"

On January 31, 1892, at long last, Pickering had the pleasure of informing her that the glasses had arrived and that he had inspected them that very day. In March all was going smoothly, the lenses were nearly ready for the great steps of grinding and polishing, and Pickering's plans were drawn for a small brick building to house the instrument during its period of testing. By early August all four lenses were ground. A month later Miss Bruce wrote (Fig. 38), "I have read in this morning's Herald of the first test examination made of your new telescope and am of course much pleased. I hold out my hand to grasp yours across all this great intervening space and want to say *Let us Rejoice.*"

For a time the correspondence slackened, but on September 10, 1893, Pickering reported, "At last a part of the Bruce telescope has been sent here . . . The bed plate weighs 3300 lbs. and required four horses and half a dozen men at work all day to get it up and in place." He hoped she was not too disturbed by the slow process of completing and installing the telescope. "As we want to get the best possible results,

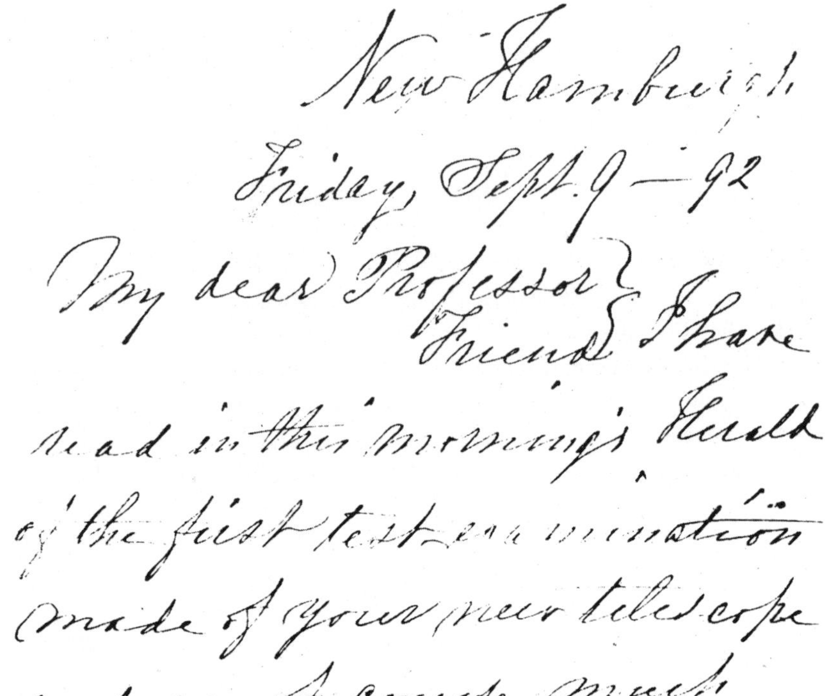

New Hamburgh
Friday, Sept. 9 — 92

*My dear Professor &
Friend I have
read in this morning's Herald
of the first test examination
made of your new telescope
and am of course much
pleased. I hold out
my hand to grasp yours
across all this green interve
ning space and want to say
"Let us rejoice". There is*

Fig. 38. Page of Miss Bruce's letter rejoicing in the first successful tests of the telescope.

The Bruce Photographic Telescope

Fɪɢ. 39. The Bruce telescope, the gift of Miss Catherine Bruce; completed 1893.

it seemed to me safer to allow whatever time is necessary, rather than to hurry the work at the risk of its being poor." Finally, on November 8 the entire apparatus was mounted without mishap—a "ponderous affair," he called it, so far revealing no flaws, but "the astronomical trials (in every sense of the word)," he cautioned, "are still before us."

During the rest of the month Clark worked strenuously to make adjustments as a result of which on the 19th Pickering wrote happily:

We have obtained some remarkable photographs. I can now safely report its assured success, and can congratulate you on having the finest photographic telescope in the world . . . You had the courage to try the experiment and it has succeeded . . . It is a great satisfaction to me . . . to feel that the telescope has justified the confidence you placed in my judgment. It was a grand experiment which could have been tried only by your most generous act.

The Harvard College Observatory

For the next year and a half the Bruce telescope (Fig. 39) underwent the most rigorous tests, and in November 1895 it was pronounced ready for its journey to the Boyden Station in Peru (Chapter VIII).

III

After the completion of the telescope and its successful mounting at Arequipa in 1896, Pickering's correspondence with Miss Bruce or her sister consisted chiefly of regular reports on progress there and of advice about the applications for aid that still poured in. Aware of her poor health and anxious to protect her from importunate demands, he offered to announce publicly that no more requests would be considered, but her sympathies had been aroused by the meager resources of many good astronomers, and she found it difficult to refuse. Some of her most lavish gifts, in fact, were made between 1895 and 1899. Pickering himself asked for nothing, but in 1897 he submitted for her consideration a different kind of proposal. "I do not know your views regarding the value of medals," he wrote, "but if you approve of them, the enclosed plan of Professor Holden seems worthy of your attention. I am sure that he would manage such a fund as he asks for, with wisdom and fidelity."

Holden's plan called for the establishment of a gold medal in Miss Bruce's name, to be administered by the Astronomical Society of the Pacific, and to be awarded, like the medal of the Royal Astronomical Society, to the astronomer who in any specified period had contributed most to astronomy. He asked for $2750. To Pickering's first letter on the subject Miss Matilda Bruce replied that her sister favored the idea but wanted more details. Would the medal be offered every year, regardless of the work, and could the officers and directors of the Society be trusted to bestow it intelligently? Could it, "in the large spirit of astronomy," go to any astronomer, regardless of country?

Pickering advised her that the terms of the medal should be perfectly understood before she made any gift, recommended specifying dates between which the work was to be done rather than any single year, and pointed out that, while a strictly national award might encourage younger American astronomers, it would be more highly prized and more widely known if astronomers of all countries were eligible. He had no doubt of the trustworthy management of Professor Holden.

On March 23 Miss Matilda Bruce wrote that her sister was ready

The Bruce Photographic Telescope

to give the $2750. She wished the medal to be awarded without respect to nationality, only when a suitable candidate could be chosen, and, as Pickering had suggested, restricted to work done between certain dates. "My sister," she added, "would like the medal to be one prized or sought for as much as those of Great Britain or France. You kindly offer to aid my sister in this matter. Would it be too much to ask you to arrange the whole affair?" Pickering readily agreed:

Your favor of March 23 has just reached me. I say favor intentionally as I highly appreciate every opportunity of aiding your sister in her great work of advancing astronomy. I write by this mail to Professor Holden regarding the medal, incorporating your suggestions, and will endeavor to arrange the whole matter as you desire. I am sure that he will do his best to insure a wise administration of the trust.

That Pickering's confidence in the administration of the award by the Astronomical Society of the Pacific[10] was fully justified is shown by the roster of distinguished astronomers who have been honored. During Pickering's lifetime there were fourteen medalists. The first, Simon Newcomb (1898), was followed by such outstanding figures as A. Auwers, David Gill, G. V. Schiaparelli, Sir William Huggins, G. W. Hill, Henri Poincaré, J. C. Kapteyn, Oskar Backlund, W. W. Campbell, George E. Hale, and E. E. Barnard. Pickering received the medal in 1908. Between 1920 and 1969 it has been awarded to 49 astronomers, of whom 28 have been American and 21 European.[11]

The Bruce telescope itself has long since been surpassed by instruments of far greater size and power, but the work it performed in recording the heavens and in providing thereby an invaluable photographic reference library for the astronomical world continues to enrich astronomy here and elsewhere. When the Arequipa station was discontinued in 1926 the telescope, after being provided with a new mounting by J. W. Fecker of Pittsburgh, was sent to the Boyden station at Mazelspoort, near Bloemfontein, South Africa, where for many years it remained in active use. Harlow Shapley wrote of it in 1931, "The 24-inch Bruce refractor has covered the whole of the southern sky in the course of the past three decades, and its many plates of long exposure are of great value in the study of the proper motions of stars, of the nature and distribution of the thousands of faint external galaxies that have left their impress, and also on the study of light variation of faint variables in the Southern

Milky Way."[12] And as late as 1954 the director, Donald Menzel, reported that 400 long-exposure plates, made with the Bruce and supplemented by photographs taken with another instrument, had provided valuable data on the distribution of faint galaxies. In 1957 the telescope was dismantled and returned to this country for possible use at the Agassiz station (previously called Oak Ridge), Harvard, Massachusetts, but with the acquisition of other instruments and the shift in emphasis to radio astronomy there, the Bruce has remained in honorable retirement.

The Bruce gold medal, however, added to the lavish gifts that Miss Bruce made to astronomers and observatories at a time when little or no support came from other sources, has kept alive the memory of one of the most generous individual donors in the history of American astronomy.

VIII

"SOME LOFTY MOUNTAIN"

Even before Miss Bruce had answered the appeal for funds to procure a large photographic telescope, and some seven years before the Bruce telescope was finally ready for use, Pickering had already taken the first steps toward establishing a Harvard observing station south of the equator, with the aid of the bequest made by Uriah Boyden.

When the Boyden Fund came to Harvard in 1887, only a handful of astronomical observatories existed in the Southern Hemisphere, and not all of them were equipped to carry on regular work. The few observatories in effective operation included one in South Africa, at the Cape of Good Hope; two in South America, at Cordoba and at Rio de Janeiro; and one in Australia, at Melbourne. The positions and magnitudes of stars in the southern skies were thus relatively unstudied, compared with those in the north. The composition of other objects of the southern skies, such as the nebulae listed by Messier for those regions, was unknown, and the true nature of the Magellanic Clouds, whose existence had first been reported to Europeans in the 15th century by the great Portuguese navigator, was still a matter of conjecture. In planning the early work for the Boyden Fund Pickering had therefore hoped to establish a mountain station below the equator, as well as one in California. The South American expedition, whose preliminary observations began about the same time as those on Wilson's Peak, was led by Solon Bailey.

Solon Irving Bailey was born in New Hampshire on December 29, 1854, and spent his youth in Concord. His interest in astronomy is said to have begun when as a boy of twelve he witnessed the Great

The Harvard College Observatory

Meteor Shower of 1866, a spectacle he never forgot.[1] After graduating from the Tilton, New Hampshire, Academy, he entered Boston University and received the B.A. degree in 1881. He then returned to Tilton as headmaster and taught a course in elementary astronomy. Hoping to continue his study of the science, in 1884 he wrote to President Eliot of Harvard, who forwarded the letter to Pickering. Like the Bonds and Winlock, Pickering regarded the Observatory as a place for research only, not as a teaching institution. He explained to President Eliot (November 3, 1884) that he could not recommend that Bailey

undertake a course of astronomical study here, since no provision has ever been made at this Observatory for the systematic instruction of students, and no lectures or recitations are carried on for their benefit. A student here has only the advantage of the library of the Observatory, and of watching the work carried out, which is usually of so special a character that it would not give that general training in practical astronomy which I presume Mr. Bailey desires.

Pickering suggested that Bailey apply for instruction to Professor C. A. Young at Princeton, or to Professor T. H. Safford who had left the University of Chicago and was now professor of astronomy at Williams College, Williamstown, Massachusetts.

In spite of this discouragement Bailey was determined to become an astronomer and in the spring of 1887 he applied again, directly to Pickering. In this period of financial plenty and expansion at the Observatory (shortly after the acquisition of the Boyden Fund and the extension of the Henry Draper Memorial) Pickering was adding to the staff, and he accepted Bailey as an unpaid assistant. Bailey moved his family to Cambridge, where he enrolled at Harvard as a candidate for the M.A. degree, and arranged to spend about 40 hours a week working at the Observatory. Within a few months he had demonstrated such unusual ability that Pickering began paying him a small salary and suggested (December 5, 1887) to James M. Peirce, Perkins Professor of mathematics and astronomy (and brother of C. S. Peirce), that Bailey's work should be "regarded as the equivalent to two courses of study" toward the degree.

This request raised a problem that seems strange today, when many graduate students receive a regular stipend from the university or a scientific foundation: since the work was paid for, could it be classed as "study"? Pickering believed it should be. In a letter (December 22, 1887) to S. M. MacVane, professor of history, he wrote:

Your letter of yesterday is at hand. Mr. Bailey's work here is decidedly of the nature of investigation. It is of course planned for him, but it is such as I should recommend to a student desirous of training himself for original researches of his own. In fact, Mr. Bailey began it entirely with the view of gaining skill in scientific work, and it is only recently that he has received any payment for it, while the character of the work remains as before. If, therefore, the mere fact of the payment does not hinder it from fulfilling the requirements of study for a degree, I am fully of the opinion that it should be accepted.

This view prevailed and Bailey received the M.A. degree in the summer of 1888. Appointed to lead the Boyden expedition for a two-years' stay in Peru, he chose as his assistant his younger brother Marshall, who was a professional photographer.

Preparations were already well advanced. During the previous year Pickering had been in correspondence with several men in Peru who kindly agreed to procure meteorological data: Victor H. MacCord, superintendent of the Arequipa, Puno, and Cuzco Railroad; W. H. Cilley of the Oroya Railroad; and Juan L. de Romaña,[2] a Peruvian gentleman with broad scientific interests. To all of these Pickering had sent thermometers, rain gauges, and sunshine recorders, with instructions for obtaining the necessary observations. They generously gave their time to the work, and in the years ahead remained enthusiastic friends of the Observatory. From the information received Pickering had selected a part of the Andes near the town of Chosica as a promising site for a temporary station.

Pickering and Bailey had planned the expedition as carefully as possible. Since wood was extremely scarce in Peru, they procured two prefabricated buildings from the Goodnow Portable House Co. of Worcester, Massachusetts, and packed them for transport. The M. A. Seed Dry Plate Co. of St. Louis agreed to make regular shipments of photographic plates. The W. R. Grace Company of New York was chosen to transport the equipment (their offices in Peru would also serve as bankers). Through President Eliot's appeal to officials in Washington, the Peruvian government agreed to admit the equipment and freight free of duty.

According to plan, Bailey (with his wife and four-year-old son Irving)[3] first traveled to California with the Harvard party to photograph the solar eclipse of New Year's Day 1889, and on February 2 embarked at San Francisco for the trip to Peru, via Panama. He recorded in his journal the incidents of the slow voyage south. As the ship called at the chief ports along the coast of Mexico and Guatemala he usually went ashore to practice his Spanish, and noted that he could now see the Southern Cross, well above the horizon.

Marshall Bailey, meanwhile, had sailed from New York to Colon and, since the Panama Canal was still unfinished, crossed the Isthmus by train to join his brother. The De Lesseps Company had recently gone into receivership, all work on the canal had come to a stop, and Solon Bailey thought that the project would probably be abandoned. In his journal he described his trip to "La Boca," the mouth of the canal, where "decay and desolation were the marked features. On the way we passed the company's hospital and burial grounds. The vastness of each bore witness to the power of the forces with which the promoters of the canal were obliged to contend."

From Panama the party sailed for Peru, with all the impedimenta necessary for an expedition to what was then a primitive country. The basic scientific equipment included the 4-inch meridian photometer and the 8-inch Bache photographic telescope, whose lenses and prisms Bailey kept in his stateroom during the journey. In addition there were nearly 100 boxes, barrels, crates, kegs, and cases, which included the sections of the portable houses, rolls of building paper, sailcloth, clocks, meteorological instruments, cans of paint, tools, lanterns, pails, photographic plates, chemicals, glassware, a ladder, furniture and bedding, as well as clothing and other personal belongings. On March 6 they disembarked at Callao, the port for Lima. Bailey recorded in his journal that the "Custom House officials made some fuss about passing some of our baggage but did so finally—under the influence of $3.00."

In Lima, the officials of the Oroya Railroad helped arrange an exploratory trip to mountain areas near the villages of Chosica, Chicla, and Matucana. Traveling on foot or on muleback, led by guides who seldom knew the country as well as they had claimed, Solon and Marshall Bailey spent several weeks in the search for a good observing site.

"Some Lofty Mountain"

The climbing demanded both skill and stamina, and the difficulties of the ascents were increased, at heights of 10,000 feet or more, by attacks of "soroche," a mountain sickness that caused dizziness, faintness, nausea, and sometimes loss of consciousness. On one such climb when Marshall suffered an attack "the patient was placed on the ground, and bruised garlic, the odor of which is thought by the natives to have great efficacy, was provided in abundance,"[4] a treatment he accepted heroically.

The results of these explorations were discouraging. Some peaks were nearly inaccessible, none had an unobstructed view of the horizon, and at no place were the skies as free from clouds as Bailey had been led to expect. Nevertheless, since rain would put an end to astronomical work during the Peruvian summer and delay in setting up the instruments would mean the loss of an entire observing season, a choice had to be made.

As the most favorable of these disappointing summits he finally selected an unnamed peak above the town of Chosica and christened it "Mt. Harvard." He had also suggested calling it "Mt. Pickering" but Pickering declined the honor, commenting (August 4, 1889) that "the name of Mt. Harvard seems very appropriate if the Peruvians do not object. Mt. Pickering might wait until I have done as good work as you have on a Peruvian mountain." With an altitude of 6500 feet, the peak had the advantages of a mild climate, clear sky, and a view to within 7 degrees of the horizon. Its other attributes were far less desirable. It was rocky and barren, and the nearest human habitation was miles away. The only access was by an inadequate path, and all the equipment—telescopes, meteorological instruments, building materials, furniture, food, and even the daily supply of water—had to be carried from Chosica to the summit, a distance of 8 miles, on the backs of mules.

The first necessity was a usable trail. Some ten or twelve peons were engaged for the work, and to hurry its completion the Baileys both supervised and shared the actual labor. The road led from the hotel in Chosica, crossed the Rimac river by a suspension bridge, continued through irrigated fields, along a ravine, and became a rough trail that then zigzagged up the mountain to the summit. Within three weeks the track was passable and the moving of equipment began on April 15, when three mule-loads of freight were taken up and dumped.

Arriving alone late that afternoon, Bailey unpacked a few boxes, stretched a piece of canvas for a makeshift tent, and prepared to spend the night. In his journal he described his feeling of satisfaction:

> Just about sunset there was a slight shower lasting a few minutes and a magnificent rainbow spanning the valley of the Rimac and resting on the mountains on either side. The fact that the nearest human being was miles away gave me a peculiar sense of proprietorship in this exhibition of nature. The impressive silence that reigned there was especially striking. By listening carefully the murmur of the Rimac rushing along in its rocky bed could be faintly but distinctly heard. The only other sound was that of a condor's wings as it occasionally swept down near me, evidently surprised at the changes.

Going to sleep at dark, he awoke about midnight and noted that although there was fog in the valley, above the mountain the skies were relatively clear. He also made a closer acquaintance with the local fauna when, picking up a stick, he "was not pleased to find a large centipede clinging to it. He paid with his life, however, the penalty of so rashly venturing to make my acquaintance."

The last of about 80 loads of freight arrived at the summit on May 7, only two months after the expedition had reached Peru. Since the installation had been planned months before in Cambridge, Bailey had only to mark the locations of the instrument shelter and the house, lay simple foundations, and assemble the framework of the buildings brought from the States. The instrument building, about 30 × 10 feet, had three rooms, one for the Bache telescope and the meridian photometer, one for the photographic darkroom, and one for the tools. Building paper nailed to the wooden frames made the walls and heavy canvas formed the roofs. Stones found on the mountain were used to build the telescope piers. On May 8 the entire family moved to Mt. Harvard, and the following day Solon Bailey took the first photograph from the new Boyden station. Two weeks later he made his first photometric measurements of southern stars.

The first two boxes of photographic plates reached Cambridge early in August. Before even examining them Pickering wrote informally to Bailey (August 4, 1889), with the promise of an official letter to follow shortly, to congratulate

you on your success. Mrs. Draper and I have both been impressed with the skill you and your brother have shown in overcoming the various obstacles in Peru and on the Isthmus. Also at the rapidity with which the work is progressing . . . I consider it one of the most important pieces of work undertaken by this Observatory, and I think that on its successful completion it will be highly appreciated. I hope that you have taken all proper precautions against illness in the way of medicines and other supplies. Bad drinking water is always a source of danger . . . I am desirous that you should all be as comfortable as possible at your station, and hope that you will go to any reasonable expense to attain this end. The outdoor life and fresh air ought to be healthful and invigorating.

After examining the photographs, Pickering wrote that the best of the plates were very good but that some of them seemed to have suffered some injury at the edges—perhaps they had been gnawed by insects? Bailey replied (September 10, 1889) with his usual dry humor:

I am inclined to think that the plates have been injured near the edges by the tanks more than by insects. The only insects that appear to be abundant here are fleas and scorpions and they have thus far evinced a prejudice in favor of us rather than dry plates—of the latter we have found specimens in our hose, pantaloons, and coats but never any on the dry plate shelf.

The life on Mt. Harvard probably was healthful and invigorating, as Pickering had hoped, but it was also lonely, monotonous, and lacking in comforts. There was no water, no electricity, no telephone, and no plumbing. For companionship the Baileys had only a Peruvian assistant, Señor Elias Vieyra, who helped with the instruments and observations, and two servants, Francisco and Vincenta. In addition to the seven human beings, the group included dogs, cats, goats, and a flock of poultry. For amusement, the family took walks on the barren mountainside, usually accompanied by the dogs, the cats, and the goats, or they played at tennis on a makeshift court they had scraped free of cacti and loose stones—a game made hazardous because a ball hit too hard was likely to roll down the mountainside and be lost forever.

The astronomical observations were directed from Cambridge. Pickering had planned them in advance but, in addition, he maintained a regular and detailed correspondence with Bailey regarding the work to be done. The most important projects were to complete the Harvard Photometry by measuring the magnitudes of the southern stars; to augment the scope of the Henry Draper Memorial by recording their spectra; and to map the sky by photographing the stars. Since the instruments employed were those that had been used in Cambridge for the same purposes—the 4-inch meridian photometer and the Bache 8-inch photographic doublet—the observations obtained in the two hemispheres would be strictly comparable. Recording meteorological data was another regular duty, in which Mrs. Bailey frequently assisted. Also, Pickering often requested observations of specific objects—a newly found variable star, a star giving a peculiar spectrum, or the satellites of a planet.

Records of the photometric determinations were mailed regularly to the Observatory in Cambridge. When a number of photographic plates had accumulated they were packed in numbered cases and shipped to the coast for the voyage to the United States. They were carefully protected, but the glass was so thin and fragile that they were sometimes broken in transit, and occasionally an entire case was lost. Still, a phenomenal amount of material began to arrive at the Observatory in Cambridge for study by Pickering, Mrs. Fleming, Miss Maury, and the computing staff.

During the first few months (autumn in Peru) weather conditions on Mt. Harvard were superb. The sky was almost cloudless so that observations were possible nearly every night. In September and October, however, thin clouds began to interfere, heralding the approach of the rainy season. Roughly a year remained of Bailey's assigned time in Peru, and it was important to find a desirable site for a permanent station, which was to be placed in charge of William Pickering. Since little astronomical work could be done during the rains, Bailey decided to use the time to resume his explorations. Leaving Señor Vieyra to carry on the work at Mt. Harvard when weather permitted, Solon Bailey and his brother Marshall packed the meridian photometer and some meteorological instruments and on November 8 set out on their search, while Mrs. Bailey and Irving settled in Lima to wait for their return.

From Lima's port of Callao the brothers sailed south to the coastal city of Mollendo, and traveled inland by train to the city of Arequipa, whose clear air and relatively cloudless sky made a very favorable impression. Continuing inland, they visited the town of Puno on the shore of Lake Titicaca, where they explored many ruins of the Incan civilization, and then continued on to La Paz in Bolivia.

After arranging to have meteorological observations made at several places, the Baileys returned to Mollendo, where they again took ship for a journey south along the coasts of Peru and Chile, as far as Valparaiso. Of the many sites inspected one of the most promising seemed to be in the desert of Atacama in Chile, a desolate area devoted entirely to nitrate mining. Early in January the Baileys became the guests of Señor Perez, superintendent of the nitrate-mining company, and set up their instruments on the flat roof of the company's office building in the town of Pampa Central. There they stayed for some six weeks of observations, and found the seeing conditions unusually fine. Late in February they left Chile and sailed again to Callao, reaching Mt. Harvard on March 5, 1890, after a journey of four months.

The situation on Mt. Harvard was discouraging. Señor Vieyra had faithfully kept the meteorological records and had carried out some photographic work, but observing conditions had been bad. In December and January fog and clouds had constantly interfered, and in February the rains had begun, causing great damage to the paper walls and roofs of some of the buildings. Vieyra had reported to Pickering in Cambridge (February 4, 1890):

Last Thursday, the fog covered all the observatory station, and the rain was so continuous day and night, that the "rain gauge recorder" retained 5½ inches of water. The rooms were all wet, the water past all them (except the one that came from the U. S.) our beds and rest of things were completely wet; but no instrument was injured—the only things that are badly hurt, as they were of paper, are my room and rooms of rest of folks, which I am fixing as well as possibly . . . This place is very solitary and musty; but I live happy with the hope that Mr. Bailey will return soon.

The damaged buildings had been repaired as well as possible by replacing the paper roofs with canvas covered by a thick coat of paint, but the human beings marooned on the peak all suffered from persistent colds.

During his journey, Bailey had mailed full reports of his explorations to Pickering. The two most favorable sites, Bailey concluded, were Arequipa in Peru and Pampa Central in the Atacama desert of Chile. At Pampa Central the seeing conditions were excellent but the isolation and barrenness of the place—there was no water, food had to be shipped from the coast, and not even a blade of grass could grow—were serious drawbacks. The second possibility, Arequipa, was a town of about 30,000 people. It was surrounded by desert, on the plain below the nearly extinct volcano of El Misti. Although the area was known to be earthquake-prone, the air was exceptionally dry, clear, and steady, and the climate would permit observations during a large part of the year. Although not on a mountain, at about 8000 feet the site was higher than on Mt. Harvard. One disadvantage was that the nearest port, Mollendo, had no harbor and ships must anchor about a mile from shore, while passengers and cargo were taken to land in a lighter. In rough weather, when the surf ran high, they had to be placed in a kind of cage suspended by ropes, lowered to a small boat for the trip to shore, and then lifted to the wharf by means of a similar cage. However, there was no real danger, Bailey wrote, and he had been informed "that no passengers life was ever lost, and no merchandise of much value."[5] Arequipa itself was on the railroad so that transporting freight there would be a much simpler procedure than at Mt. Harvard. Water would be available and food and supplies could easily be obtained.

Having sent all this information to Pickering, Bailey resumed work at Mt. Harvard but was hampered by unexpected bad weather—the skies were cloudy in May and worse in June. July and August brought some improvement, but by then Pickering had decided to move the station to Arequipa.

Work came to a close on Mt. Harvard. The buildings and instruments were dismantled and packed for the hazardous journey down the trail, and the residents finally left on October 15. Bailey wrote to Pickering (October 17. 1890) :

We were fortunate in getting down the mountain with out damage to ourselves or the apparatus—Four mules loaded with lumber rolled down the mountain some 20–30 feet and lodged among the rocks, but by some kind fate which seems to watch over observatories & mules they escaped with out broken bones—How they escaped broken legs & necks I fail to understand.

Bailey recorded that although "we were pleased to go to a site nearer civilization and where more of the comforts of life would be found, we, nevertheless, felt sorry to leave the mountain that had served us as a home for a year and a half."[6]

"Some Lofty Mountain"

They reached Arequipa late in the month. Bailey rented a house in Carmen Alto, a village about three miles outside the city. On the flat roof of the house he installed the meridian photometer and settled down to complete its scheduled work. Cloudy weather, and problems connected with the arrival of the new expedition, caused some delay but on May 9 he made the last of the planned measures. A few days later he and his family left Peru, via the Straits of Magellan, for a vacation in Europe before their return to Cambridge.

The accomplishments of this first, exploratory expedition were impressive. In two years, almost single-handed, Bailey had measured the magnitudes of about 8000 stars from −30° to the South Pole, to complete the photometry of the entire sky.[7] With the assistance of Marshall Bailey and Señor Vieyra, he had obtained some 2500 photographs that recorded the spectra of southern stars brighter than the eighth magnitude, for the Henry Draper Memorial, and provided charts of the stars brighter than the fourteenth magnitude. He had arranged for regular meteorological observations at four sites, ranging in altitude from sea level to around 14,000 feet. He had also found what promised to be an ideal site for a permanent high-altitude observatory, as Uriah Boyden had wished.

II

The second Peruvian expedition reached Arequipa on January 17, 1891, under the leadership of William Pickering. He had brought with him two assistants, George F. Vickers and Andrew Ellicott Douglass. In looking for candidates for these posts, which paid $500 a year and expenses, William had demanded only modest qualifications, and had stated that he did "not care especially for a brilliant man, but one who is fairly bright, and who would be an agreeable companion. Skill in physics and mathematics desirable but not necessary."[8] In the end he had selected Vickers, a young man recommended by the president of the Boston YMCA as a well-traveled gentleman, and Douglass, who had joined the Observatory staff in 1889 after graduating from Trinity College in Connecticut and who proved to be an excellent choice.

The expedition carried a large amount of new equipment: the 13-inch Boyden refractor, which was designed for photography and had been used at Wilson's Peak; a 20-inch reflector that Edward Pickering had bought from the English astronomer A. A. Common; and a 2½-inch Voigtlander photographic doublet. These, with the Bache telescope, pro-

297

vided the means for a wide variety of astronomical work. Since the cloudy season had begun, however, regular observations were not possible and William turned to the immediate problems of selecting a place to mount the instruments and finding a residence for the members of his party. In addition to the three persons employed by the Observatory, the group included William's wife Anne and their two children, his wife's mother, and a nurse.

Before William left Cambridge, Edward Pickering had discussed with President Eliot the question of acquiring property in Peru. Eliot had suggested that instead of buying land for what might prove to be a temporary installation, as both the Wilson's Peak and the Mt. Harvard stations had been, it would be wiser to lease a tract of ground, unless a barren hilltop could be found that would have so little value for other purposes that it might be had for a small sum. In the end, they had left the decision to William's discretion, on the understanding that he must not spend more than $500, whether for a long-term lease or for an outright purchase of land. As for living quarters, both Edward Pickering and President Eliot had assumed that the members of the new expedition would occupy the large house already rented for that purpose by the Baileys.

The first intimation of a change in these plans reached Cambridge less than a month after William arrived in Peru. The ensuing events plunged the Harvard Observatory into a financial crisis that for a time threatened the continued existence of the Arequipa station. Early in February, Edward Pickering received a cable from his brother that read "Send four thousand more." Since the expedition had been supplied with funds to cover the estimated expenses of the first period, this demand for extra money was wholly unexpected. The sums already spent for new equipment and for the costs of travel had been great, and very little money remained in the current Boyden account. Edward Pickering had received no letters from his brother as yet (mail sent from Peru required about a month to reach Cambridge) and, lacking an explanation, he replied "Account overdrawn. Economy necessary. Name minimum. State object." William's cabled response was confusing but insistent. After consulting with Mr. Hooper, the College Treasurer, Pickering therefore dispatched the $4000 along with the admonition "Use care. Money scarce. Write fully."[9]

When letters at last arrived from Peru they brought disturbing news.

298

"Some Lofty Mountain"

For the observing station, Bailey had recommended the neighborhood of Carmen Alto. William Pickering had agreed with this choice and selected a site on the crest of a ridge about 300 feet above the village and surrounded by irrigated fields. Without consulting Edward Pickering, who as director had sole authority to make major decisions, William had bought the land outright for a sum much greater than the $500 allowed for the purpose and had begun construction on a lavish scale. In addition to the telescope shelters and dome, instrument sheds, and a road to provide access from the village, he had laid out plans for a large residence, which was being erected on the grounds of the observatory, together with servants' quarters and a stable. In addition to the $4000 already received, William asked for $2000 to cover current expenses, plus an estimated $7000 for the new house.

Edward Pickering was dismayed. The total annual income from the Boyden Fund was only about $11,000; the Baileys, living and working in their paper buildings on the summit of Mt. Harvard, had kept their expenditures to half that sum, including the costs of travel and freight shipments. To accede to William's requests, and to continue paying the normal running expenses of the station, the Observatory would incur a deficit of more than $7000 for the year. To William, however, the problem seemed easily solved: Edward should simply dip into the principal of the Boyden Fund.

The terms of the Boyden bequest did not in fact prohibit the spending of the principal. Theoretically, with the consent of the President and Fellows, Edward Pickering could have used the entire sum but neither he nor the Corporation would have approved of this course. When the fund first came to Harvard the Corporation had agreed to allow the expenditure of $20,000 from the capital. Most of that had already been used for the cost of the Wilson's Peak expedition, the Boyden telescope, and other new equipment, but to diminish the principal still further seemed unwise.

Faced with this anxious situation, Edward Pickering for a time considered altogether abandoning his plans for a southern station and recalling the expedition immediately. He consulted with Mr. Hooper and with President Eliot, however, and in the end recommended instead that the necessary money be borrowed from the principal of the Boyden fund. By the exercise of severe economies, he believed, it could be repaid from income in the years ahead. President Eliot concurred in this decision, although with evident reluctance, and wrote to Pickering (March 24, 1891):

The Corporation yesterday made provision for the erection of the house for your brother at the expense of the Boyden Fund; but they felt, as I doubt not you feel, that the expedition was involving them in heavy, unforseen expenditures. They regret to mortgage the future income of their funds; but in this case they did not see what else was to be done. We seem to have embarked on an enterprise the cost of which was really unknown to us. Having embarked we must go on. I do not doubt that you have urged your brother to avoid all unnecessary expenses, and to bear in mind the temporary quality of the whole undertaking . . . The Corporation of course remembers that they are at liberty to use the principal of the Boyden Fund; but they always use principal with extreme reluctance, feeling that their successors will have a right to complain if they reduce the legitimate resources of the future. The general rule of the Corporation is to spend every cent of their income, but nothing of the principal.

This decision made, Pickering wrote to William, giving his consent to the construction of the house, which was finished late in September. Made of the local white volcanic stone called "sillar," it was an impressive structure on the hill above the cultivated fields, with the snow-covered crest of El Misti rising beyond (Figs. 40 and 41). A flight of stone steps flanked by large flower urns led up to the broad veranda and a roofed balcony on the second floor extended across the entire front. Shrubs, hedges, and a rose garden were added later. (Both land and house proved in the end to be a wise investment, and served as a base for Harvard astronomers until 1927, when the Boyden Station was moved to South Africa.)

In his letter to William (March 28, 1891), Pickering urged him to start some photographic work immediately with the Bache telescope, without waiting for the Boyden refractor to be mounted, and requested him to limit running expenses of the station to $500 a month. He added:

Sending an expedition, at least on this scale, was doubtless a bad mistake, for which I am primarily and you secondarily responsible. Whatever the urgency we cannot escape the blame. More importance will be attached to a good financial than to a good scientific condition of the department . . . We get no scientific results from Peru. What is Mr. Bailey doing with the meridian photometer? Is the Bache telescope at work? . . . Also, how about the meteorological work?

"Some Lofty Mountain"

Bailey was of course finishing his photometric measures and before he left Peru early in May he set up the Bache telescope, which was put to work in the middle of April. The Boyden telescope was mounted early in July but a portion of the driving gear had been broken during the journey from Boston so that, until new castings could be made, it could be used only for visual observations.

William wrote enthusiastically about the advantages of the Arequipa site. He reported that he had secured a large number of plates of stellar spectra, made with the Bache; described the excellence of his visual observations of the moon, Mars, and the other planets; and urged a further expansion of the station—he had already hired another assistant, a young Peruvian named Luis Duncker. William also suggested that Edward Pickering send down the 12-inch telescope, and appropriate still more of the Boyden capital to erect additional buildings and employ a larger number of assistants.

The fact remained, however, that this costly new expedition had been in Peru for several months and had not yet sent any photographic material to Cambridge. Edward Pickering wrote (August 8, 1891) that he strongly doubted the Corporation's willingness to approve further use of capital, particularly without something to show for what had already been spent.

I fear that your mistake throughout is attempting to do things on too large a scale. Everything should be sacrificed to attaining results . . . I am very glad that you have 500 plates but very sorry that they are not here. I am very anxious lest some mistake regarding instructions may make them worthless. I would give up everything to keep the telescopes running all night, the plates developed, and sent on promptly. This seems to settle the case of the 12 inch telescope. You evidently cannot use more instruments at present. When the photographs come your case will be much stronger.

After an interval of nearly nine months, during which no photographs from Peru had reached Cambridge, Pickering cabled the direct order: "Photograph with thirteen inch." He supplemented this demand with a letter to William (October 8, 1891):

Your visual work is valuable and doubtless you would prefer it to photographic work. But the instrument was sent to fulfill a plan which should have preference. I have no objection to a certain amount of visual work especially during the opposition of Mars but I want photographs charts

Fig. 40. The Arequipa station in Peru, photographed February 17, 1892; El Misti in the background.

FIG. 41. Parlor of station residence, Arequipa.

and spectra of the principal objects as soon as possible. If the spectra are not taken I cannot charge so large a part of the expense to the Henry Draper Memorial and this would cut down the available income.

On November 11 the Boyden took its first photograph of the southern skies, but it continued such work only sporadically.

III

This surprising neglect of Edward Pickering's photographic program apparently stemmed from a hitherto unrecognized difference in the basic interests of the two brothers. Edward hoped to accumulate large numbers of stellar spectra and astronomical photographs that could be used by all astronomers to help determine the chemical composition and structure of the various types of stars. William, however, had developed a strong curiosity about the members of the solar system and believed that for their study the methods of spectroscopy and photography were of less use than the human eye.[10] He himself had unusually keen eyesight. When atmospheric conditions were at their best, with the naked eye he could easily distinguish 13 stars in the Pleiades cluster, while the average person sees only 6 or 7, and he was fascinated by the changes he perceived in the color and form of surface features of the moon and the planets, particularly Mars.

Mars had attracted unusual attention since the period of the favorable opposition of 1877 when, on the nights of August 11 and 16, Asaph Hall at the Naval Observatory had discovered two Martian moons. Detailed computations had convinced Hall that, if Mars possessed any satellites, they would have to move in orbits very close to the planet itself. Knowing that Mars would be making an unusually close approach to the earth in 1877, he had planned a careful search, which he later described in a letter to Pickering (February 14, 1888):

A little trial, and the analogy of other planetary systems, led me to search very near the planet . . . In the case of the Mars satellites there was a practical difficulty of which I could not speak in an official Report. It was to get rid of my assistant [E. S. Holden]. It was natural that I should wish to be alone; and by the greatest good luck Dr. Henry Draper invited him to Dobb's Ferry at the very nick of time. He could not have gone much farther than Baltimore when I had the first satellite nearly in hand.

304

"Some Lofty Mountain"

Hall kept the discovery a secret for six days. He then "bubbled over" and broke the news to Simon Newcomb, who at once notified the press, much to Hall's displeasure.[11] A telegram announcing the discovery reached the Harvard Observatory on the morning of August 18. That same evening Pickering and his assistants were able to locate the outer moon, and in the weeks following they made many measures of the position, distance, diameter, and brightness of the small satellites. Their periods of revolution and distances from the planet proved to be astonishingly similar to those Jonathan Swift had imagined and described 150 years earlier in *Gulliver's Travels*.

Other astronomers had at once joined the search for additional Martian moons and, in the excitement of the hunt, had reported several more, none of which proved to be real. At the Draper Observatory at Hastings, New York, Henry Draper, his brother Daniel, and E. S. Holden for a time thought they had found a third satellite, as Holden reported in a letter to Alvan Clark (September 1, 1877):

On August 26, Henry Draper & I found what we supposed to be a 3d Sat. of Mars. It was about ½ as bright as the outer satellite & a little less distant revolving in about 23+ hours . . . On Aug. 27, D & I followed the new satellite, x, four hours & *proved* it to be no *eyepiece* ghost by several tests. On Aug. 28, Dan'l Draper & Henry followed it two hours. We could not reverse for fear of throwing the spectroscope or the reflector out of adjustment. Hall has had 2 clear nights & has not seen x, & from the period (24 hours) I fear it may have been an object glass ghost—tho' I can hardly see how it should travel round the planet. I hope you will have time to look for this & settle it one way or another.

In writing Edward Pickering a few weeks later (October 30, 1877), Hall complained that he was getting behind in his double-star work because he was having to spend so much time looking for "fictitious" Martian satellites:

The object which the Drapers and Holden followed all the night of Aug 27 was my outer satellite, which they took for a new one. No doubt they saw a fixed star on Aug 26; and the Drapers saw another star on Aug 28. Also their observed distances will not admit a period of "24 hours or less," as they say.

The object which Professor Holden found here about Sept. 20 not only violates Kepler's third law but also the simplest rules of geometry. Its existence therefore is a mathematical impossibility . . . Besides these several other moons of Mars were reported, but they gave me but little trouble. If I were to go through this experience again other people would verify their own moons.

During this same opposition of 1877, G. V. Schiaparelli, director of the Royal Observatory in Milan, had announced his observation of dark streaks on the planet's surface, which he called "canali," a term that was unfortunately translated into English as "canals." Schiaparelli had not meant to imply that the markings represented artificial structures. He thought they might result from natural geographical changes in the planet's surface and were not necessarily the work of intelligent beings. Other astronomers confirmed the presence of the markings, and some two years later Schiaparelli reported the phenomenon of "gemination," the occasional doubling of a canal.

William Pickering's interest in Mars had been stimulated by two fine photographs taken by E. S. King from Wilson's Peak, on the nights of April 9 and 10, 1890. They showed that in the single day's interval the white area around the south pole, which William interpreted as snow, had greatly enlarged.[12] In Cambridge, he spent some time in observing the changing colors and system of markings on the planet. In Arequipa, where the seeing conditions were far better than in Cambridge, the images of Mars and the moon in the telescope showed unexpectedly sharp detail, far more than was visible in photographs. William therefore chose to neglect photography in favor of the old-fashioned methods of visual observation and drawing, a work that eventually led to his becoming a recognized authority on the planet Mars, along with Schiaparelli and Percival Lowell.

At the opposition of August 4, 1892, Mars was somewhat less than 35,000,000 miles away. Although it was not ideally placed for observation in the Northern Hemisphere, it was easily visible from Peru and William observed it every night but one between July 9 and September 24. He distinguished an astonishing amount of detail, which he described in a series of incautious cables to the New York *Herald*. Without qualifying his remarks as being only a possible interpretation of the surface markings, he stated explicitly that there were two mountain ranges on Mars near the south pole and that water from melted snow had collected between them and was flowing northward. He reported that a yellowish transparent cloud cover over the mountains was breaking up, and that there were forty lakes whose areas varied from 80 by 100 miles to 40 by 40 miles. William's accounts sent during the same period for publication in *Astronomy and Astro-Physics* were more conservative (he mentioned seeing more than forty "minute black points" that "for convenience" he called "lakes") but they included a report of clouds 20 miles above the Martian surface.[13]

After the first telegram was published, Edward Pickering sounded a warning note. He wrote to William (August 24, 1892) that the

telegram to the N. Y. Herald has given you a colossal newspaper reputation. A flood of cuttings have appeared, forty nine coming this morning. In my own case I should have restricted myself more distinctly to the facts in this as in other cases. You would have rendered yourself less liable to criticism if you had stated that your interpretations were probable instead of implying that they were certain. A reexamination of your telegram shows that it is more guarded than some newspaper versions of it.

These reports on Mars were indeed subjected to criticism, particularly by astronomers at the Lick Observatory who were also observing Mars. The seeing conditions on Mt. Hamilton were excellent and the 36-inch telescope was far more powerful than the 13-inch Boyden.[14] Yet the Lick observers could not perceive the details described by William and they received his reports in the newspapers and the journals "with a kind of amazement."[15]

That autumn, Jupiter was also in a favorable position for study and William made many observations of the planet and its satellites. Using the 13-inch to measure the diameters of the satellites, he was astonished to find that they did not always show a circular disk but periodically became elliptical. This seemed to be particularly true of Io, the first satellite, and he cabled the New York *Herald* on October 14: "The first satellite is egg-shaped and revolves end over end, and nearly in the orbital plane. Its period is twelve hours and fifty-five minutes."

This claim caused consternation among other astronomers.[16] Holden, who had been observing the satellites from Lick, wrote to Edward Pickering (November 2, 1892), "We think your brother is hardly right about Jupiter's Sat. I. It is egg-shaped, no doubt, as was found here—but I doubt if it revolves as the telegrams make him say it does; but very likely the telegrams are wrong?"

The telegram was transmitted correctly but the information it contained was wrong. No other astronomer could confirm the alleged periodic changes in shape of the satellites. The problem of their appearing as ellipsoids, however, was not settled for some time. Schaeberle and W. W. Campbell at Lick believed only the first satellite to be an ellipsoid but Barnard, using the same telescope, insisted that all of them were in fact spherical,[17] an opinion later shown to be correct. The dusky markings on the surface of Io, seen against the bright surface of Jupiter, sometimes made the satellite's disk appear elongated.

IV

By the spring of 1892, when William had been in Peru a little more than a year, it had become apparent to Edward Pickering that a change in the management of the station was imperative. Although a large sum of money had been spent, there were few results to show for it. William Pickering was a gifted man, enthusiastically devoted to planetary astronomy, but he had little talent as an administrator. Douglass had proved to be a hard-working and able assistant in all phases of the work, and had obtained some fine pictures of Swift's comet, a1892.[18] The other two, Vickers and Duncker, needed more supervision than they received and their work was not dependable. William himself had published a number of articles describing the Arequipa station and its great possibilities for astronomical work[19] but most of that work still lay in the future. He had maintained the meteorological stations established by Bailey and had added another, but the observations were made irregularly and the records were not always carefully kept. He had set up instruments for recording earthquakes and had kept a record of their frequency. He had also collected material for a catalogue of double stars in the Southern Hemisphere. His several papers describing the surface markings of the moon and Mars displayed the keenness of his observation but his interpretations were of doubtful value and had exposed the Observatory to severe criticism. Of the many plates of spectra photographed for the Henry Draper Memorial only a few had arrived in Cambridge. Useful pictures of the Magellanic Clouds had been made with the 2½-inch doublet but the Boyden telescope, which should have been devoted to photography of the Magellanic Clouds, nebulae, and other celestial objects visible from the Southern Hemisphere, had been used almost entirely for visual observations of members of the solar system.

"Some Lofty Mountain"

To prevent the failure of his cherished plan, Edward Pickering decided to recall his brother at the end of the customary two-year term, and to send down Solon Bailey to replace him. Pickering announced these plans in a letter to William written May 29, 1892, and added:

As the routine work with the 13 inch cannot advance very far this year and must be mainly done by Mr. Bailey, there is no reason why you should not make rather more visual observations than formerly proposed . . . If you want to return to Peru, before Mr. Bailey's term expires, it can perhaps be arranged, if satisfactory accommodations can be provided for you. As Mr. Bailey will take your place in charge of the station, he would probably wish to occupy the present house. If a large telescope is sent down, perhaps a second house can be built. I do not understand about your account with Grace and Co. Have you a large deficit on which you are paying interest? I am not sure that the Corporation will authorize such interest to be charged to the Observatory.

William had known from the correspondence of the preceding months that the work of the station had not been satisfactory to Edward Pickering. Nevertheless, this letter came as a shock. It evoked an angry defense, which also revealed the muddled financial situation at Arequipa. The house had cost more than the original estimates, William wrote (June 27, 1892), and the Observatory must find the money to make up the deficit. Running expenses had been heavier than estimated; remittances from Cambridge had sometimes been late in arriving; he had therefore been forced to borrow money from W. R. Grace and Co., which must be repaid and, with interest, would probably amount to about $8000 by the end of the year. Furthermore, routine expenses for the coming year would be about double the amount that Edward Pickering had allowed. After stating these facts, William continued:

Without being boastful, I think I've accomplished a pretty big thing, and if the authorities could see it they would say I had got them a great deal for their money. Now it turns out that the running expenses will be about equal to the annual income [of the Boyden Fund], or have been so the past six months. Perhaps they may be less in future. I shall make them as small as possible, but we must live, and live decently . . . Now what are you going to do about it? That's what I want to know. If the Corporation is willing, now that the Observatory is an established success, to make an appropriation from the principal and pay it off, well and good. You remember I recommended a 10,000 appropriation once before.

If the Corporation is not willing, we must pay it off slowly, little by little every year . . . I shall be glad to hear your views on the matter.

As to our coming down here again to Peru and living in a small hut, while the Baileys occupy the Director's house, it is out of the question. I planned and built that house, and while I am in Peru I expect to live in it. I don't choose to live in a shanty while one of my subordinates occupies the house I built.

I like Peru, and I like to live and work here, but when I am here this is *my* Observatory, as a department of Harvard College Observatory, and it is not Bailey's Observatory nor anybody else's. If Bailey and I are here together again, I want him to distinctly understand as he did before when here, and also in California, that as you said to me when I first came to Cambridge, that under you, *I* am at the head of this Department. I have no equal in authority, and only one superior, and that is yourself. If Bailey and I are here together again, that is to be understood, and he must understand that he is only at the head of this Observatory because I am away. This Observatory is not to be run by a Council while I am in it . . . As you like Cambridge and have a talent for publicity, and as I like Peru and have I think a talent for overcoming photographic and mechanical difficulties, I think we can run this establishment very well between us.

This intemperate response, with its lack of respect for the Harvard Corporation and the director of the Observatory, and its evident jealousy of Bailey, did not receive an immediate answer. First, Edward Pickering sent President Eliot a long account of the troublesome state of affairs in Arequipa, including the fact that no detailed financial statement had been received for the operations of the last six months or more. He had supposed that once the construction period was over the expenses would have diminished to only $7000 or $8000 a year, including salaries. William's estimate, however, was $12,000 (more than the entire annual income of the Boyden Fund), which included a sum of $8000 for "running expenses" and "sundries." President Eliot commented (August 8, 1892) that it might be well to get "some quite distinct ideas" of what these items covered:

This seems to be a large sum for the number of working persons in the party, considering that no printing is done and that there can be no considerable expenditures on the spot for supplies and apparatus . . . Are you sufficiently informed about the scientific results of the expedition to make up your mind whether the results are worth what they are costing? I have heretofore understood that the astronomical results were excellent and hardly to be procured in any other manner.

310

"Some Lofty Mountain"

Pickering had already replied to William (August 7, 1892) in a letter notable for its restraint as well as its sternness:

Your letters of June 10 and 27 duly reached me. The last of these was not a pleasant one to receive either as your superior officer or as your brother. In future semi-official communications of this kind please use forms of expression which cannot be used to your disadvantage if seen by others. The President and Fellows of Harvard College are not easily bullied and had your letter been presented to them the results might have been disastrous to you. As soon as possible after learning the unfortunate financial condition of your account with Grace and Co., I sent all the facts to President Eliot, with a copy of a portion of your letter, omitting all the objectionable portions. With regard to the deficit you have incurred you write "Now what are you going to do about it? That's what I want to know." Under the letter of the Bursar, the Corporation does not appear to be responsible for this deficit. If this view is insisted on, "what I am going to do" is to help you out of this difficulty by paying a portion of the deficit myself. I do not think that the Corporation will insist, and I have already written to the President asking authority to send you two thousand dollars in addition to two thousand just sent as part of your usual remittance. Please use the most rigid economy during the remainder of your stay.

Your position as regards the Boyden Fund appears to need explanation. When I first spoke with you about coming to Cambridge [from MIT], I told you that your position would not be an independent one. Your case is like that of Mr. Wendell who has charge of the 15 inch equatorial. In each case you carry out my plans and I hold you responsible for their execution . . . The Corporation having granted me almost absolute authority in these matters, it is not my intention or policy to delegate it to another. In the future, as in the past, I expect to ask your advice freely, adopt your suggestions and carry out your plans so far as they seem to me good, taking care to give you full credit for them. If you suppose that I gave you more authority than this, it arose from a misunderstanding and I now withdraw it.

Neither William nor his wife wanted to leave Peru. They both wrote Pickering, earnestly requesting permission to stay longer than the scheduled two-year period, and assuring him that they would try to be very economical and to produce more photographic work. He replied (August 24, 1892) that there was no alternative to their return, together with Douglass and Vickers, immediately after the solar eclipse of April 16, 1893. He added his own belief

311

that in Peru as in Cambridge you have done your utmost, and that any mistakes you may have made are not due to lack of energy, enthusiasm, or hard work on your part. Your visual observations have been excellent of their kind, though I fear that many astronomers will not place as high a value on them as I do.

During the remainder of his stay in Peru William did increase the amount of work done. He and his assistants took, altogether, nearly 2000 plates of stellar spectra with the Bache for the Henry Draper Memorial, and he established a new meteorological station on Mt. Chachani, at 16,650 feet.

The advent of the cloudy season in January 1893 put a stop to further astronomical work, and William and Douglass (Vickers had resigned and left after a bitter disagreement with William) began preparations to go to Chile to observe the solar eclipse of April 16, on their return journey to Cambridge.

v

Solon Bailey, with his wife and their son Irving, reached Arequipa again on February 25, 1893. Bailey had recently been appointed to an assistant professorship. Under the terms of his agreement with Pickering, he was to remain in Arequipa for a period of five years and was to receive a salary of $2500 a year in addition to traveling expenses, board, and lodging. The costs of running the station, which would include his salary and those of his assistants, must be limited to $7500 a year, part of which would come from the Henry Draper Memorial. Thus the Observatory's debt to the Boyden Fund would gradually be paid from unused income.

Marshall Bailey[20] had stayed at home to enter medical school but an older brother, Hinman C. Bailey, took his place as assistant. A second assistant, George A. Waterbury, reached Peru a few days later, traveling from California with Professor J. W. Schaeberle of the Lick Observatory, who visited the station on his way to Chile to view the coming eclipse.

The rooms at the station residence were all occupied by William Pickering's family, his assistants, and several guests, so that members of the new expedition had to take rooms in a local hotel for the weeks remaining before the house would be free. While waiting to resume astronomical work, Bailey called on officials and old friends in Arequipa, made the ascent to the new meteorological station at Chachani Ravine with Douglass and Schaeberle, and occasionally visited the observing station to begin checking the equipment.

Since William and his wife believed that their recall was to be blamed at least in part on Bailey,[21] relations between the old and the new heads of the station were polite rather than warm. Some friction was inevitable but the shift in administration took place smoothly, on the whole. By the end of March, William's party, which included Douglass, Schaeberle, and A. Lawrence Rotch, director of the Blue Hill Meteorological Observatory, had left for Chile, taking with them the Common and the 5-inch telescopes. A few days later, when Anne Pickering and her family had also left to return to the United States, Bailey and his group moved into the residence.

His first task was to organize the astronomical work. In addition to Hinman Bailey and Waterbury, he had two assistants, Luis Duncker and H. Mechelhof, a German resident of Arequipa who had replaced Vickers. Thus Bailey was well supplied with help, as he had not been on Mt. Harvard. All of the instruments, however, were in need of attention. He recorded in his journal (April 10–24, 1893) that the small Voigtlander doublet was

encumbered with old iron and stones whose use must have been to balance it but all this was cleared away and the instrument put in good condition. No possible use for this rubbish could be found. The instrument promises to work well though considerable trouble was found in getting it into good order.

The Bache itself, Edward Pickering's pride, was in bad condition: it was very dirty, screws were loose, and the brass frame was bent out of shape so that the instrument could not maintain a constant focus. The Boyden photographic telescope had been adapted for visual observations and needed much work before it could again be used for photography. There were loose screws in the finder, and the resolution was so poor that the images were double or much elongated. The clocks were unreliable and stopped frequently. One control clock had been taken to pieces and some of the parts were missing. The earthquake recorders had been installed in the clock room and were not stably mounted, so that the mere opening or closing of the door produced a false record of a quake. Some of the meteorological instruments brought from Cambridge had never been set up.

The Harvard College Observatory

In addition to these technical problems, Bailey had also to

systemize work & house service. Heretofore, to judge by present customs, servants have been accustomed to obey everyone—assistants have ordered in the city articles according to their wishes without control either for the Observatory or for other purposes. Assistants (& servants?) have been accustomed in town to draw money without authority from the grocer and have it charged against them. Private bills of all sorts have been run by the assistants and the accounts settled by the Observatory. Of course these accounts were charged back to the assistants but this has involved much extra work and trouble—An attempt has been made to forbid any assistant or servant to draw money in town on Observatory account, and to get money only from Mr. Mechelhof, who keeps the accounts of the station . . . These radical changes seem to cause some friction with the old assistants but will be of great advantage to the Observatory & finally to all concerned.

Some of these problems were easily solved. The earthquake apparatus was moved to a small stone building in a quiet part of the grounds. The Bache and the Boyden continued to give trouble for some time, but by mid-May Bailey had put them in repair and had begun to photograph stellar clusters, while Hinman Bailey and Duncker were photographing the Milky Way. A long period of great productivity had begun at the Arequipa Station, and regular letters to Cambridge kept Pickering informed of every detail.

After these months of hard work, Bailey took advantage of a spell of cloudy weather early in September to indulge in a little archaeology. With the help of a guide, Francisco, he did some excavating in an ancient Incan burial ground across the river. He wrote Pickering (September 8, 1893) that, although the site had been pretty well dug over, he had been lucky enough to find an untouched grave and had opened it carefully, according to instructions given by Professor F. W. Putnam of the Peabody Museum at Harvard.

The grave contained one skeleton and three pots or jars. Curiously I hit upon this grave the first time and Francisco was so impressed with my skill that he lifted his hat to me four times while digging and thought I must have done it by Astronomy.

The most difficult undertaking of the year was the construction of a new meteorological station. Pickering believed such work to be second

314

in importance only to astronomical research. In Colorado, in California, and in Peru, the Observatory had spent much time and money in procuring systematic records of rainfall, temperature extremes, barometric pressure, and winds. After visiting the meteorological station set up by William Pickering at Chachani Ravine, Bailey began to doubt the value of the records made there, since the site was sheltered from the weather on three sides—an opinion that Rotch had concurred in. With Pickering's permission, Bailey therefore began exploring for a better site. During the spring and summer of 1893, he studied the possibilities of a number of places and discussed their merits in his correspondence with Pickering. In one letter (August 16, 1893) he remarked that "for an ideal mountain station, however, my thoughts keep turning to the MISTI. For a permanent lofty station it seems to be the best place in this region."

El Misti is a nearly extinct volcano that towers above the plain behind the station, about 11 air miles away. An isolated, symmetrical cone whose peak is rarely free of snow, usually quiet but occasionally emitting clouds of smoke, the Misti reaches a height of 19,200 feet. Although the ascent was extremely difficult, in 1784 a party of priests had succeeded in reaching the summit where, a few feet from the lip of the crater, they had erected an iron cross, which many of the local people regarded as the protector of the city. In the century that followed at least 20 climbers had reached the summit, and Señor de Romaña had made the climb three times.

An enthusiastic mountain climber and a member of the Appalachian Mountain Club, William Pickering had made the attempt in 1891, soon after his arrival in Peru. Of the party, which included Douglass, Vickers, Solon Bailey, and Indian guides, only William and some of the Indians had been able to climb all the way to the cross—the others had been forced to turn back by the severe nausea and headache that characterized "soroche." Nevertheless, Bailey was convinced that the climb was not impossible. The symptoms of mountain sickness were greatly increased by physical activity, he noticed; but if he could be carried most of the way on the back of a mule and thus avoid the exertion of climbing on foot, he believed he could reach the crater. He therefore determined to try making a path for mules—a plan that the citizens of Arequipa regarded as wild and impractical.

During the late summer Bailey and his brother Hinman, with two guides, spent three days on the mountain, to determine how high the mules could go without a path, to choose the best route, and to select sites for the stations. The survey convinced him that with a path and fresh mules a man could ride most or perhaps all of the way to the summit. Having received Pickering's approval, he began carrying out his plan. A crew of Indian workmen were employed to construct a way station at about 16,000 feet, which was appropriately christened Mt. Blanc. Stones in the area were used to build a hut where the observers could spend the night before and after the final ascent. To make a path above the Mt. Blanc station, the workmen moved aside the boulders and with shovels cleared away lava chunks and sand in a narrow track whose width varied between one and two feet. (The Indians at first thought the work a mere blind to the "real" purpose of the trip which, they thought, was to search for mines or buried treasure.)

During an eight-day expedition to the Misti late in September, Bailey, Waterbury, their two mules, and a group of six Indian workmen did at last reach the cross at the crater. Describing their success to Pickering, Bailey wrote (October 2, 1893) that the Indians

all embraced me in turn and afterward embraced G. A. W. Then each removed his hat and kissed first the base and then the arm of the cross. Then they drank our health and afterward or before dug a little hole and each placed therein a little coca and poured on it a little wine and covered it over . . . A little of the romance of mountain climbing is perhaps lost by using mules legs, in part, instead of depending entirely upon ones own: but what is lost to sentiment is many times made up to science, for one arrives at the summit with quiet nerves and in good condition for exact work. I think I may claim to have taken mules higher than they ever went before in any well authenticated instance. Perhaps this may prove to be a step in an evolutionary process, from the primary dependence on personal endurance to the time when scientists will regularly ascend to great heights for meteorological study, by means of captive balloons or flying machines. I hope our mules on our next expedition will appreciate their honorable position as connecting links, and exert themselves accordingly.

Just how seriously Bailey meant this half-jocular prophecy is impossible to know. In our own space age it has of course been amply fulfilled.

On October 10 the final work of placing the station began. The two Baileys and Duncker, together with 12 mule drivers and workmen and a caravan of 21 mules and horses, left Arequipa for the mountain.

Fig. 42. Shelter on top of El Misti, October 12, 1893; standing at right foreground, Solon I. Bailey; at left, Luis Duncker.

The freight included a portable hut for the observer, self-recording meteorological instruments (which ran by clockwork and would give a continuous record for 10 days without rewinding) for both the Mt. Blanc and the Misti stations, and shelters for the instruments. Three days later, after an extremely difficult ascent, they reached the summit. There they erected the observer's hut and placed the instruments in their shelters a few feet above the iron cross (Fig. 42).

After his return to Arequipa, Bailey cabled to the New York *Herald* the news that Harvard had now established the highest meteorological station in the world. Pickering (October 28, 1893) sent his hearty congratulations

on the successful establishment of a meteorological station on the Misti. It will be worth considerable expenditure to maintain observations on it. I await anxiously your account of your experiences in establishing the station. I sent for a New York Herald reporter and gave him additional facts.

Fig. 43. Ascent of El Misti, at about 18,000 feet (January 5, 1894).

Thereafter, someone from the Arequipa station (usually Waterbury or Duncker) made the ascent at least once a month (Fig. 43), when snow conditions permitted, to check the instruments and collect the records.

The new station evoked much interest among the townsmen. Bailey wrote Pickering (November 15, 1893) that the secretary of the Bishop of Arequipa had asked permission to go along on the next trip to celebrate High Mass on the summit and also to give his benediction to the meteorological station.

I think I shall go with him myself, in order to be sure that he takes no offense at anything. Our station is rather close to the cross and a trifle higher but I think this padre is a reasonable man and will find no fault.

I imagine the ceremony up there will be interesting and I am sure that it will be the 'Highest' Mass ever celebrated. When he gets us properly blessed, I think he would be willing to go to Cambridge for a reasonable price and finish the [Harvard] observatory.

Fɪɢ. 44. Blessing of El Misti station and reconsecration of the cross, November 23, 1893; priest and gentlemen of Arequipa.

The ceremony took place on November 23, and lasted about an hour (Fig. 44).

With the establishment of the Misti stations, Harvard now had a line of meteorological installations extending from roughly sea level, at Mollendo, to 19,200 feet, near the crest of the western Andes. To complete the chain, Bailey determined to install meteorological instruments and shelters at a site in the valley between the western and eastern cordilleras of the Andes and at another beyond the eastern range. For these. he chose Cuzco, the ancient capital of the Incas, where he placed the instruments in the enclosure belonging to a brewery and arranged for a German employee to make the observations, and Santa Ana, a group of sugar and cocoa estates beyond the eastern cordillera. In the years following, the locations had to be changed from time to time because reliable men could not be found to check and collect the records, or because of civil disorders in the country.

The Harvard College Observatory

The astronomical work of the Arequipa station went well, much to Pickering's satisfaction. Boxes of photographic plates—of stellar spectra, star charts, nebulae, the Magellanic Clouds—were regularly shipped from Peru. Before packing the plates, Bailey or one of his assistants looked them over for suspected variable stars, but he had neither the facilities nor the time to carry out the labor of confirming them and determining their periods. That work was done in Cambridge, where Pickering, Mrs. Fleming, Miss Leland, Miss Wells, and others studied each spectrum in detail and found many variable stars and objects with peculiar spectra.

Late in October, examining a plate made in Arequipa on July 10, Mrs. Fleming discovered the spectrum of a new star in the constellation Norma,[22] and thus brought to roughly a dozen the number of such objects then recorded in astronomical history. A plate of the region taken three weeks earlier showed no trace of the nova. Pickering cabled the news to Bailey, who obtained photographs and additional spectra of the object. In the next few years Mrs. Fleming discovered other novae from the Arequipa spectral plates, in the constellations of Carina, Centaurus, and Sagittarius.[23]

In addition to the meteorological and photographic programs, the routine research at Arequipa also included attempts to observe the zodiacal light, and some visual work on the planets and their satellites. During the summer of 1894, when Mars was in favorable opposition, Bailey made a few observations but left close study of the planet to other astronomers. In what was obviously an oblique reference to the Martian observations of William Pickering, Percival Lowell, Schiaparelli, and others, he wrote to Pickering (October 9, 1894):

> I have given very brief attention to Mars. I have concluded that my time will in general be better spent in other lines, especially as the northern observatories will doubtless give so much attention to it and see all there is to be seen, if not more, and also I fear that I have not the creative faculty sufficiently developed to make a mark as an observer of Mars.

Soon after his return to Peru in 1893 Bailey had begun a pioneer study, which developed into the major research project of his life, of the relatively rare celestial objects, once thought to be nebulae, that the telescope reveals as magnificent globular clusters of stars. One of the most spectacular of these is ω Centauri, which Bailey often referred to as "wonderful"; in writing to Pickering (September 8, 1893), he remarked that ω Centauri was "the finest cluster in the southern if indeed not in the whole sky," but later he reported that 47 Tucanae might rival it in beauty.

To the naked eye, ω Centauri looks like a hazy star, fainter than fourth magnitude; in the telescope it appears as a compact, slightly flattened ball of many hundreds of faint stars, so numerous toward the center that they seem only a luminous blur. Deciding to attempt a star count, Bailey used the Boyden telescope and a 2-hour exposure to photograph a square of sky, 30 minutes of arc on a side, whose center coincided with the center of the cluster. He then photographed a ruled grid of 400 squares, on glass, which he placed over the glass negative. Using a microscope he and Mrs. Bailey, independently, undertook the immense labor of counting the stars in each square and adding the counts. Since the two totals did not agree exactly he averaged them and found that this single cluster, for which Gould's catalogues of the southern zones had listed only 7 stars, contained at least 6389. The actual number, he wrote, must be far greater, since at the center the images of the stars were too crowded to be resolved as distinct objects.[24] (Photographs made with the larger telescopes of our own day have shown that ω Centauri and other rich globular clusters may contain as many as 100,000 stars.)

Bailey also made longer exposures (from 4 to 12 hours) of other clusters, of the nebula in Orion, and of the Magellanic Clouds, but his two favorites, 47 Tucanae and ω Centauri, he photographed so often that he became familiar with the appearance of their chief stars. While studying a newly made photograph of 47 Tucanae, he noticed that one of the stars had changed in brightness. Variable stars were still relatively uncommon objects. In 1863, when George Bond discovered the first variable reported from Harvard, only slightly more than 100 were known and in 1893 the number was still less than 400, of which fewer than half a dozen belonged to clusters. (Perhaps the first known cluster variable was that discovered by Pickering in 1889, from a plate of Messier 3 made by King at Wilson's Peak.) Looking for others on plates of 47 Tucanae, Bailey found two more. Meanwhile, in studying the plates of ω Centauri that had arrived in Cambridge, Pickering and

The Harvard College Observatory

Mrs. Fleming had each discovered a variable star. Thus in his report to the Visiting Committee for the year 1893 Pickering was able to note the discovery of five of the first Harvard variables, now numbered in the thousands, that have been found in the globular clusters.

With Pickering's permission, Bailey began to concentrate on the study of clusters, when routine work permitted. Making a preliminary examination of each plate before packing it for shipment, he marked the stars that seemed to be possible variables. The computing staff in Cambridge carried out the final tasks of comparing the images of a suspected variable on a series of plates, measuring its changes in magnitude, and determining its period. As Bailey continued his survey of the southern clusters, he found new variables not only in ω Centauri and 47 Tucanae, but also in the clusters Messier 3 and Messier 5, of which he wrote to Pickering (July 30, 1895):

> Of hardly less interest to me than the great number of variables found in these two clusters, is the apparent complete or almost complete lack of variables in many other clusters, in other respects finer objects—such as the cluster in Hercules . . . I believe that the number of these variable star clusters, if I may call thus these clusters in which the ratio of variable stars is much greater than in the general sky, will not be so great as to cheapen them: nevertheless it seems to me improbable that there are no more and it may be well for us to keep on photographing these clusters as fast as is consistent with other work . . .

Pickering approved of this project, as long as it did not interfere with the routine work scheduled for the station.

One difficulty in carrying out so many types of research was the lack of a competent staff. From the beginning of the Arequipa station, getting and keeping reliable assistants had been a problem. They were required to work long and irregular hours, their salaries were low, they received little public recognition, and, somewhat isolated by the language barrier, they found few amusements in the town. For dedicated astronomers like Pickering and Bailey, the astronomical material accumulating in Cambridge and the growing reputation of the Harvard Observatory were sufficient reward. Except for Douglass, however, few of the assistants who went to Peru were so devoted to astronomy for its own sake, nor were they able to carry out projects over a long period of time without supervision.

"Some Lofty Mountain"

Although Mechelhof had been a valuable helper to both William Pickering and Bailey, he was in frail health and unable to do night work. He became seriously ill in the spring of 1894 and resigned and returned to Germany, where he died a few months later. At about the same time Luis Duncker, who had become less and less dependable, resigned to go into business. Thus Bailey was left with insufficient help: in addition to Hinman Bailey and Waterbury, whose terms would expire in December 1895, he had only the assistance of Juan E. Muñiz, a young Arequipan with a knack for handling machinery but with little scientific training. Faced with this problem, Bailey wrote to ask Pickering's plans for the future of the station; if Bailey was to remain in charge he would need one or two assistants sent from Cambridge.

Pickering responded (September 28, 1894) not only with an outline of his general plan and the assurance that assistants would be provided, but also with warm appreciation of Bailey's abilities and accomplishments:

In reply to your letter pages 241 to 244, I give below my views with regard to the future running of the Arequipa Station. The length of time required to get an answer to any question is a sufficient reason for making plans so far in advance. You of course understand that any plans I may make are liable to be changed with changed conditions, and are in all cases subject to the approval of the President and Fellows of Harvard College.

First, as regards your own connection with the Observatory. The work you have done at Cambridge, at Mt. Harvard, and at Arequipa has been very satisfactory to me. I hope, therefore, that you will look forward, as I do, to your connection with the Observatory being permanent, not temporary. You are more familiar with the work of the Observatory in general than anyone else, and as you have the necessary executive ability I want to make your position one of increasing responsibility. How would the following plan strike you? That you should finish your term of five years at Arequipa, then come here for a period of one, two, or three years during which you should prepare your Arequipa work for publication and undertake more and more of the executive work here. By that time I may want to take a Sabbatical Year, or shorter vacation abroad. Of course Professor Searle represents the Observatory when I am away, but by that time he also may want a vacation, as he is much older than either you or I. Moreover, he is not familiar with a large part of the work, including that relating to photography and meteorology. After this, the question would arise, whether you had better take another term in Peru, or remain here, relieving me of a large part of the general management of the Institution. This cannot well be decided now as meanwhile conditions may greatly alter.

323

The Harvard College Observatory

I feel that you are more familiar than anyone else with my policy regarding the Observatory work, and that you sympathise with it. I should like to have you, year by year, become more closely identified with it.

Pleased with this expression of confidence, Bailey replied that he was willing to complete the five-year term, and that the work of the Observatory would command his best efforts. The first of the new assistants, William B. Clymer, arrived in March 1895, a few months before Hinman Bailey and Waterbury returned to the states. Dr. De Lisle Stewart, who had studied astronomy under Professor Payne at the Carleton College Observatory, arrived in April 1896 to complete the replacement.

IX

"HARVARD OBSERVATORY
ARRAIGNED"

For the first fifteen years of Edward Pickering's directorship, the Observatory in Cambridge followed a generally satisfactory course. Pickering had been successful in augmenting the financial resources, increasing the staff, adding to the equipment, and contributing to our knowledge of the brightness and classification of stars. In the years 1894 and 1895, however, a series of critical problems developed that for a time checked this steady progress. A plan to establish another Harvard station, in the Territory of Arizona, had to be abandoned. E. S. Holden and other astronomers challenged the validity of William Pickering's work on Mars and the satellites of Jupiter. The accuracy of the Harvard Photometry came under attack, an attack that threatened to destroy the scientific reputation of the Observatory. The Bruce telescope proved to need major adjustments that delayed its planned journey to Peru. Civil war in Peru endangered the Arequipa station and its staff. The first of these difficulties, that of the Arizona station, proved to be the least serious.

I

William Pickering and his assistant, A. E. Douglass, had observed the eclipse of April 16, 1893, from Chile. They had had good weather and obtained photographs of the corona and its spectrum. After returning to Cambridge in the autumn, William had started preparing some of his Arequipa work for publication and, at the same time, urged the forming of a new expedition to test the seeing conditions in the Territory of Arizona. Another favorable opposition of Mars would occur in the summer of 1894, and he believed that the high dry air of Arizona would allow detailed study of features on the Martian surface.

325

The Harvard College Observatory

Edward Pickering did not oppose the plan. He had become convinced that all observatories must eventually be moved away from the cities, where an atmosphere perturbed by the results of industrial activities, as well as the increased use of electric lighting, were already interfering with astronomical work. The idea of establishing another Harvard station in the Southwest appealed to him but he had no way to finance the venture. The Observatory was still in debt to the Boyden Fund, and had also lost the income formerly derived from the sale of time signals (about $3000 a year), a service that the Naval Observatory had recently taken over for the entire country. The full income of the Paine Fund had now become available but it was being used to support the Paine Professorship of Practical Astronomy (held by the director) and for routine expenses. A possible means of carrying out the Arizona project appeared early in January 1894 when Percival Lowell of Boston offered to collaborate in a joint expedition to observe the planet Mars.

Percival Lowell was a member of a long-established New England family (his brother Abbott Lawrence later became the president of Harvard, serving from 1909 to 1934, and his sister Amy became a well-known poet). A man of independent wealth, Percival Lowell could afford to indulge his wide interests, which ranged from business and polo playing to astronomy and Eastern religions. After graduating from Harvard in 1876 (where he studied mathematics under Benjamin Peirce) he had spent some years in the Orient, where he learned Japanese and wrote several brilliant books dealing with the people of the Far East and their religions.[1] In the autumn of 1893 Lowell returned to Boston from his last visit to Japan. His boyhood interest in astronomy had reawakened and, having learned (probably from his cousin A. L. Rotch, who had accompanied William Pickering on the eclipse expedition to Chile) of William's plans to observe the coming opposition of Mars, Lowell suggested that they join forces.

At first, Edward Pickering treated the proposal with some caution. Lowell was known to be a strong individualist, whom some acquaintances regarded as charming but others characterized as an "intensely egoistic and unreasonable person."[2] Edward Pickering therefore wrote (January 24, 1894) to thank Lowell for his kind offer, which would allow William "to carry out the plan formed by him some time ago, for the observation in Arizona of the next opposition of Mars. I am greatly obliged to you for the personal and pecuniary cooperation which you propose to supply, for I have much confidence that the expedition will result in very interesting additions to our present knowledge." However, Pickering asked for a clearer statement of Lowell's views on printing the observations obtained, and suggested that the final publications resulting from the expedition should be reserved for the Harvard *Annals*.

Lowell replied (January 27, 1894) that the official publication

would be as joint as the expedition and would be so stated—It would appear officially both in the *Annals* of the Observatory and in any account I determined to publish—Prior publication of any matters of interest would only be made by one with the consent of the other. I am thus specific because a corporation, Harvard College, is the other party and needs more formality than another man—

Although this phraseology was ambiguous and implied some distrust of the Observatory, Pickering gave his consent to the plan. He agreed to contribute the professional services of William Pickering and A. E. Douglass, and to provide the telescope for the expedition. Lowell agreed to pay the costs of the journey and the other expenses involved in mounting the instruments, and in return would share the credit for the astronomical data obtained. All went well until Pickering released the news to the press, as was his custom when some interesting event occurred at the Observatory. Then a major misunderstanding became apparent.

The story appeared in the Boston *Herald* on February 13. Under the headline "Another Harvard Observatory" and the subhead "Funds for the Work Provided by Mr. Percival Lowell, Who Will Also Take Part in the Expedition," the article stated that "by the generosity of a citizen of Boston, Mr. Percival Lowell, Professor Pickering is enabled to carry out his long-cherished hope of establishing an observing station in Arizona, which has not previously been possible on account of lack of funds, and this donation of Mr. Lowell came like a benediction." The article went on to say that Mr. William Pickering "has been placed at the head of the explorers," and that Mr. Lowell would "not only furnish the necessary financial support for these experiments but he will go in person as a regular member of the party."

This version of the agreement, which gave the primary credit to the Harvard Observatory and relegated Lowell to the position of a subordinate who merely supplied the money, evoked an outraged protest from

The Harvard College Observatory

Lowell, who threatened to withdraw his support. He wrote to Pickering (February 14):

> I see by the papers that such serious misstatements in regard to our expedition to observe Mars have eminated [sic] from the Observatory that unless they are officially corrected I fear I shall be unable to proceed further in the matter.
>
> I had no intention, as I clearly explained to you, of aiding the work of the Harvard Observatories but proposed to you an expedition on joint account. I cannot see how I can obtain the recognition promised me, after such misleading statements, unless the Expedition or the Observatory bears my name.

It was obvious, however, that each party to the dispute needed the other. Edward Pickering believed that a station in the clear air of the Southwest would be able to make important contributions to astronomy, but the Harvard Observatory could not finance it. While Lowell could well afford to build an observatory of his own, he had not yet gained the experience necessary for its construction and was still an amateur astronomer. Conferences and telephone calls followed. In the end Pickering chose to sever Harvard's official connection with the expedition, which thus became a Lowell project. At the same time, Pickering granted a year's leave of absence without pay to William Pickering and to Douglass, who were thus free to accompany Lowell as his employees. Pickering also leased the 12-inch telescope to Lowell until his own (an 18-inch being made by Brashear)[3] should be ready. Thus the trouble was smoothed over. (Lowell's distrust of the Harvard Observatory remained so great, however, that relations between him and Edward Pickering were always somewhat distant thereafter, and in his will Lowell stipulated that "the Lowell Observatory shall at no time be merged or joined with any other institution.")

Douglass left Cambridge early in March. After testing the observing conditions at several places he selected a site (later known as Mars Hill) near the village of Flagstaff, at an altitude of about 7200 feet, where Lowell and William Pickering soon joined him. Construction of the Lowell Observatory began at once and in the last week of May observations of Mars commenced. The atmospheric conditions were perfect, William wrote his brother—second only to those at Arequipa.

In August the first paper from the Lowell Observatory appeared in *Astronomy and Astro-Physics* and was soon followed by others.[4] The surface features and colors of Mars, Percival Lowell reported, could be seen with great clarity. Geographically the planet seemed to be not unlike the Earth. The markings indicated the existence of an atmosphere, sky-blue seas, islands, a continent whose coastline was indented by bays, land that bore vegetation as well as very steep snow-covered hills, and an orderly network of canals that strongly suggested the work of intelligent beings.

Schiaparelli, who in 1877 had started all the excitement over the canals, had not thought it necessary to interpret them as artificial structures, but preferred to believe that they were natural geological formations produced during the evolution of the planet.[5] Lowell's speculations exhibited no such restraint. He continued to publish observations and inferences so sensational that Edward Pickering commented in a letter to President Eliot (October 13, 1894) that "it is probably just as well that this Observatory is not responsible for all of Mr. Lowell's personal views about Mars."

William Pickering returned to Cambridge at the end of the year but Douglass stayed on in Arizona.[6] The Lowell Observatory, which thus began as a kind of stepchild of the Harvard Observatory, continued to concentrate on planetary astronomy throughout Lowell's lifetime.[7] During the winter of 1894–95 snow fell at Flagstaff and the seeing was disappointingly bad. Hoping to find a better site, Lowell sent Douglass to Mexico, where he built an observatory in Tacucabaya, a suburb of Mexico City (not far from where the Mexican National Observatory now stands) and made observations for some months. In the end, however, they gave up the Mexican site and the Lowell Observatory at Flagstaff became permanent.

Lowell was hard-working, energetic, and enthusiastic. In the intervals between periods of travel and other periods of ill health, he tirelessly observed his favorite planet. He described what he saw in colorful, poetic language, and stated his conclusions explicitly. He thought it probable that life of a higher kind existed on Mars and wrote, "That Mars is inhabited by beings of some sort or other we may consider as certain as it is uncertain what these beings may be."[8] Few astronomers could see all the detail he reported, and even fewer agreed with his interpretations. The photographs taken by Mariners IV, VI, and VII have of course confirmed these doubts and have shown unmistakably that Mars is a barren, cratered sphere only less desolate than the moon, and wholly without seas, rivers, or canals.

Douglass later came to believe that many of the reported Martian features derived from defects in the eyes of the human observers. He once wrote to Edward Pickering (January 27, 1907) that he attributed

a good many of the very faint canals of Mars to rays in the eye from black spots. These rays can be made visible at any time by screening a part of the pupil of the eye,—showing that they are always on the retina. I think that in very close work they are seen and mistaken for canals.

A later preoccupation of both Lowell and William Pickering was the search for a planet beyond Neptune. A number of astronomers had pursued this quest at intervals ever since it became evident that Neptune did not wholly account for certain residual irregularities in the motion of Uranus. Lowell and William, independently, spent years in calculating the probable orbit of the hypothetical planet, and each published a succession of revised positions. After Lowell died in 1916 the search continued at his observatory at Flagstaff. In February 1930, Clyde Tombaugh, a young assistant at the Lowell Observatory, finally detected on a photographic plate the image of the tiny planet, but announcement of the discovery was delayed for several weeks, until the Lowell astronomers could confirm the reality of the new planet and calculate its exact path. The existence of this new member of the sun's family was finally revealed to the world on Percival Lowell's birthday, March 13.

At once the problem of selecting a name arose. Among those suggested were "Lowell," "Minerva," and "Chronos." A member of the Visiting Committee of the Harvard Observatory commented that

a favorite indoor sport around Boston is naming the new planet. My own suggestion would be Postumus in allusion to the fact that it was discovered after Percival Lowell died,—as a postumous child or book.[9]

Harlow Shapley, who had succeeded Edward Pickering as director of the Observatory, asked in return, "What kind of symbol would you use for the planet—a skull and crossbones?"[10]

William Pickering, then working at his observatory in Jamaica, was delighted that the planet had been found and that it was so close to the position he had predicted. In a letter to Shapley (March 27, 1930) he wrote:

As soon as the existence and identity as a planet of this object has been fully confirmed, a name of universal acceptance must be selected for it.

330

"Harvard Observatory Arraigned"

Unless someone can suggest a more appropriate one, the writer would propose that it should be called Pluto, after the Greek god of darkness who was also able at times, when he so desired, to make himself invisible.

William also sent (April 10) the suggestion in an article that appeared in the May issue of *Popular Astronomy*. (Some 30 years earlier, when the asteroid eventually called Eros was discovered, S. C. Chandler had suggested to Edward Pickering (January 16, 1899) that it ought to be called Pluto, "for malignity alone.")

The privilege of naming the new planet, however, belonged to the Lowell Observatory at Flagstaff, then directed by Professor V. M. Slipher. The administration of the observatory, independently, chose the name Pluto and devised the symbol ⯓, a monogram containing the initials of Percival Lowell. The symbol seemed entirely appropriate to William Pickering, since the initials composing it also stood for the two names "Pickering" and "Lowell." William wrote to Shapley (October 21, 1931): "The Lowellian suggestion of PL for Pluto I took as a rather delicate compliment, not to be disregarded."

Lowell and William Pickering had used different mathematical approaches in their years of searching for the trans-Neptunian planet and had evolved different orbits. Nevertheless, Pluto's position, at the time it was found, was within a degree or two of that calculated on both the Lowell and the Pickering orbits. Later, however, the planet was shown to have a mass too small to account for the observed perturbations of Uranus and Neptune, and most astronomers agreed that the closeness of the actual and the calculated positions was only a lucky accident. (Similarly, some astronomers still doubt, as did Benjamin Peirce at the time Neptune was discovered, that the calculations of Leverrier and Adams led logically to its finding.)[11]

II

A problem far more serious than the managing of the Arizona expedition arose early in 1894, when Edward Pickering confronted a sustained attack, which continued for more than a year, on the scientific work of the Observatory and on his own competence as director. The impeachment came not from rival institutions in the United States or in Europe but from two men who had once served on the staff of the Observatory, John Ritchie, Jr., and S. C. Chandler, and, less openly, from their close friend B. A. Gould, editor of the *Astronomical Journal*. The conflict stemmed largely from their distrust of the methods of the "new astronomy" employed at the Observatory, which relied more on the techniques of physics than on visual observations made with a telescope.

During his years at the Observatory Chandler had performed brilliant work in computing cometary orbits and finding variable stars. To make a more accurate determination of star positions he had devised an ingenious instrument he called the almucantar, a telescope mounted horizontally on a base that floated on a pool of mercury. It could be rotated around an imaginary vertical axis, but the force of gravity kept it always at the same angle with the vertical. He had also used the almucantar to determine the exact latitude of the Observatory dome at Cambridge, and had found what seemed to be a progressive variation in the position of the North Pole. Some European astronomers had suspected such a shifting, but it was generally regarded as an impossibility because mathematicians had supposedly proved that the earth was as rigid as a steel sphere. Although Chandler had confidence in his data, he regarded his conclusion as too bold an inference, and merely published his computations, leaving them to speak for themselves. (Later, after German astronomers had detected an actual variation in the latitude of Berlin, Chandler resumed his work and in 1901 was able to publish the exact laws of the Pole's variation, which occurred systematically over a period of 14 months and showed that the earth was not a rigid body.)

Like some other astronomers of the old school, Chandler distrusted the use of photography and spectroscopy and placed his confidence in visual observation, particularly for the study of variable stars. In 1886 he had severed his connection with the Observatory, to concentrate on preparing his excellent catalogues of variable stars and to assist his friend Dr. Gould with the *Astronomical Journal,* which had just resumed publication after an interval of more than 20 years. However, Chandler had continued to assist Ritchie with the Harvard announcement service. Using the Chandler-Ritchie code and its later revisions, this bureau had functioned efficiently for several years.

Sometime in 1890, however, Pickering had begun to receive complaints from American astronomers that important announcements occasionally arrived late or not at all. Early in January 1892, when French astronomers rediscovered Comet Brooks (1890 II), Holden at the Lick

"Harvard Observatory Arraigned"

Observatory did not learn the news until a month after the event, and then only by reading a notice in the *Astronomische Nachrichten*. Another irritating situation occurred in relation to the discovery of a nova. On February 2 the bureau at Kiel cabled Harvard that a new star had appeared in the constellation Auriga, but Chandler failed to telegraph the news to American observatories so that several days of invaluable observing time were lost. When Pickering asked for an explanation Chandler replied (February 6, 1892) that he had not been able to decipher the fifth word of the cable and that he had not had time to procure a copy of the original message. He added impatiently that the object was

probably of little importance. They have the fashion of finding an ordinary variable and calling it a new star, as in the case of U Orionis a few years ago. If you desire I will telegraph (cable) to find out. Perhaps you had better success in deciphering your copy. If so, I would be glad to learn the result.

The "result" was contained in Pickering's reply of the same date:

The fifth word of the telegram was "chiaurigae," meaning Chi Aurigae. The object is not an ordinary variable, but a star with a peculiar spectrum, and apparently temporary. As it may be inconvenient to you to send the messages I will have them despatched unless I hear from you that you would prefer to attend to the matter.

Chandler insisted on sending them himself, but the problem did not end there. A few days later, when Nova Aurigae had been photographed at the Observatory, Pickering released the news to the press but did not communicate it to the *Astronomical Journal*. In the resulting exchange of letters, Gould made clear his resentment of what he considered Pickering's lack of patriotism in continuing to send his papers to the *Astronomische Nachrichten* rather than to the *Astronomical Journal*.

Almost at once a further difficulty arose. In distributing astronomical news to the Associated Press, Ritchie stopped acknowledging the role of the Observatory so that the service seemed to be solely a function of the *Science Observer*. Pickering objected to this omission. A formal correspondence ensued in which Ritchie (March 30, 1892) made clear his belief that the announcement service belonged to him and Chandler alone, although he acknowledged the cooperation, the "pecuniary aid," and the "moral support" of the Observatory. Since the bureau had been formally transferred by the Smithsonian Institution to the Harvard College Observatory, Pickering could not agree with this view and, when Ritchie then resigned from the Observatory staff, accepted the resignation. Holden and others urged Pickering to appoint someone else to manage the bureau but, to avoid unnecessary friction, Pickering allowed Ritchie to continue in charge. (In 1905, however, after further difficulties with Ritchie, Pickering at last took over the bureau completely. Staff members of the Observatory carried on the work thereafter until 1966.)

Following the disagreement with Ritchie, the Observatory began to suffer a series of small harrassments. As old-line observational astronomers, both Chandler and Ritchie seemed to doubt any astronomical fact discovered by spectroscopic methods alone, and demanded confirmation by the human eye at the telescope. When Chandler published his "Second Catalogue of Variable Stars" he referred to "the profuse announcements of variability during the past few years," spoke of the facility with which "rubbish" collected around the subject, and emphasized that in his catalogue "no star should be inserted, no matter how high the authority on which its variability is declared, without independent verification on undoubted authority and evidence. Otherwise the result will be chaos."[12]

Following this principle, he omitted almost all of the variables recently reported from the Observatory. Of one such star that he did include, he noted that it had been "suspected" by Mrs. Fleming, and confirmed by others. His supplementary list of alleged but unconfirmed variables included 15 such stars reported by Mrs. Fleming and one by Pickering. A particularly irritating incident occurred in the autumn of 1893 when Mrs. Fleming, studying stellar spectra photographed in Arequipa, discovered a nova in the constellation Norma. Ritchie and Chandler delayed telegraphing the announcement, and then transmitted it in a form that gave the wrong position. Hence, until a correction was sent, astronomers were unable to locate the nova.

Such events, though annoying, were scarcely more damaging than mosquito bites. Then, suddenly, a whole hornet's nest erupted. On March 17, 1894, the *Boston Transcript* published a letter from "Student" under the headline "Harvard Observatory Arraigned." In the February issue of the *Astronomische Nachrichten*, the letter said, S. C.

"Harvard Observatory Arraigned"

Chandler had published an article denouncing the Harvard Photometry as being all but worthless. Chandler stated that 15 out of 86 variable stars listed in Volume 24 of the Harvard *Annals* were wrongly identified, and that many of the estimates of brightness for ordinary stars were wrong by several magnitudes. The work had been badly planned and carelessly executed, Chandler said. His "somewhat desultory" examination of the volume made him doubt whether any of the observations were reliable, and justified the inference that "mis-identification prevails on so liberal a scale in the work with the meridian photometers . . . as to deprive the photometric catalogues executed in this manner, of any scientific value whatever."[13]

After reviewing Chandler's allegations in detail, "Student" (who was probably John Ritchie, Jr.) reached his conclusion:

The position is plainly this: Harvard College Observatory has received from our wealthy citizens large and important contributions of money which have amounted to at least half a million of dollars. It has recently been before the public asking for about a hundred thousand dollars more for a great southern telescope and for further funds with which to endow it. It has recently received aid in the construction of a building in which to store its photographic plates and its last report suggests the need of still more money. In return for this investment of their money, the donors and the public as well have the right to know whether the quality of the work is what it should be, and on this point, the evidence in the Nachrichten has the strongest application. Adverse statements so sweeping and from so well known an authority as Dr. Chandler call for an explanation which shall be satisfactory to scientific men.

This was a devastating indictment. It appeared not only in a journal that would be read by all astronomers, but also in an influential newspaper read by thousands of citizens who had no way of evaluating the scientific facts at issue. Potentially the damage was incalculable. The strength of the Observatory depended on the confidence of patrons such as Mrs. Draper and Miss Bruce, as well as on its standing among astronomers, and a general loss of confidence could bring disaster.

Pickering published a brief reply on March 20:

To the Editor of the Transcript: Unwarranted attacks have recently appeared against the observatory of Harvard College. The questions raised are scientific in their character, and unsuited to a discussion in a daily journal. Reply will be made through the proper channels to all criticisms.

The Harvard College Observatory

The *Transcript* thereupon dropped the story but other newspapers in Boston and New York gave it full publicity for the next two weeks. Mrs. Draper immediately wrote to Pickering (March 29, 1894) to express her confidence in him:

> Professor Chandler's attack on your photometric work was in part in the "Evening Post"—I[t] seems to be prejudicing people very much against Chandler & Co. Mr. Rood[14] came here last evening, full of indignation about it—He said Prof Reese of Columbia had expressed himself very strongly against Chandler and said among the people who understood the subject, he was injuring himself, as it was evidently the result of jealousy at your success, for what he said amounted to nothing—I am glad to find the scientific men accept your results and are not impressed by the attack.

Before writing a formal defense of the photometric work Pickering consulted with Professor Searle, who wrote a detailed statement of his views (April 2, 1894). Searle found himself in a difficult position, he wrote, for he regarded both Pickering and Chandler as his friends. Also, unlike Pickering, he believed

> that Mr. Chandler has no personal motive for his criticism, in this sense: that he has no conscious antipathy to you as a man, and no conscious resentment against you arising from any incidents in your acquaintance with him. My belief is that he supposes himself to be simply opposing your system of managing the affairs of the Observatory, in the interest of astronomical science. Of course, since so large a part of all our mental activity is unconsciously carried on, neither he nor any one else can say whether this supposition of his is correct or not. . .
>
> When any charge is brought against a man affecting his reputation for morality or honor, he cannot be too quick in denying it and challenging proof. When the evidence is brought forward, if it is too weak to sustain the charge, he has merely to say so; if it has any plausibility, he must meet it by contradictory testimony.
>
> But when, as in your case, the charge is merely one relating to capacity or good judgement, there is no need of any defence at all. The question will settle itself in due time. I do not mean that there is any impropriety in attempting a defence, but merely that it may not be most judicious to do so; and when any explanation is called for, it should ordinarily be given. I think that dignity is best preserved, and that no permanent loss of reputation can occur, by leaving actions to speak for themselves to those who understand the questions at issue, and leaving the defence to the unsolicited testimony, private or public, of these professional

judges. . . In the present instance, as Mr. Chandler practically demands explanations by his article in the Astronomische Nachrichten, I think such explanations ought to be given in the same periodical.

Searle went on to express his own strong approval of Pickering's management of the Observatory, and his confidence in the accuracy and the value of the photometric work. He added that "my own very high opinion of Mr. Chandler as an investigator does not induce me to think him equally eminent in criticism. In that field, I consider that his conclusions are often too hastily drawn and too sweeping."

Pickering's reply appeared in the May issue of the *Astronomische Nachrichten*. He referred briefly (against the advice of both Searle and President Eliot) to the possibility that Chandler's attack might have been inspired by unconscious animus. He agreed that 15 variable stars had indeed been wrongly identified, but pointed out that they were unusually faint and difficult objects. Such isolated errors, which crept into every catalogue, did not justify the sweeping conclusion, made from an admittedly cursory examination, that the whole catalogue of more than 20,000 ordinary stars was similarly riddled with error.

Every astronomer is of course aware that the difficulties of identification in the case of variable stars, especially when they are faint, are much greater than with stars photometrically constant. It is somewhat as though it should be argued from a physician's losing twenty per cent of his cholera patients that he had been equally unfortunate in his general practice.

He had full confidence, he said, in both the method he had used and the accuracy of its results.

In spite of the preoccupying worries of this very critical period, Pickering found time to send a letter (May 18, 1894) of congratulation and appreciation to President Eliot (Fig. 45) on the twenty-fifth anniversary of his presidency of Harvard:

My dear Mr. Eliot,

Congratulations will be coming to you from all quarters on the important anniversary of tomorrow. I hope that it will be pleasant to you to recall the influence you have had upon the life of one person during the last thirty-two years. In 1862, I asked your advice and followed it when entering the Scientific School, in 1863 when beginning as I supposed a career in mathematics rather than chemistry, in 1867 when becoming a physicist, and in 1877 when undertaking astronomy. In every case your opinion had

Fig. 45. President Charles W. Eliot, from a painting by John Singer Sargent. (Harvard University Portrait Collection.)

great weight in my decision and in every case had I not taken your advice I should always have regretted it.

I was pained to hear you say one evening at the Club that your last twenty-five years in Cambridge had been stormy, rather than peaceful. You may not appreciate how very large a following you have among those who watch and approve of your work, rather than take a public part in a discussion.

I wish also to express my appreciation of the aid and support you have always shown me in my official connection with the Observatory and of your unfailing thoughtfulness, and kindness to me personally. Hoping that for many more years the work of the University may be continued under your guidance, I remain

Faithfully yours,
Edward C. Pickering

In his warm reply (May 19) President Eliot took the opportunity to express his full confidence in the director of the Observatory:

Dear Mr. Pickering:
Your note of yesterday gives me a great deal of pleasure. I am glad that your experience has made you content with decisions to which I contributed a little. As to my own experience during the past twenty-five years, you must not suppose that it has all been combat and storm. There have been great satisfactions and rewards also. But there has been incessant discussion, much difference of opinion, and a deal of criticism. These things were necessary accompaniments of the work, & doubtless improved the result; but they inevitably produced on my mind the impression of struggle. You know that with us New Englanders approval is often silent, while disapproval is loud.

The progress of the Observatory during your administration has been a lively satisfaction to me, as to all friends of the University and of science.

I am, with many thanks for your note,

Very truly yours
Charles W. Eliot

When the issue of the *Astronomische Nachrichten* containing Pickering's defense reached Boston, President Eliot wrote that the answer to Chandler was

excellent throughout, with the possible exception of the two sentences on animus. I am not sure that these might not better have been omitted.

What is the present state of volume 34? Will it be possible to issue it shortly, on the principle of giving the experts something new and better to think about?

The Harvard College Observatory

Again the newspapers turned the spotlight on the Observatory. Angry counterattacks appeared in the Boston *Commonwealth* over the signature of John Ritchie, Jr., and more or less informed discussions of the controversy appeared in the Boston *Globe, Post,* and *Transcript,* as well as in New York papers. Besieged by reporters, Pickering made a single brief statement (Boston *Globe,* June 25):

My position in the matter is precisely this. If you seek the facts in the dispute, you can find them in my article. If you want personal denials or contradictions, or want to get me into a newspaper controversy, I must decline to be used.

President Eliot was not so reticent. He undertook to defend the Observatory in an interview reported June 24 in the *Post* and June 25 in the *Globe* under the headlines "Here's a Mess" and "Astronomers at War." Although he had questioned Pickering's reference to possible unconscious "animus" in the *Astronomische Nachrichten,* Eliot himself used even stronger language. He was quoted as saying that "undoubtedly Dr. Gould is at the back of" the attack, that Chandler was "influenced by animus," and that "there are three persons in this, who have combined against the credit of the university. One of them is Mr. John Ritchie, who has the newspapers well in hand." The arraignment of the Harvard Photometry would have no effect on the work of the Observatory, Eliot said, except to make the staff more careful than ever to avoid mistakes.

Dr. Gould, approached by a reporter, repudiated the charge but made his views obvious by his choice of words when he added that he had absolutely no part in "the exposure" published by Dr. Chandler. Gould embodied this denial in a letter to President Eliot, who did not reply. Gould then published the letter in the Boston *Transcript* (June 30).

All through the summer the newspapers waged the astronomical war, but they got no further ammunition from Eliot or Pickering, who resolutely refused to speak for publication. President Eliot apparently regarded the subject as closed, and he wrote to Pickering (July 31):

I think on the whole that the newspaper appeal made by Messrs. Gould, Chandler, and Ritchie is doing very well without any added word on my part. I shall say nothing more; for I am persuaded that the original interview with me gives, on the whole, quite an accurate picture of the real state of affairs. My only interest in the subject is that the Observatory should get every possible advantage from the disclosures now made by the trio; first, advantages in the way of precautions to secure all possible exactness in the published work of the Observatory; and secondly, the advantage of having to deal with an open criticism traceable to certain individuals rather than with incessant insinuations and derogatory remarks for which nobody can be held responsible.

You need not fear for a moment that the public will believe that the work of the Harvard Observatory is invalidated, or that any sensible person will suppose that "Mr. Chandler's article has ruined forever the astronomical reputation of the Harvard Observatory." Such a statement as that refutes itself and predisposes every sensible reader to believe that the whole attack is baseless.

As I have said to you before, the best way of meeting this and all other criticisms is to issue more fresh good work, and this I doubt not that you are bent on doing. My chief anxiety in connection with this matter is that it should not disturb your peace of mind or impair your scientific activity. At first it had to a little; but I hope the temporary effect is wearing off. If it does not, I beg to repeat what I said to you at our last conversation—you ought to take a good vacation.

Other letters of encouragement came to Pickering, from members of the Visiting Committee, from Bailey in Peru, from William Pickering in Arizona. Mrs. Draper, too, wrote him warmly (July 2, 1894):

I am glad that you are still intending to take a longer holiday than usual, and hope on my return [from Europe] to find you are well rested, and worrying no more about the buzzing of such insignificant flies as Chandler and his contingent. Your letter of reply seems to meet with universal approval.

Pickering took a long vacation in the White Mountains, and then returned to work. In the autumn, vindication came with the publication of the first volumes of the *Potsdamer Photometrische Durchmusterung,* compiled by G. Müller and P. Kempe. This great work, carried out with deliberation, included a comparison of the magnitudes determined for the many stars that were listed in the catalogues of Harvard, Oxford, and Potsdam. The magnitudes found for most of the stars were identical in all three catalogues and in only a few cases did they differ by more than half a magnitude.

Understandably eager to rehabilitate the reputation of the Observatory completely, Pickering made further comparisons of the three catalogues and found evidence that the average difference between the Harvard

341

estimates and those of Potsdam was less than two-tenths of a magnitude, and that the Harvard measures were slightly more accurate than those made by Pritchard at Oxford. He embodied these conclusions in his annual report to the Visiting Committee and published them in the *Astronomische Nachrichten*. The Boston *Morning Journal* seized on the information, which it printed (January 9, 1895) under the headline, "AHEAD OF OXFORD, Harvard Photometry Vindicated From Charges."

The article in the *Astronomische Nachrichten* inspired a new series of attacks in the Boston *Commonwealth*, which appeared sporadically throughout the spring. And, far worse, it evoked protests from the astronomers at both Potsdam and Oxford, so that this unhappy controversy seemed likely to have no end. The Potsdam astronomers replied that their magnitudes were more accurate than those of the Harvard Photometry (volumes 14 and 24 of the *Annals*), partly because the Harvard staff had worked at a speed incompatible with exact results—a charge that Pickering denied. They also added:

We only desire here to again state, and with even stronger emphasis, that in our opinion the Cambridge catalogue will always have a high value because it gave for the first time a systematic catalogue of a large number of stars, but that on the other hand it has not that degree of accuracy which must be striven for with the instrumental means of today, and which can in fact be obtained.[15]

H. H. Turner, who had replaced Pritchard as Savilian Professor of Astronomy at Oxford, published a sharp rejoinder defending the Oxford work, which moved Pickering to explain himself to Turner (February 23, 1895):

Your article in the Astronomische Nachrichten was received this morning. I regret that you should regard my comparison of photometric magnitudes as an attack upon the excellent photometric work done at Oxford. This was far from my intention. You may not be aware of how vindictive an attack has been made upon this Observatory in the local newspapers by two of my former assistants and their friends. For instance, the statement has appeared that the catalogues of the meridian photometer had no scientific value whatever. A few days later a second statement appeared that this view had been universally accepted by astronomers, and then a third statement that we shall now hear no more of the photometric work of the Harvard Observatory. As I do not indulge in newspaper controversies, and some friends of the Observatory may not understand the true state of the case, a comparison of results obtained elsewhere did not seem to me out of place.

Turner replied (March 6, 1895) with entire friendliness:

I fear we are but selfish, we mortals. All *I* thought of was the credit of this Observatory, and that it must not suffer or our funds would be endangered. Perhaps you will forgive my blindness as regards Harvard (I did not realize your motive until I read your Report in which something of what you said was repeated) when I remind you that Oxford has similar reasons to your own for not wishing its work to appear unfavourably before the public. I am very sorry indeed however to hear from you how gross the statements about Harvard have been: and could wish that after all I had let the matter pass: your remarks were of course not at all of the same kind, though I think you will still see that *in effect* they tended to the same result,—possible discredit of the work here: and will perhaps forgive me for playing rather for my own hand.

Frankly, though confidentially of course, I should be very sorry if serious occasion arose to defend the photometric work [of Professor Pritchard] in detail. I fancy the scale is a bit wrong: this could of course be corrected but it is not a gracious attitude towards a predecessor to criticize his work.

Fortunately, the Observatory emerged from this controversy with reputation untarnished. A letter written by Mrs. Draper on February 3, 1895, expressed the feelings of many friends of Pickering's:

I hope that both Mrs. Pickering and yourself are quite well, and that you are having no further annoyance from the attacks of your envious enemies, how they do wish they were in your place.

The sounds of battle slowly faded and even the breach with Chandler eventually healed. The improved relation began, appropriately, with an astronomical problem that could not be solved without cooperation. On August 13, 1898, the German astronomer G. Witt of the Urania Observatory at Berlin discovered an unusual minor planet, first known as Witt's planet or Planet DQ, and later called Eros. The 433rd asteroid to be found since the first (Ceres) had been discovered on January 1, 1801, Eros had a highly elliptical orbit. Preliminary computations suggested that at its minimum distance Eros approached the earth more closely than any similar body, except the moon. The planetoid could therefore be used to get a more accurate determination of the sun's distance, which was known to be in the neighborhood of 93,000,000

The Harvard College Observatory

miles. To obtain more accurate estimates, some fifty expeditions had been sent to far parts of the earth to observe the transit of Venus in 1874, but the results had been extremely disappointing. The newly discovered Eros now seemed to offer a better means of finding the true distance. The first essential was a determination of its exact orbit, which required knowledge of the planet's position in the years immediately preceding its discovery.

Professor Chandler (who after the death of Gould in 1896 had become the editor of the *Astronomical Journal*) made some rough calculations of the object's earlier path. Its actual movements in recent years could be verified from only one source: the Harvard collection of photographic plates. A major reason for starting the plate collection had of course been to provide a permanent history of the sky, which could be consulted when necessary. Simon Newcomb, in his address at the dedication of the Yerkes Observatory on October 22, 1897, had remarked[16] that Harvard's photographic program was

so organized that a new star can not appear in any part of the heavens nor a known star undergo any noteworthy change without immediate detection by the photographic eye of one or more little telescopes, all-seeing and never sleeping policemen that scan the heavens unceasingly while the astronomer may sleep, and report in the morning every case of irregularity in the proceedings of the heavenly bodies.

Faced with the critical problem of Eros, Professor Chandler, whatever the cost to his pride may have been, sent his calculations to Pickering with a request for help (November 3, 1898). He wrote formally:

I deem it my duty, in the interest of science, to send you the inclosed ephemeris of the planet DQ which I have computed in behalf of the attempt, which all astronomers will be interested in, in common, for the recovery of any previous observations of this most important body . . . I suppose of course you will agree with me on the importance of recovery of such observations of a body which is destined to play an important part in our astronomy, in various ways, and will appreciate that it is this general scientific interest that leads me to write you.

Eros would have been at very favorable opposition in 1894; could Pickering search the appropriate photographs, try to locate the object, and provide measures of its position at the time? Chandler closed his appeal by adding, "I feel sure you will estimate my motive rightly."

344

Pickering agreed to Chandler's request, and turned the search over to Mrs. Fleming. The rough positions given were so uncertain that a wide segment of sky had to be covered. After a long and discouraging hunt, however, she finally identified the elongated image of Eros among the stars on plates made in 1893. When the positions were sent to Chandler, he expressed (January 16, 1899) his surprise and gratitude for "this inspired find," and calculated a more exact orbit, which Mrs. Fleming used to find Eros on other plates made in 1894 and 1896. (The amicable relation between Pickering and Chandler thus resumed did not again lapse. Later, Pickering was able to offer financial support for some of Chandler's research projects, which Chandler appreciated but could not accept, for lack of time.)

Calculations of the asteroid's orbit, by Chandler and others, showed that at its closest approach Eros would be only some 15,000,000 miles from the earth. In the autumn of 1900 it would still be relatively near, only 30,000,000 miles away. An "Eros Campaign" was therefore organized, in which some fifty observatories in all parts of the globe measured the position of the asteroid to determine its parallax. The results of this campaign did provide a new measure of the sun's distance—about 92,890,000 miles—which is very close to the figure accepted today.

III

While the astronomical battle raged in Cambridge in 1894 and 1895, civil war broke out in Peru and for a time threatened the existence of the station in Arequipa. Several times during the summer and autumn of 1893, and at intervals throughout the following year, Bailey's letters mentioned rumors of rebellion. At first he seemed to take the danger lightly, and in writing to Pickering (July 10, 1893) remarked that if a revolution occurred he might "have to remove the lenses and use the telescope tubes for cannon." Later (September 30, 1893) he wrote more seriously that he planned to have some heavy wooden shutters made for all the windows on the ground floor, to protect the station.

We have two or three revolvers and with the addition of a few good clubs, I think we should be able to keep off any drunken rabble, who would probably be the only serious source of danger. If a really serious armed party demanded the surrender of the house we should have to "hands up" and rely on the government for indemnity. We shall also lay in a special supply of provisions as at such times no marketing can be done.

I do not really expect any trouble but yet I think it well to be on the safe side. The extra expense will probably be in the vicinity of $100.00 and the shutters for the windows and doors will possibly be useful in later years, unless the Peruvians learn wisdom, and the food we can eat in any case.

A few weeks later rioting broke out in Arequipa, several persons were shot, and troops were called in. Bailey wrote that the shutters were nearly ready for use and that if the station should seem to be endangered he would take up the floor of a closet, sink a deep hole in the ground, and bury the telescope lenses.

After the Peruvian elections in the spring of 1894 the civil disorders increased. The observing station remained safe but Bailey had an adobe wall built along the west side of the grounds, between the station and the road, and requested permission (which Pickering granted) to buy a small piece of land on the north side, bordering the village of Carmen Alto, and build another protecting wall. Another worry developed early in September when Waterbury, on his regular trip to the summit of El Misti, found that thieves had broken into the shelters and had taken away tools and meteorological instruments. The robbers were probably Indians, Bailey wrote, but they would be difficult to trace because of the unsettled condition of the country.[17] Troops of mounted rebels were overrunning the smaller towns and had already occupied the region around the first station, on Mt. Harvard.

During October the disorders came nearer to Arequipa. The revolt was spreading, revolutionary and government troops were battling in several towns, and Lima was surrounded by rebels. Alarmed by Bailey's reports, Pickering consulted President Eliot, who at once wrote to Mr. Gresham, the Secretary of State in Washington. A cipher telegram was dispatched to the American Minister in Lima (December 4, 1894): "It is feared Arequipa Observatory endangered by present war. Spare no efforts for its protection." This telegram was little more than a gesture. The Minister had no soldiers at his command and the Peruvian government was too insecure to spare troops to protect the property of a foreign observatory. The situation remained tense, but the Baileys celebrated the holidays as usual and their young son Irving entertained the observatory staff by giving a show with his magic lantern, a Christmas gift from Pickering.

Early in January the Baileys made their first real contact with revolu-

346

tion. Professor Bailey had gone to Mollendo to straighten out some problems relating to the meteorological station there, and had taken Mrs. Bailey and Irving with him. On the return trip, their train was captured by rebels. Bailey described the experience in a letter to Pickering (January 14, 1895):

> I was reading, when we heard a tremendous shout of "Viva Pierola". I looked out only to see a crowd of men armed with rifles and revolvers come rushing around the train and into the car. The car was at once filled with cries of "Jesus Maria" and "Por Dios," by the ladies and children. The natives really seemed much more frightened than we were. I advised Mrs. Bailey and Irving to keep quiet and there would be no harm done and so it turned out. The revolutionists behaved with great moderation and offered us no indignity whatever. We were sent back to Mollendo however while the men followed us in another train which they had captured. When near the town they left us locked in the car and forming in line marched in and took the place in a few minutes. Mollendo is said to have a population of about 3000 but there were only 15 soldiers and they surrendered after about a hundred shots were fired. After the affair was over we were allowed to enter the town and found everything quiet.

The Baileys spent an anxious night (along with more than 100 other persons) at the house belonging to Mr. Golding, the agent of the steamboat lines—Golding armed with a revolver and Bailey with a heavy club. However, government troops recaptured the town after the rebels had fled, and the Baileys returned to Arequipa, where they found the town in a state of excitement. The buildings at the station had not been molested, although, as a precaution, Hinman Bailey had removed and buried the telescope lenses. Less than two weeks later the rebels attacked Arequipa and cut the telegraph line; savage fighting took place in the city. Bailey again secreted the lenses, put the wooden shutters in place, and raised the American flag over the residence. For more than two weeks Arequipa was cut off from the rest of the world. On January 27 Bailey began a letter to Pickering describing the events of each day; he continued the account until February 12, when at last the siege was lifted and mail service was restored. The rebels had been victorious throughout Peru, and the revolutionary government assumed power.

Even in the midst of these dangers Bailey did not forget his astronomical responsibilities. Each evening, with the sounds of rifle fire only 50 feet away, Hinman Bailey ventured into the grounds with a dark lantern to collect the day's meteorological records. Deprived of their lenses, the telescopes could not be used, but Solon Bailey commented on how fortunate it was that the battle for Arequipa was taking place "in the cloudy season as otherwise it would sadly interfere with our night work."

When this long letter reached Cambridge with its play-by-play description of the revolution, Pickering responded (March 10, 1895) with vigor. He suggested explicit ways to defend the station and gave advice that was amazingly warlike, coming as it did from a man who disliked controversy. The chief danger, he wrote, would probably be

from attacks of drunken or disorderly stragglers rather than from regular troops. Perhaps it would be well, in view of future disorders, to put the Observatory or at least a portion of it in a condition in which it could be safely defended from stragglers. A stone house which cannot be burned from the outside is not readily captured by a mob. The greatest danger is that using a stick of timber as a battering ram, any ordinary door can be broken in. The scarcity of wood in Peru diminishes this danger. Loop holes through which you can see what is going on, and if necessary fire without danger might prove very useful. During the troubles in Ireland a form of country house was proposed as shown in the annexed sketch. The circular bay windows in the corners have steel shutters with loopholes commanding all four sides of the house. It was claimed that four determined men, armed with repeating rifles, could hold such a house against a mob of any size, unless it was provided with artillery . . . I have an improvement to suggest in your proposed line of action in case of an attack, but I wish that your position might be behind a steel lined loophole. Can you conveniently pour buckets of water (preferably hot) on persons attempting to force an entrance. The most dangerous mobs are sometimes dispersed by a thunder shower.

But by the time this letter arrived in Peru the fighting had died down and the rebels were in control. Bailey hoped for two or three years of peace in which to continue astronomical work, and he wrote regretfully (March 8, 1895) to Pickering, "What with robbery, revolution, and the snows of the rainy season, the work on the Misti has not gone as I hoped."

Early in April the Baileys held a reception at the residence. To establish friendly relations with the new government, the Arequipa station entertained the successful leader of the revolt and other men prominent in the new regime and gave them a tour of the observatory. Bailey wrote (April 15, 1895): "The expense was moderate, about twenty dollars, and as Pierola is sure to be the next president, if he lives, I think it was a wise act."

IV

The years 1893–1895 had been a critical time in the history of the Observatory but this anxious interval closed with a new triumph: the successful performance of the Bruce telescope, the 24-inch photographic doublet. Pickering's design for the Bruce had been a bold experiment in photographic instrumentation and in the period of construction many astronomers had remained skeptical of its success. During the early photographic trials in Cambridge the instrument had indeed offered some unexpected problems. The optical system was superb but some flexure in the tube at times caused the star images to be elongated. Pickering and the Clarks eventually made the necessary adjustments, and it was in good working order by autumn, when H. H. Turner with a group of other British astronomers visited the Observatory in Cambridge. They were on their way home from Japan, where they had hoped to observe the solar eclipse of August 9, 1896 (clouds interfered). Turner later reported his experience at the November meeting of the Royal Astronomical Society. He had been very curious to see what the Bruce photographic doublet could do, he said, and to compare the results with those of other instruments:

I was very much struck with the character of the images given by this object-glass and the way in which a large field was covered satisfactorily—a field a good deal larger than one can get with the ordinary achromatic object-glass. But what struck us more than anything else was the man who was directing all this work and the wonderful energy and skill he showed in organizing it and carrying it out in such a successful way. It seemed extremely difficult to imagine how any one man could manage to keep so many different branches of work going satisfactorily; and I am sure the Fellows of this Society will feel that Professor Pickering deserves the thanks of astronomers for the enormous amount of work he is turning out from that observatory.[18]

The Bruce was now ready for its task in the Southern Hemisphere. To ship the instrument by the shortest route, over the Isthmus of Panama, would have involved unloading it at Colon, taking it overland by train to Panama, and again loading it on board a ship for Peru.

The Harvard College Observatory

Each transshipment would have added to the risk of damage. Pickering therefore chose the safer, though longer and more expensive, Atlantic route by way of the Straits of Magellan and up the west coast of South America.

Professor Gerrish accompanied the telescope to New York to oversee its loading and to place the lenses in the ship's treasure room. When the ship reached Mollendo on February 13, 1896, Professor Bailey was waiting to supervise every step of the unloading. The heavy pieces of the mounting, the framework of the dome, and the 11-foot tube made their hazardous journey across the rough water from ship to land and were placed on the train for Arequipa. From there a cart pulled by a team of oxen conveyed them over the last few miles to the station. Bailey wrote Pickering immediately (February 18, 1896) to assure him that everything had arrived in good order.

The Captain of the Condor thought it a great joke that Mr. Gerrish waited for the tide to change in New York so that the steamer would be on an exact level with the wharf. In Mollendo where the ship and launch roll about so heavily even in the best weather, it did seem very funny. I was on the steamer nearly all one day and everything was done with the utmost care and nothing got a bad bump on landing. It does look rather risky however to see the heavy pieces roll up and down over the heads of the boatmen.

Bailey erected a new building to shelter the Bruce, and a heavy stone pier on which to mount it. By April he was ready to take the first photographs, but the results were disappointing. The mounting was still not satisfactory, he found. In some positions a degree of torque caused unsteadiness so that star images on the plates were not circular. For a time Bailey was discouraged by what seemed the temperamental behavior of the instrument, but by May 26 he was able to write Pickering that "in the Bruce I have got some plates in the way of clusters which are more than I feared and nearly what I hoped."

A few weeks later the occurrence of a severe earthquake (which preceded by about 20 hours a great tidal wave that killed thousands of persons in Japan) threatened to interfere with the work of the Bruce. Bailey wrote Pickering (June 15, 1896):

Yesterday we had the strongest earthquake which I have ever experienced. It came at 10.05 A.M. I could distinctly see the ground move, something which I never saw before. I was in the laboratory. I rushed into the Bruce building which was near to see the effect. The whole mass of the castings etc swayed visibly and the tube shook violently. The main house shook considerably as if it were being shaken and twisted. It lasted about one minute. The seismograph record gave a twisting mark or marks like a lot of figure eights. The apparatus was thrown out of position after the first.

Fortunately, the Bruce and the other instruments at the station escaped damage.

The more Bailey worked with the instrument, the more pleased he was with its performance—its light-gathering power was about 3½ times that of the Boyden, and with long exposures it could record even very faint nebulae. The plates of the cluster Messier 3, which he had earlier photographed with the Boyden telescope, distinctly showed the 95 variables he had previously discovered there.

When the first plates made by the new instrument reached Cambrige, Pickering wrote to its donor, Miss Catherine Bruce (September 13, 1896):

You will be as much interested as I am to hear the good news from Peru. We have just received a large number of photographs taken with your grand telescope which establishes the entire success of that instrument. They were taken by Prof. Bailey who is now in charge of our Peruvian station and their excellence is largely due to his skill and perseverance . . . Some of them are very fine indeed, as fine as we could expect, so that I consider the entire success of the instrument as at last established.

Toward the end of the year (December 1, 1896) Bailey wrote to Pickering that they had been getting

some interesting long runs recently in the Bruce, from four to six hours. The Magellan clouds are very interesting and I am endeavoring to get a sufficient number of plates of different exposures so that you will have the best plates possible for a thorough study of these objects should you wish to make it. They are really wonderful objects. I did not appreciate the fact so well before.

Also we have made some good plates of the Pleiades in the same instrument, of one, four and six hours. I did not know before that the nebulosity was so extensive. It is very beautiful and in part of wisp like filiments.

In a Circular (No. 15) issued December 30, 1896, Pickering announced the unqualified success of the Bruce telescope, which vindicated his choice of a large photographic doublet to map the sky. As evidence he included a plate, made with a 3-hour exposure, which showed the Trifid nebula in a field of thousands of sharp, distinct images of stars. In the Circular he called attention to the courage of Miss Bruce,

who permitted an experiment to be tried on a scale never before attempted, and whose liberality, both in the amount of her gift and in the terms on which it was made, rendered every aid to secure success. It is a great satisfaction to be able to show by these photographs that the results obtained are exactly as expected, and that no unforseen difficulty interfered with the success of the experiment.

During its months in Peru, he said, the Bruce had mapped the entire southern sky in a series of such photographs. Each plate measured 14 by 17 inches, and each one covered a section of sky 5 degrees square. Thus the plan he had proposed 10 years before to the astrographic congress had been carried out. However, to avoid duplicating the work of the congress, he had decided not to continue his original plan to map the entire sky. Contact prints on paper of the existing plates would be made available to astronomers who wanted them.

A copy of this Circular was sent to H. H. Turner at Oxford, who had criticized Pickering so severely at the time Miss Bruce made her donation. In the intervening years Turner had modified his views, and he wrote (February 25, 1897) that he had read the Bruce Circular with great interest:

I have gradually been coming round to your views about the Chart—the doublet with a large field is surely the way to do it, at least in the first instance.

Can you not reconsider the decision? If you could & would publish a complete chart on this scale it would be a grand thing. I don't think an International Chart *will* be published. You can see from the reports how it hangs fire. The Catalogue plates will I hope be taken & measured; but the long exposure plates I fear will not, at any rate for some time. I think the interests of Astronomy would be far better served by your plan. Is it still possible?

In 1903, Pickering did issue a "Photographic Map of the Entire Sky"—the first such chart ever published—but it was not made by the Bruce. The 2½-inch doublets were used in Arequipa and in Cambridge to cover sections of sky about 30 degrees square, on 8-by-10-inch plates, which showed stars down to the 12th magnitude. Thus 55 plates covered the entire sky. Glass contact prints were made, and the set was offered for sale to astronomers at less than cost—$15.00.

v

During the final two years of Bailey's term at Arequipa the station played host to a number of visiting astronomers. Late in July 1896 Winslow Upton (who in his days in Cambridge had written the *Observatory Pinafore*) arrived with his wife and three children for a year's work. Now director of the Ladd Observatory at Brown University, he was on sabbatical leave and intended to make meteorological studies in Peru. He brought with him a large amount of equipment—portable transits, chronometers, and surveying instruments—and during his stay he determined the exact latitude, longitude, and altitude of the Arequipa station.

Soon after the Uptons had gone, the Harvard meteorologist Robert de Courcy Ward[19] and his wife arrived for a three-months' visit, bringing a supply of kites (which proved difficult to assemble and to fly) for studies of the winds at various altitudes.

These and other visitors were welcome, but they were not allowed to interfere with the scheduled production of photographic plates. The number of variables found by the staff at Arequipa and in Cambridge had been increasing so rapidly that Mrs. Draper had commented in a letter to Pickering (February 20, 1894) that "the discovery of variable stars, seems now to be a matter of very frequent occurrence. It does not look as if we would leave many for those who came after us," but new variables continued to appear. By the spring of 1897, 293 in globular clusters had been found on the Boyden photographs—285 by Bailey and 8 by Pickering and the Cambridge staff. When Pickering expressed his pleasure in this achievement, and the hope that Bailey would soon find seven more to complete an even 300, Bailey replied (July 13, 1897) that "the seven variables you ask for have been found but unfortunately (?) a number more, so that I fear we shall now have to try and complete *four* hundred."

During these years in Arequipa, Bailey had become as much devoted to the study of cluster variables as was Pickering to the study of stellar spectra. In the last year of his term at Peru, he sent to Pickering (March 30, 1897) a formal request

that you will intrust to me the study, determination of periods, and discussion of results for publication, subject as usual to your approval, of the variable stars in clusters (or elsewhere) which I have or may discover. I have become much interested in this subject, and in so far as consistent with your plans and my other duties, should like to make this something of a specialty and make myself thoroughly familiar with every thing known on the subject and add something thereto. The amount of work involved is of course very great and might occupy my spare time perhaps for many years. Any advice and assistance would of course be most acceptable.

Pickering replied that he would be most happy to accede to this request.

(Bailey later exerted an important influence on the career of Harlow Shapley, who in 1921 became the fifth director of the Observatory. In 1914 as a graduate student at Princeton and about to go to Mt. Wilson, Shapley visited the Observatory in Cambridge and for the first time met Pickering, Miss Cannon, and Professor Bailey. Bailey urged him to use the 100-inch Hooker reflector at Mt. Wilson to study the variable stars in globular clusters. Shapley did take up the work and later recorded that "within a month or two after I got to Mount Wilson, Shapley and the globular clusters were synonyms. Bailey showed me that this was a rich field to get into.")[20]

The hunt for variable stars was an absorbing one for the Observatory staff, and success brought welcome praise. Unfortunately, the hunt also involved a competitive spirit among the searchers that was not always easy to deal with. To stimulate interest in the work as a whole, Bailey had encouraged his assistants to look over the photographic plates before sending them to Cambridge, and to mark suspected variables. This practice displeased Mrs. Fleming, who was thus deprived of both the joy and the credit of original discovery. Pickering explained the situation to Bailey (September 29, 1897):

Mrs. Fleming's work is now partly duplicated by the examination of the spectra and marking of peculiar objects, She feels that in these cases the credit goes to the Peruvian observers while a large amount of work falls upon her. She is obliged to measure the positions, the variations in brightness, if any, and to identify the individual lines, classify the object and see if it is a catalogue star. She also has to reexamine the plates since the fainter objects, including about half of the peculiar objects, and as many more having slight peculiarities are omitted. All of this is part of her regular routine work and has been for the past ten years, and much of it could not be done in Peru. On the other hand Dr. Stewart would doubtless feel aggrieved if after all the labor especially in following the Bruce plates, he is not allowed to examine them. The delay might also prevent the early discovery of a new star or other object of special interest.

When Pickering asked whether Bailey could suggest any way of removing this "source of friction," Bailey replied (November 2) with a defense of his assistants. The plates had to be examined before they were shipped to Cambridge; it scarcely seemed fair that the man who had made the photograph and inspected the result should be forbidden to note the presence of any unusual object on the plate. Also, Bailey had supposed that such evidence of interest in the work would win Pickering's approval and some public recognition:

Personally I think that the ability to make first class plates is greater than that required in the mere picking up of new objects by certain well known characteristics, as it has been done generally by the assistants here. But as the latter has been publicly recognized and the former seldom or never, it is perhaps not strange that an ambitious assistant should desire to try also the latter . . . Mrs. Fleming is not the only one who has felt vexed at times.

Since Bailey was to leave within a few weeks to return to Cambridge, no direct reply to this letter was sent. But after Clymer had taken charge of the station, Pickering wrote (December 9, 1897), expressing his own perhaps utopian views on how an assistant should regard his daily work:

As regards hours of work, the Observatory expects to pay for the entire working time and energy of each assistant. In view of the great competition for positions here no one is likely to rise high who does not devote his entire energy to the work. A minimum time of seven hours a day has been named, since it is supposed that in fatiguing work like numerical computations and some forms of observing this is as much as could be done satisfactorily by the average man allowing proper time for rest and recreation. It is for the best interests of the institution as well as of each individual that the best possible work, both as regards quality and amount, should be done. I wish to aid this in every way and to properly recognize the deserving work of each assistant. It has not been customary to name in print the assistant by whom each photograph is taken, but where special skill or care is required this should evidently be done.

I entirely approve of assigning special investigations to any assistant who desires it if it will increase his interest and pleasure in his work. There ought to be no distinction in the interests of the Observatory and of each assistant. What benefits one, should benefit the other. Evidently it is for the mutual interest to avoid duplication and there surely should be some way by which justice should be done to all. In any case, especially for those at the distant station of Arequipa it is much better that they should tell me frankly of any grievance they have and I will certainly endeavor to correct it if possible. Of course, if any striking object, as a Nova, a bright variable, or a comet appears, numerous photographs under varying conditions should be taken at once.

On January 1, 1898, Bailey formally turned over to Stewart and Clymer the responsibility for the Arequipa station. A few days later he started the journey home, by way of Mexico. Before leaving, he received a letter of warm commendation from Pickering (undated, November 1897):

Your term in Arequipa has been a great success and I congratulate you heartily on the results. We are looking forward with great pleasure to your visit to us on your return here. I wish you were coming back by way of Europe so that you might see how highly your work is appreciated at the Observatories there.

X

FRUITS OF THE ENTERPRISE

I

On January 31, 1902, in a letter to President Eliot, Pickering (Fig. 46) wrote:

Today I complete twenty five years of my connection with the Observatory. I have been very happy in it, and have been greatly aided and relieved whenever difficulties have presented themselves by your unfailing help, good counsel and kindness.

With this letter he enclosed a deed of gift of $50,000 to the Observatory and directed that, after his death and that of his wife,[1] the income should be used to support astronomical research. Believing as he did that scientific advance was more important than individual prestige, he stipulated that this aid should be available to astronomers in any part of the world. Mr. Eliot replied (February 19, 1902):

Your Deed of Gift was suitably accepted by the Corporation. It was doubtless one of the ways in which you celebrated the completion of twenty five years of invaluable service.

The staff of the Observatory (Fig. 47) celebrated the event by presenting to the director a silver loving cup bearing the inscription, ". . . with the cordial congratulations of his associates." Among the many letters of felicitation was one from Mrs. Draper (January 30, 1902):

You can certainly look back with satisfaction upon the superb work which has been accomplished under your supervision, and which I hope may be continued for many years to come.

357

FIG. 46. Edward C. Pickering in the later years of his administration.

FIG. 47. Gathering to celebrate Pickering's twenty-fifth anniversary as director of the Observatory.

The Harvard College Observatory

With my congratulations and good wishes, will you accept a clock I have had forwarded to you by express today; it will, I think, interest you, in the fact that it will run for 400 days without rewinding—Please place it in your room in the brick building,[2] and let it remind you occasionally of your sincere friend, who is so deeply indebted to you.

During this first quarter-century of Pickering's directorship, fundamental changes had occurred in the research methods used by most astronomers and only a few continued to rely chiefly on visual observations. The spectroscope and the camera, working in combination with the optical telescope, had produced an increase in our knowledge of the stellar universe that, as Arthur Searle remarked in his own letter of congratulations (January 31, 1902), had "no parallel, I think, since the time when Galileo applied the telescope to the work of astronomy." In summarizing these advances, Searle wrote:

We have seen the beginning of accurate astronomical photometry on a large scale, to set against Bradley's beginning of accurate observations of position; the final detection of the long suspected variation of latitude, to set against his work in aberration and nutation; the discovery of the satellites of Mars and the planet Eros, small but interesting members of the solar system, to compare with Herschel's discovery of Uranus. His discoveries in the field of stellar astronomy are recalled and indeed surpassed by what has lately been done. It is only in theoretical astronomy that we cannot compete with our predecessors, so far as striking and brilliant results are concerned. But, in fact, it is difficult to suppose anything resembling Newton's work to appear again, although that may come too.

A few omens had in fact already appeared, presaging a revolution in theoretical physics that would indeed compare with the one brought about by Newton's discoveries. In the 1890's Norman Lockyer and his colleagues had shown in laboratory studies that the character of the spectrum from a given element would change at extremely high temperatures (an effect we now recognize as the result of ionization, or the successive removal of electrons from an atom). Becquerel had discovered radioactivity and Roentgen had produced X-rays. In 1901 Max Planck had proposed his quantum theory, that light was emitted or absorbed as discrete units, not as a continuous flow of waves. The significance of these and other developments was not yet apparent, but theories of atomic structure proposed by Ernest Rutherford and by Niels Bohr, which

360

would greatly expand the science of stellar spectroscopy, lay only a decade in the future.

During these first 25 years of Pickering's administration, astronomers had also witnessed a phenomenal proliferation of observatories. When the nineteenth century began, no permanent observatory had existed on the North American continent or in the Southern Hemisphere.[3] Now there were nearly fifty in the United States and Mexico, and more than a dozen south of the equator. A steadily growing number of American colleges had established active observatories in the last quarter of a century. They included the Chamberlin Observatory of the University of Denver (1894), the Carleton College Observatory at Northfield, Minnesota (1878), those at the state universities of Wisconsin (1878), Virginia (1883), Indiana (1900), Pennsylvania (1896), Minnesota (1892), Illinois (1896), and those at two women's colleges—Smith (1886) and Wellesley (1900).

Other new institutions that began operating during this period were the Lick Observatory (1888) on Mt. Hamilton, California, under the directorship of E. S. Holden; the Ladd Observatory (1890) of Brown University at Providence, Rhode Island, directed by Winslow Upton; the Astrophysical Observatory (1890) of the Smithsonian Institution, directed by S. P. Langley; Harvard's southern station at Arequipa, Peru (1891); the Lowell Observatory (1894) at Flagstaff, Arizona, directed by Percival Lowell; and the Yerkes Observatory (1897) of the University of Chicago, directed by George E. Hale. Soon to come into existence were the Mills Observatory (1903) of Lick's southern station near Santiago, Chile, and the Solar Observatory of the Carnegie Institution (1904) on Mt. Wilson, California, directed by Hale. With the building of the 60-inch instrument for Mt. Wilson began the age of large reflectors—a 72-inch for the Dominion Observatory at Victoria, British Columbia, and Mt. Wilson's 100-inch Hooker telescope, both completed in 1918; and, the culmination of these optical giants, the 200-inch Hale telescope, completed in 1948, on Mt. Palomar.

The large number of active observatories and the steadily widening scope of research based on physics had emphasized the need for new journals. Since the early days of the Civil War, when Gould's *Astronomical Journal* had stopped publication, the United States had not been able to boast even one periodical devoted solely to astronomical work. Then in 1882 the *Sidereal Messenger* (not related to the journal of the same name that had flourished for a short time in the late 1840's) began publication under the editorship of W. W. Payne, director of the Goodsell Observatory of Carleton College, and provided an outlet for some astronomical news and reports. In 1886, Benjamin Gould revived his *Astronomical Journal.* More technical in nature than the *Sidereal Messenger,* it encouraged the submission of original research papers.[4] Three years later the newly organized Astronomical Society of the Pacific began including some research material in its *Publications.*

In the closing years of the century came an innovation in astronomical publications, a journal designed specifically to report research based on the methods of the "new astronomy." The full development of the idea required several years and was largely the work of George Hale.

After finishing his astronomical apprenticeship at the Harvard Observatory in 1890, Hale had returned to his home in Chicago, where he enlarged his privately owned Kenwood Physical Observatory to continue his study of the sun. He planned also to start a new journal devoted to spectroscopy and astrophysics. In a letter to Pickering (October 14, 1891), he wrote:

I have returned from Europe rather sooner than I expected and have resumed my work of photographing the spectrum of the [solar] prominences. After consulting with Prof. Young, Dr. Huggins, Dr. Vogel, and others, I have decided to publish "The Astro-Physical Journal" in place of ordinary publications from this observatory. I had hoped to have an opportunity of talking over the matter with you before returning home, but my time was too short to allow me to go to Boston. I propose to publish the journal at irregular intervals, but as often as sufficient material is at hand, and it seems desirable to translate a good many articles from foreign journals, in order to collect under a single cover a large proportion of the literature relating to spectroscopy and allied subjects.

However, some astronomers felt that so specialized a publication was not necessary, and Professor Payne suggested that the *Astro-Physical Journal* could be incorporated in the *Sidereal Messenger* as a separate section, to be edited by Hale. Since the success of the new venture was so uncertain, Hale accepted this suggestion.

The first number in the new series, whose name was changed to *Astronomy and Astro-Physics,* appeared in January 1892, but the union lasted only three years. From it emerged two new journals: *Popular Astronomy,* edited by Payne, and *The Astrophysical Journal, An Inter-*

Fruits of the Enterprise

national Review of Spectroscopy and Astronomical Physics, edited by Hale with the assistance of Professor James E. Keeler of the Allegheny Observatory. Volume I, Number 1 appeared in January 1895, and contained articles by A. A. Michelson, on the "spectro-photography" of the sun; by Henry A. Rowland and R. T. Tatnall, on the arc spectra of boron and beryllium; and by E. C. Pickering, on detecting variable stars from their spectra. There was no paper on purely observational astronomy.

Professional groups of American astronomers also formed during this period. The first, the Astronomical Society of the Pacific, was organized in 1889 with E. S. Holden as president. The second, the American Astronomical and Astrophysical Society (the word "astrophysical" was later dropped), developed from discussions between Hale and the associate editors of the *Astrophysical Journal*—Keeler, Michelson, Hastings, Young, Rowland, and Pickering. Plans took final shape in 1897 at the dedication of the Yerkes Observatory, and the society was formally organized in March 1899. The members elected Simon Newcomb as their first president. The first international astronomical society came into being in 1904 with the formation of the International Union for Cooperation in Solar Research, whose first president was George Hale. The more inclusive International Astronomical Union was not officially organized until July 1919, a few months after Pickering's death.[5]

At the time of Pickering's 25th anniversary, the Harvard College Observatory itself had reached perhaps the highest point of development it would attain during his administration. When Charles W. Eliot had become president of Harvard he remarked, in his inaugural address of October 19, 1869, that "with the exception of the endowments of the Observatory, the University does not hold a single fund primarily intended to secure to men of learning the leisure to prosecute original researches." Now the University was liberally supplied with such funds, and during Pickering's years the Observatory itself had increased its endowment from less than $200,000 to more than $900,000. The annual income had grown from less than $20,000 to some $50,000.[6] The number of observing instruments had multiplied, a fireproof building had been erected to house the collection of photographic plates, and the old wooden structures had been modernized. A large staff of men and women assisted in the research both in Arequipa and in Cambridge, and the work of the Observatory had received universal recognition.

The Harvard College Observatory

In 1901 Pickering had again been honored with the award of the gold medal of the Royal Astronomical Society, this time for his work in stellar photography and spectroscopy. His old friend Arthur Ranyard, who had helped procure the first medal even though he objected to the entire system of awards, had died in 1894, but congratulations came from many others. H. H. Turner wrote (January 13, 1901):

> We hope you will appreciate our Medal, which was awarded you for the second time last Friday quite unanimously . . . I should have opened this letter with my hearty congratulations. You are one of a very select few who have had the Medal twice; and there is a thorough conviction in the minds of the Council that it is well deserved. We really should like to see you on the occasion if you can spare the time and energy to come.

Pickering was not able to make the trip to England. The British astronomer William Shackleton, who was about to accompany Ernest Henry Shackleton's expedition to the Antarctic, in writing (April 12, 1901) to Pickering for advice as to what spectroscope to take for photographing the spectrum of the aurora, remarked:

> I was delighted to be present at the meeting of the Roy. Ast. Soc. when the American Ambassador [Joseph Choate] received the Gold Medal on your behalf. I am sure all of us on this side who have taken spectroscopy as our work feel sure that it has gone to the right place & no one more than us at South Kensington (i.e. speaking for the assistants) appreciate the quality of the work put out from the Harvard Observatory.

II

Some of the most important material for this work was of course the stream of photographs coming from the Boyden station in Arequipa.

The Bruce telescope, with its ability to record very faint celestial objects, had produced an exciting discovery in April 1899 when William Pickering in Cambridge, studying photographs made at Arequipa some seven months earlier, detected the image of a ninth moon of Saturn. (The eighth moon, Hyperion, had been discovered visually half a century earlier by George Bond and, two days later, by William Lassell.)[7] The new satellite, which received the name Phoebe, was so faint (magnitude 14.5) that other telescopes were not able to locate it. Phoebe's existence therefore remained in doubt until 1904, when William Pickering again found its image on more than 40 Bruce plates. Knowing where to look for it, Professor Barnard at the Yerkes Observatory was able to find it, visually, and thus confirm its reality. (A few months later William Pickering reported finding a tenth moon, which he called Themis, but its existence has never been confirmed.)

Unlike the eight other moons of Saturn, this outermost satellite moved in a clockwise direction, a fact that impelled an unidentified versifier to compose an address:

> Phoebe, Phoebe whirling high
>   In our neatly plotted sky,
> Phoebe listen to my lay,
>   Won't you swirl the other way?
> Never mind what God has said,
>   We have made a law instead.
> Have you never heard of this
>   Nebular hypothesis?
> It prescribes, in terms exact,
>   Just how every star should act.
> Tells each little satellite
>   Where to go and whirl at night.
> [Disobedience incurs
>   Anger of astronomers
> Who—you musn't think it odd—
>   Are more finicky than God.]
> And so, my dear, you'd better change;
>   Really we can't rearrange
> All our charts from Mars to Hebe
>   Just to fit a chit like Phoebe.[8]

The publicity given the first report of Phoebe may have been responsible for a suggestion sent to Edward Pickering, probably during this period:[9]

95 Irving Street March 29.

Dear Pickering,

My friend F. W. H. Myers writes to me:

"I wonder if you can help me in getting some member of our solar system called *Jeanne*. The adoration, or the timidity, of astronomers has given the names of Sappho and Xanthippe to two minor planets. But it is surely Jeanne d'Arc and not Xanthippe, who divides with Sappho the primacy of her sex, and now that she is to be canonized on Earth

365

The Harvard College Observatory

I think she should be commemorated in Heaven and her name stamped upon our system, to endure until that expires. I dare say some of Prof. Pickering's young men can find a minor planet when they want to!"

What think you of the suggestion? It seems to me an admirable one, bringing some human life and suggestiveness into that cold classic nomenclature (?)—Jeanne's is a touching figure and this would be some amends for burning her as a witch.

Always truly yours
Wm James

The Bruce photographs revealed the existence of another interesting though minor member of the solar system when on a plate made August 14, 1901, Dr. Stewart at Arequipa found a new asteroid, the 475th such object detected in the century after the discovery of the first one, Ceres. Deciding that a name from Peruvian mythology would be appropriate, Stewart called the newcomer Ocllo, after the sister-wife of Manco, the first Inca, both of whom were supposed to have descended from the Sun. Ocllo moved in an orbit more eccentric than that of any previously known asteroid.[10]

Mrs. Draper, who of course was kept informed of the progress of work at the Observatory, had occasionally questioned the wisdom of a plan that involved photographing the sky over and over again. She had been thinking of revising her will, she wrote Pickering (June 15, 1903):

Looking to the future—how long do you think it will be advisable to continue to take photographs, with the eight inch telescope, of the entire sky, night after night? The accumulation of plates will become unmanageable in a few more years, and the handling and examination of them will involve an outlay of means which might be more wisely expended for the purposes of this Memorial. Will you think over this? . . . In providing for the continuance of the work in my Will, I have to remember that in a relatively short time, you and I will no longer be living, and I have to guard against the possibility that your successor may not be interested in this branch of research, and may prefer to use the fund in some other direction, if he can. In this he may be wise, but I am not disposed to trust his judgement alone, or that of the trustees of Harvard University, which would be practically the same thing . . . Circumstances compel me to assume the duty and responsibility of deciding as to the disposal of a certain amount of money, some of the questions involved are difficult, and I hope you will be kind enough to aid me with your advice concerning them.

Fruits of the Enterprise

At the time Pickering was able to convince Mrs. Draper that continued photography of the sky "night after night" was indeed worth while. Such photographs had made it possible to determine the orbit of Eros and to reveal the existence of Phoebe and Ocllo. But these discoveries, while interesting, were of minor importance compared with another fact that gradually emerged when, in 1904, the Bruce began to concentrate on photographing the Magellanic Clouds: that the Clouds were as rich in variable stars as some of the globular clusters had proved to be, and that these stars offered a way of measuring the true distance of celestial objects outside our own Galaxy.

The search for variable stars in the Magellanic Clouds became the special province of Miss Henrietta Leavitt, a young member of the Observatory staff who had graduated from Radcliffe in 1892. The method most commonly used by her (and by Bailey) was that of superposition. A negative of a photograph taken on one date was placed on top of a positive made from a plate of the same region of the sky, taken on a different date. If the white and the black image of a star did not exactly coincide, the star was suspected of variability, which could be confirmed by further comparisons with plates made at other times.

By this laborious process of inspection, in 1904 Miss Leavitt found 152 variables in the Large and 59 in the Small Magellanic Cloud. The next year she reported 843 new variables in the Small Cloud. These discoveries, announced through the Harvard Circulars, led Professor Young at Princeton to comment in a letter to Pickering (March 1, 1905), "What a variable-star 'fiend' Miss Leavitt is—One can't keep up with the roll of the new discoveries." Mrs. Draper, who in spite of advancing age maintained her concern for the work of the Observatory, wrote Pickering (May 7, 1905) that she was "greatly interested in Miss Leavitt's discovery of the large number of variables in the small Magellanic Cloud. It is certainly strange, that so many of them should be found apparently so close together. Will you please congratulate Miss Leavitt for me, and may I also ask you to congratulate your brother upon the discovery of the tenth satellite [Themis, which was never confirmed] of Saturn, he is now the proprietor of two of the planet's attendants." One correspondent even suggested that Miss Leavitt's name should be proposed for the Nobel Prize.

Most of the variables in the Clouds belonged to the Cepheid class, which was named for δ Cephei, one of the brightest stars of the type.

The Harvard College Observatory

At the time, a Cepheid was generally believed to be a binary system—a star that varied in brightness because one member of the revolving pair was periodically eclipsed by its companion, which was too faint to be seen. Later work showed that Cepheids are not eclipsing systems but are single stars that pulsate at regular intervals; as the star expands, it brightens; as it contracts, it grows dim. Up until 1895 only 33 Cepheids were known, with periods (the interval between times of maximum brightness) ranging from 2 to 39 days. Then Bailey at Arequipa had begun to find large numbers of short-period Cepheids, many with periods less than a day, in ω Centauri and other globular clusters.

Miss Leavitt continued to add to the list. Despite long interruptions when poor health made work impossible, in 1908 she was able to publish a list of 1777 variable stars she had found in the two Magellanic Clouds. She included a table that gave the carefully measured periods and magnitudes of 16 variables in the Small Cloud, and commented briefly, "It is worthy of notice that in Table VI the brighter variables have the longer periods."[11] This statement was apparently the first enunciation of the fundamental discovery that later became known as the period-luminosity relation.

Neither the table nor Miss Leavitt's cautiously worded inference seems to have attracted the attention of astronomers, perhaps because she had arranged the stars not according to their magnitudes or periods but according to their identifying numbers, which indicated merely the order of their discovery (818, 821, . . . , 1742). Hence the real correlation between their periods and luminosities was not obvious at a casual glance. Four years later, however, she issued a report on a somewhat larger number of stars and firmly restated her earlier conclusion: "A remarkable relation between the brightness of these variables and the length of their periods will be noticed."[12] Miss Leavitt had measured the periods and magnitudes of 25 variables (including the 16 reported earlier), and had tabulated them according to the length of their periods, which ranged from 2 to 120 days. The measurements given in the table clearly showed that the fainter stars had the shorter periods, and that the brighter the star, the longer its period. To display the relation graphically, she plotted the magnitudes of the 25 variables, at both maximum and minimum brightness, against the logarithm of their periods; the resulting straight lines showed unmistakably that the longer the period of a star, the greater its brightness at maximum.

Fruits of the Enterprise

Miss Leavitt obviously recognized some of the far-reaching implications of the relation, for she commented, "Since the variables are probably of nearly the same distance from the Earth [since all were in the Small Magellanic Cloud], their periods are apparently associated with their actual emission of light, as determined by their mass, density, and surface brightness." She herself did not follow the paths opened by this brilliant discovery, perhaps because of Pickering's conviction that the Observatory's function was to collect data, not to try to explain them, but others pursued the question.

Ejnar Hertzsprung, who some years earlier had quickly grasped the significance of Miss Maury's c characteristic, was perhaps the first to appreciate that the period-luminosity relation offered a way of measuring celestial distances. Sir John Herschel, during his stay at the Cape of Good Hope in 1834–1838, had made the earliest detailed study of the Magellanic Clouds, and he, unlike some other astronomers of the time, correctly believed them to to be stellar systems outside our own. How far away they were was a question that, at the time, had seemed unanswerable. Hertzsprung realized that if we knew the period of a Cepheid and could determine its actual brightness in comparison with the sun's brightness, simple calculations would give its distance from the Earth.

He made some measurements on 13 Cepheids in our own Galaxy, to determine a scale of absolute magnitude, and applied the scale to the apparent magnitudes of Miss Leavitt's 25 Cepheids. From these calculations he made the first theoretically valid estimate of the distance of the Small Magellanic Cloud: 30,000 light-years.

In 1922 Shapley and Menzel, who had found a number of globular star clusters in the Large Cloud, used the apparent diameters of these clusters to estimate a distance of 110,000 light-years.[13] A few years later Shapley, having determined a more accurate zero point and magnitude scale, measured a larger number of Cepheids and revised the earlier estimate to about 100,000 light-years. (Various later revisions reduced the figure to about 85,000 light-years.) On the basis of this work, Shapley proceeded to use the period-luminosity relation to estimate the distance of the globular clusters, and eventually to formulate a standard by which he could measure the dimensions of our Galaxy and estimate the scale of the universe.

In 1918, when Shapley had begun a study of the light curves of short-period variables, he commented on the importance of continuing Miss Leavitt's work, when he wrote to Pickering (July 20, 1918):

I believe the most important photometric work that can be done on Cepheid variables at the present time is a study of the Harvard plates of the Magellanic clouds. Probably Miss Leavitt's many other problems have interrupted and delayed her work on the variables of the clouds for the interval of six or seven years since her preliminary work was published. I wrote last year something about the importance of these variables in my investigations of cluster parallax. The theory of stellar variation, the laws of stellar luminosities, the arrangement of objects throughout the whole galactic system, the structure of the clouds—all these problems will benefit directly or indirectly from a further knowledge of the Cepheid variables.

Miss Leavitt, meanwhile, had continued her routine tasks. Her "Record of Progress" for the time (a notebook begun August 5, 1912) lists the projects she was working on: a study of Algol variables undertaken for Henry Norris Russell, measuring the luminosities of stars in Kapteyn's selected areas, experiments to determine the colors of faint stars, and methods of transforming photographic to visual magnitudes. One of the most important of these undertakings, carried out in collaboration with astronomers at Mt. Wilson, was determining the exact photographic magnitudes of a sequence of stars (ranging from very bright to very faint) near the north celestial pole. This "North Polar Sequence" later became the basis for an internationally accepted system of stellar magnitudes.

Miss Leavitt never abandoned the search for variable stars, however, and by the time of her death in 1921 she had discovered some 2400—about half the number then known to exist.

III

During these very productive years the Observatory established yet another observing station outside Cambridge, at Jamaica in the British West Indies, under the supervision of William Pickering. This was the last such venture carried out during Edward Pickering's administration, and the least productive of basic astrophysical data. Finally set up in 1911, the station had its beginnings more than a decade earlier when on a brief visit to the island in 1899 William Pickering was favorably impressed with its clear skies and excellent seeing conditions.

Edward Pickering had long wished to try to photograph stars with a new type of telescope that would use a very long focal length. When a gift from "two friends of the Observatory" made it possible for him to procure the instrument in 1900, William Pickering was assigned the task of testing it. With the aid of the Boyden Fund, he returned to Jamaica and set up the instrument in the grounds of a rented house in Mandeville. The new telescope had a 12-inch lens and a focal length of 135 feet. As a recorder of star positions it was a disappointment, for it could not photograph stars fainter than about the 7th magnitude, but it did reveal the surface of the moon with great clarity. William soon became convinced that the moon's surface closely resembled that of Mars in having canals (which sometimes doubled) and changing features that strongly suggested the existence of an atmosphere and vegetation. The suggestion of snow caps that expanded or receded with the season and apparent changes in the craters Eratosthenes, Linné, Messier, and Plato stimulated William's earlier enthusiasm for studying the moon. Other astronomers in the past had reported seeing changes at intervals on the moon's surface and in these craters, especially Linné and Eratosthenes, but none had before reported seeing canals. William wrote to Douglass in Arizona, asking him to try to confirm the existence of the lunar canals, and described them to Edward Pickering, adding (July 23, 1901) that he "did not intend to publish at present. I don't think any other astronomer will discover them, but if he did extracts from my letter to you would apparently, if published, establish my claim to priority." (Brief changes, including flashes of light and short-lived glows in the craters are still occasionally reported, but they have not been confirmed by photography. Astronomers today generally agree that transitory events do occur on the moon, but that they do not include major structural changes. The seismographs placed on the moon by Apollo 11 and Apollo 14 may yield information on this point.)

During more than a year's residence in Jamaica, William systematically observed, sketched, and photographed the moon, and after his return to Cambridge was able in 1903 to publish, as Volume 51 of the *Annals*, the first complete "Photographic Atlas of the Moon," a compilation of some eighty photographs. In this volume he presented the evidence suggesting seasonal changes. He added a word of caution, however, against accepting the evidence of photography and, in so doing, expressed what he continued to believe throughout his life—that the eye was more trustworthy than the camera. In commenting on changes

371

in Eratosthenes, apparently confirmed by photographs taken at different times, he wrote:

We must be careful, however, not to put too much confidence in results obtained from a mere comparison of photographs, as slight changes in exposure and development will sometimes produce results that are very misleading. The only safe course to follow is to confirm all suspected changes by a careful visual study of the formation at different ages of the Moon.[14]

Although astronomers in general could not confirm and did not accept his conclusions, William did not lose confidence in his observations, being convinced that the seeing conditions in Jamaica were among the best in the world and that, although he was now in his mid-forties, his eyesight was still better than that of most men.

In 1904 William spent some months at the Lowe Observatory on Echo Mountain, near Wilson's Peak in California, where he continued his study of the moon and of Jupiter's satellites. Meanwhile he continued to urge that Harvard send him back to Jamaica, where in his opinion the seeing conditions were far superior to those in Arequipa. The following year he visited Hawaii, as well as Alaska, Canada, and the Azores, to study the formation of volcanic craters and calderas, which he regarded as similar to those he saw on the moon. In 1905 he had the satisfaction of being awarded the Lalande Prize of the Paris Academy of Sciences for his work on the satellites of Saturn.

Not until 1911, however, did Edward Pickering consent to setting up the new station, to be supported in part by the Boyden Fund and in part by William's own funds. William then returned to Mandeville, taking with him the 11-inch Draper and a smaller telescope, to carry out a program of photographing double stars and sections of the Milky Way and to measure the photographic magnitudes of stars in Kapteyn's selected areas.

From the beginning, however, the Mandeville station functioned somewhat outside the main stream of the Observatory's research, in that it made very little use of photography and concentrated not on stars but on the solar system. Just as had happened earlier at Arequipa, the lure of visually observing the bodies near the earth proved to be overpowering. When after some months Edward Pickering wrote to express his disappointment because no photographs had yet arrived in Cambridge, William explained (September 2, 1912) that he had in fact taken very few pictures. "You see I adjusted all my instruments visually instead of photographically, which is I believe a quicker and more accurate method, and saves plates."

In response to William's continuing reports of lunar canals, snow caps, and vegetation, Edward replied (October 10, 1912), "Your results on the Moon are interesting. Can you not get some other astronomer to confirm them?" A few months later William wrote (January 7, 1913): "I am getting a great deal of evidence in regard to the lunar atmosphere, & wish to make it so complete that astronomers will have to accept it. I have about a dozen different proofs of its existence, and after they accept that, it will be less difficult for them to accept the evidence of vegetation."

His new studies of lunar phenomena caused him to write to Edward (September 12, 1912) with great enthusiasm that he had been obtaining really startling results on the moon:

Whatever reputation as an astronomer I lost when I published my former observations, will be nothing to the destruction produced when these get into print, & especially the drawings. I have seen everything practically except the selenites themselves running round with spades to turn off the water into other channels!

He had seen marked changes on the floor of Erathosthenes, and concluded that "since the dark areas in many cases increase in size while the Sun is rising, & diminish while it is setting, the reverse of what happens to shadows, I do not see what they can be but vegetation." On March 1, 1913, he wrote: "The latest addition to my lunar investigations is the finding of certain small snow craters b & c [sketched in the letter], which early in the lunation each contain a small black spot at their centers. As the lunation progresses these spots enlarge as shown. I believe this is due to the melting of snow, and is analogous to the melting snow caps of Mars."

Except for occasional visits to Cambridge and to Europe, William continued to live in Mandeville until his death in 1938. After he retired in 1924, the Harvard Observatory severed its connection with the station and William thereafter maintained it as a private observatory, supported by his own funds. Between 1913 and 1920 he published 44 "Mars Reports." During the latter part of his life he published dozens of articles, mostly in *Popular Astronomy,* on the moon, the planets, and the satellites of Jupiter, but he never succeeded in convincing professional astronomers in general that his interpretations of the observations were correct. When,

The Harvard College Observatory

after Edward Pickering's death, Professor Bailey was asked for his opinion of William Pickering's views, he replied that they had not been confirmed by photography, but that he did not wish

to enter into any public controversy with another member of the Harvard Observatory staff. Moreover, it is only fair to Professor W. H. Pickering to state that no other member of this Observatory has made any special study of the lunar surface . . . It seems to me that one can safely accept most of his observed phenomena. The difficulty comes in the interpretation. For changes in surface brightness in special areas and spots, he assumes snow, cloud, and vegetation. Others would assign a different explanation.[15]

Unfortunately, the sensationalism of William Pickering's views, and the often incautious terms in which he expressed them, have tended to obscure the sometimes brilliant theoretical work that he did produce. Unlike Edward Pickering, who wished to amass facts but was reluctant to construct hypotheses, William had an innately speculative mind, and was always ready to devise a tentative explanation of any puzzling fact. After the discovery of Pluto, and the general realization that its mass was too small to account for the perturbations of Uranus and Neptune, William concluded that one or more other unknown planets remained to be found beyond the orbit of Neptune, and carried on the search throughout his life.

His published papers ranged over a wide variety of subjects. He made studies of spiral nebulae and attempted to calculate the distances of globular clusters. In a letter to Edward Pickering (June 5, 1918) he wrote, "I see that Shapley has just published an article on Globular Clusters giving the distribution and general arrangement, which contains a good deal of the material that I was going to send you. I presume he has one or two computers to help him, so that he is able to work faster. I am sorry, but it can't be helped."

One of the most striking of William Pickering's astronomical achievements, one that was long overlooked,[16] was the first valid physical explanation of the shifting dark absorption bands and bright lines that characterize the spectrum of a nova. He suggested that, when a star exploded, a series of "prominences" erupted, ejecting a shell of gases that surrounded the star and rapidly expanded in all directions:

In a few hours the gases first emitted have receded to a considerable distance from the star, and have cooled down owing to the rapid expansion

374

involved by their recession, which causes them to fill a very much larger sphere than that originally occupied by them. Thus the star very soon becomes enveloped in an atmosphere whose outer regions are comparatively cool . . . This cold advancing atmosphere produces a series of dark absorption lines . . . The hot receding prominences, however, extending away for perhaps millions of miles behind the limb of the star, give out a light whose wave-length cannot be absorbed by the cold advancing atmosphere. They therefore shine with their full brilliancy.[17]

IV

When Solon Bailey had returned to Peru in 1899, the Arequipa station was beginning its highest state of productivity. The staff included four competent astronomical assistants—Bailey, Dr. De Lisle Stewart, Royal H. Frost, and W. B. Clymer, in addition to Juan E. Muñiz, who had been employed in 1894 to help establish the meteorological station on El Misti and to collect observations from there and from others in the chain of stations. All the meteorological installations except the one at Arequipa were given up in 1900 but Muñiz stayed on, to help with instrumental problems and keep the telescopes in repair.

Solon Bailey returned to Cambridge at the end of his five-year term, leaving Frost in charge at Arequipa. As had been true from the beginning, the relative isolation of the site, in what was then a primitive country, plus the severe physical demands of the observing schedule, made it difficult to keep assistants for a long period of time.[18] In spite of frequent changes in the staff, the system of work had been so well established that the station functioned smoothly on the whole and except during the cloudy season there was rarely any interruption in the flow of plates from the Boyden, the Bruce, the Bache, and the smaller telescopes.

Extracurricular activities occasionally varied the routine of celestial photography. An unusual interruption in the spring of 1903 was the collecting of wild guinea pigs (*Cavia cutleri*). In sending his financial statement for March, Bailey pointed out to Pickering (April 6) that there was

also a charge in these accounts for guinea pigs. This may impress you as something new in the astronomical world, but Professor (or Mr.) W. E. Castle, of the Zoological Laboratory, of Harvard University wrote me, requesting me to send him some guinea pigs for some studies he is making in heredity.

375

The Harvard College Observatory

Bailey had collected the specimens, as requested, but unfortunately the health authorities in Peru "saw a resemblance in these animals to rats"[19] and refused to release them. Subsequent attempts were successful, however. Frank Hinkley, who was in charge of the station in 1909, and Leon Campbell, who took charge in 1911, were able to collect numbers of the desirable guinea pigs and help dispatch them to the laboratory of the Museum of Comparative Zoology at Harvard.[20]

Excellent though the Arequipa site was for more than half the year, it had the disadvantage that its cloudy season sometimes lasted nearly four months, and Pickering therefore began a search for an even better astronomical climate. Sir David Gill and others had strongly recommended South Africa. In 1908 Pickering sent Bailey, equipped with two visual telescopes, a photographic telescope, and several meteorological instruments, to explore the possibilities. After trying various sites in the Cape Colony, the Transvaal, and Rhodesia, Bailey returned to Cambridge after nearly a year's observations. He reported that the skies at Bloemfontein, in the Orange Free State, had less cloudiness than those at Arequipa and the seeing was excellent, although dust storms and violent thunderstorms were occasionally a problem.

The tentative plan for transferring the Arequipa station to South Africa had to be abandoned,[21] however, and all the work of the Observatory suffered an abrupt check when early in 1909 Mrs. Draper informed Pickering that she could no longer contribute so generously toward the support of the research. She wrote that after considering her financial situation and prospects she had found, to her great regret, that after August 1, 1909, she would have to reduce her contribution to a maximum of $400 a month—less than half the sum she had been providing for more than 20 years. Pickering replied (January 26, 1909):

Your letter of January 24 has just reached me. I am truly sorry on your account, as well as on that of the Observatory, that your finances are not what you wish. I had supposed that your income permitted you to have everything that you desired. If this is not so, of course the first saving should be made on your good deeds to others. We will do the very best we can with reduced income. The change will, of course, cause delay, and I am sorry for the assistants we shall have to discharge.

The blow was the more severe because for several years the expenses of the Observatory had been greater than the income, and the deficit had been growing larger. Interest rates had fallen from more than 6 to about 4 percent, so that the Observatory's invested funds earned less than before. Efforts to meet the deficit by raising a subscription of $50,000 in 1906 had been a failure. The sudden loss of more than $5,000 in yearly income meant that several assistants had to be discharged, and that the research must be greatly curtailed.

Even though she was forced to reduce her contribution, and was now elderly and in poor health, Mrs. Draper never lost her interest in the Observatory. For several years she had served as a member of the Visiting Committee, the first woman to be so appointed.

In one of her last letters to Pickering (July 12, 1914), she wrote:

Thanks for your letter, and for the photographs, received yesterday,—they are wonderfully fine, I am not surprised they are wanted for exhibition, they are most interesting.

I have not been well since I came here, I do not seem to recover from the effects of the run-away, which shook me up, and has left me very shaky, they say it will come right with time, I hope so.

It is wonderful what Miss Cannon has accomplished, she is to be congratulated.

I wish I felt able to go to Chicago,[22] it would be most interesting and I should enjoy it.

Mrs. Draper died on December 8, 1914, leaving a large estate. Her will provided for generous bequests to the New York Public Library, several hospitals and charitable institutions, the National Academy of Sciences, and $150,000 to the Harvard College Observatory, the income to be used "for the purpose of caring for, preserving, studying and using the photographic plates of the Henry Draper Memorial." To avoid any interruption to the stellar research, she had provided that her executors should pay to the Observatory, on account, the sum of $5000 the first year after her death, and $4000 each succeeding year until the final settlement of her estate.

In an obituary, Miss Cannon paid tribute to Mrs. Draper as one whose name would "always be honorably associated with the science of astrophysics."[23] Indeed, the Observatory's great contributions to our knowledge of the classes and structure of stars would not have been possible without the generous, sustained support of Mrs. Henry Draper.

The effects of the suddenly diminished income were even more severe in Arequipa than in Cambridge, for the Peruvian station was made to suffer most of the loss. The two assistants, H. E. Blackett and C. J. G. Vogel, were given notice, and Hinkley, then in charge, was instructed to keep the expenses to about $2300 a year, even though this meant drastically curtailing the photographic work.

Hinkley's term expired in 1911, and he was replaced by Leon Campbell, who since 1899 had served as an assistant in Cambridge. During much of his five-year term he had only one helper. Nevertheless, he was able to send photographs to Cambridge, to carry out his own visual observations of variable stars, and to cooperate with visiting astronomers in their projects. He gave valuable assistance to the Yale Peruvian expedition and, at the request of C. G. Abbott of the Smithsonian Institution, set up a pyrheliometer in the garden of the residence to measure the sun's radiation and study how the earth's temperature depended on it. In particular, Abbott wanted to find out how volcanic eruptions affected the transmission of sunlight through the earth's atmosphere. After the eruption of Mt. Katmai in Alaska in 1912, when astronomers in North America and Europe found the sky so turbid with volcanic dust that the amount of sunlight reaching the earth's surface was 10 to 15 percent below normal, the skies in Peru remained unaffected, and throughout the year 1913 as well.

After Campbell returned to Cambridge in 1915, the Arequipa station was again in Hinkley's charge for about two years. The problems of communication and transport posed by the advent of the first World War, as well as the lack of money, greatly diminished the amount of work produced.

L. C. Blanchard, the last of the supervisors during the Pickering regime, resigned in November 1918 to join the armed forces, and the once-flourishing station was shut down. Just before leaving, Blanchard wrote to Pickering (November 10, 1918) that "the building[s] are all closed and lens[es] covered so that no harm will come to them." The doors were locked and a notice was posted to inform the public that the place was no longer open to visitors. Juan Muñiz, who for years had served as a caretaker and repairman, was left in sole charge until someone could be sent from Cambridge.[24]

v

As Pickering approached his mid-sixties he was still well and vigorous. Nevertheless, he began considering what would happen to the Observatory if he should become unable to serve. Early in 1910 he wrote to President Lowell, requesting that Professor Bailey be appointed assistant director and that his salary be increased from $2800 to $3000. Pickering himself was growing older, he pointed out, he wished to be freed from some of his administrative duties so that he could spend more time on astronomical projects, and he had complete confidence in Bailey's judgment. The Corporation were willing to grant the raise in salary, Lowell replied, but they were unwilling to remove any of Pickering's responsibilities by officially appointing an assistant. A few weeks later Pickering again approached the subject and in a letter to President Lowell (March 17, 1910) he wrote:

Referring to our conversation of last evening, perhaps the following statement of my views regarding the administration of the Observatory may be of service to you.

While it is generally assumed that a man should not attempt to influence the choice of his successor, it seems to me that you should be in possession of such facts as you might hereafter find useful. As in the case of making a will, it is best to do this when in good health and strength rather than wait till it might be too late.

In the case of my permanent disability two courses would be open to the Corporation. First, the appointment of a Director, who would continue the general plan of work of the Observatory along its present line. In this case Professor Bailey would be the only person here who would be competent to do this. I am now taking care to consult with him on all important matters, and he is familiar with the numerous details in the various departments of the Observatory. In case of my temporary disability I should delegate to him my duties, under the authority conferred in your letter of January 18, 1910. If unable to do so, I hope the Corporation would take this action, and if my disability became permanent, before appointing my successor, would give Mr. Bailey the opportunity of showing whether he could fill the position satisfactorily.

Lowell replied that, fortunately, there was no need to consider the question at present, and the matter was dropped. Unofficially, however, Bailey did act as an assistant director. In 1912 he was made Phillips Professor of Astronomy, to fill the chair left vacant by Professor Searle's retirement.[25]

One of the most important researches carried out during Pickering's final years was Bailey's work on the magnitudes, periods, and light curves of the variable stars in the globular clusters Messier 3, Messier 5, and

Messier 15.[26] Shapley, then at Mt. Wilson, had begun similar studies of cluster variables, for which a liberal exchange of photographic plates took place between Harvard and Mt. Wilson.

The last period of Pickering's administration was one of solid accomplishment. The pattern of research chosen at the beginning of his directorship had undergone no basic changes. The photometers, telescopes, and cameras produced their data, the photographic plates accumulated, the assistants measured and catalogued. When astronomers at other observatories needed pictures of a particular section of the sky, or wished to know the spectral type or magnitude of a certain star, they had only to write to Pickering, who gladly found and sent the required material. That had been his purpose from the start, to create a reservoir of astronomical facts that others could draw on.

Two important new instruments acquired during this period were a 10-inch photographic triplet, which Shapley later called "one of the most industrious telescopes on the planet"[27] and a 16-inch photographic doublet. This instrument introduced a new principle, in that it employed a vacuum to bend the photographic plate so that it fitted more closely to the curved focal surface. Both telescopes were made for the Observatory, without charge, by the Reverend Joel Hastings Metcalf, an able amateur astronomer and an expert in applied optics. He maintained a private observatory at his home in Winchester, Massachusetts, cooperated closely with Pickering in the photographic programs of the Observatory, and was himself the discoverer of several comets and a number of asteroids. The 16-inch Metcalf was used chiefly to supplement, for the northern stars, the work done by the Bruce for the southern stars.

Pickering never lost interest in stellar photometry, the subject of his first major research project. In 1908 he published the Revised Harvard Photometry in volume 50 of the *Annals*. It included all stars of magnitude 6.5 and brighter, and became the standard reference catalogue for visual magnitudes.

Photographic photometry developed under the direction of Professor E. S. King, who after his return from Wilson's Peak (Chapter VI) had been put in charge of the photographic work in Cambridge. He devised methods for using photography to measure the magnitudes of bright stars by throwing the image of the star out of focus by a known amount, measuring the enlarged image, and then determining the amount of silver precipitated on the plate. Working chiefly with the

11-inch Draper telescope, he tested the properties of various photographic emulsions and determined how temperature and humidity affected the image on the film and hence the apparent magnitude of the star. One of his special interests, which occupied many years, was determining the photographic and visual brightnesses of the sun and the moon. King also worked out the numerical relations between photographic magnitudes, visual magnitudes, and spectral class, so that when the spectral class of a star was known its photographic magnitude could easily be converted to the visual scale, and vice versa.[28]

Another of Pickering's earliest interests, variable stars, also continued. As early as 1880, when there were only about 200 variables known, he had devised a classification into five groups: novae, long-period variables, irregular variables, short-period variables (Cepheids), and Algol variables. When Miss Cannon began preparing a catalogue of variable stars, which was published in 1903, the number of known variables had grown to 1227, and Pickering's classification was still basically valid. In 1907 she published a second catalogue of variable stars.[29] It contained 3748 known variables, 2909 of which had been found by workers at the Observatory. (In July 1939, Harvard's 10,000th variable was detected by Miss Cannon.)

The visual study of variables was research that required neither the resources of an observatory nor great astronomical and mathematical knowledge. Pickering had long believed that amateur astronomers could make valuable contributions to the science by observing variable stars. He offered full cooperation when in 1911 the American Association of Variable Star Observers was formed with headquarters at the Observatory. William Tyler Olcott, a gifted amateur, became its president and received help and advice from both Pickering and Leon Campbell.[30]

In 1911 Miss Cannon began the colossal project of compiling a new Henry Draper Catalogue of stellar spectra. Her goal was to classify the spectra of all stars brighter than the 8th magnitude in all parts of the sky and, in the process, to reclassify all those previously studied by herself, Miss Maury, and Mrs. Fleming.

The work was virtually completed within four years but publication was a long and expensive undertaking. The first volume, whose costs were paid by Pickering from his own funds, appeared in 1918 and, like the succeeding volumes, was printed on special rag paper that would not deteriorate with time. Only three volumes had been completed at the time of Pickering's death, but by 1924 the entire work was in print, occupying volumes 91–99 of the *Annals*. They included the spectra of more than a quarter of a million stars, all classified by one person—Miss Cannon. Without her great patience and devotion, the Henry Draper Catalogue, which is still a fundamental work, would never have come into being.

XI

A FIELD FOR WOMEN

To the Bruce telescope and the Henry Draper Memorial, the gifts of two remarkably generous women, the Observatory owes the means of achieving its greatest triumphs during the Pickering years. Though they were the most important donors among women, Miss Bruce and Mrs. Draper were both preceded and followed by others whose subscriptions, legacies, or outright gifts of funds demonstrated their special interest in both science and Harvard.[1] Less spectacular or familiar, yet of enormous significance, were the scientific contributions of a large corps of women on whose infinite patience and unflagging industry depended the ultimate results of their sisters' munificence. As observers, computers, and discoverers they made possible the interpretation of the stellar and other phenomena that the great photographic sweeps of the sky revealed.

In the period when visual observation demanded a physique that could endure constant night-long vigils in a frigid dome, astronomy remained a strictly masculine world. Abroad, to be sure, England boasted three women who had entered that exclusive domain—Caroline Herschel, famous for her domestic and scientific devotion to her brother William, as well as in her own right as an observer and cataloguer of stars; Mary Somerville, to whom, among her other accomplishments, John Couch Adams is said to have owed the idea of making the calculations that led to his prediction of the existence of Neptune; and Lady Huggins, Sir William's indispensable collaborator at Tulse Hill. On the continent, too, a number of women, often in association with their husbands, enjoyed a certain fame in mathematics, biology, and philosophy. A French work published in 1897 by Alphonse Rebière, *Les Femmes*

FIG. 48. Maria Mitchell at her telescope in Nantucket, from a painting by Hermione Dassel (1851). (By permission of Miss Virginia Barney.)

dans la Science, lists or describes the work of 610 women, 100 of whom he associated with astronomy.

In the United States, at least until the last quarter of the nineteenth century, Maria Mitchell (Fig. 48) stood alone.[2] On October 1, 1847, she discovered a comet, but, naturally shy and unwilling to believe that her little telescope mounted on a Nantucket bank building could have

384

revealed a wholly unknown celestial object, she refused to allow her father to announce it publicly. Fortunately, she did consent to his sending the news to William Bond on October 3: "Maria discovered a telescopic comet at half past ten on the evening of the first instant . . . Pray tell me whether it is one of George's . . . Maria supposed it may be an old story."

A storm on the island held back the mails, and the Cambridge post office added to the delay, so that, as the Bond diary notes, the letter did not arrive at the Observatory until October 7. On the 8th and 9th the Bonds and their guests observed the comet, on the 14th George Bond computed its elements, and on October 30 he and his father sent to President Everett a complete report on the comet. The dates are important because at stake were a gold medal and world-wide fame, and Maria Mitchell's priority in the discovery was in dispute. On October 3 Father Francesco de Vico, director of the Observatory at the Collegio Romano, a well-known comet-hunter, had reported his observation of the comet, and, without any competing claim, he was adjudged the winner of the medal. This award had been established by the King of Denmark to be given to "the first discoverer of any comet, which, at the time of its discovery, is invisible to the naked eye, and whose periodic time is unknown."

President Everett, who believed that George Bond had been unjustly deprived of the medal in 1846, determined that an American astronomer should not be outdone a second time. With characteristic vigor he pressed the indisputable claim of "a lady industrious, vigilant, a good astronomer and mathematician."[3] He wrote letters to the judges pointing out the dates provided by the Bonds as well as the fact that Father de Vico had himself acknowledged Miss Mitchell's right to the medal. He alerted the American consul in Copenhagen. That astute diplomat, in his direct appeal to the King, referred to the "pleasurable coincidence" that "the trophy of science . . . was achieved on the very coast where, as far back as the tenth century, the intrepidity and enterprise of your Majesty's Scandinavian ancestors first discovered and planted a colony upon the great western continent." Thanks largely to the evidence of the Bond diaries, the campaign succeeded in bringing to the young Nantucket astronomer probably the most ornate medal ever awarded the discoverer of a new celestial body.

In his first *Annual Report* of the Smithsonian Institution in 1848

The Harvard College Observatory

Joseph Henry announced a forthcoming "account of a new comet, the discovery of which . . . is one of the first additions to science . . . ever made in this country." Perhaps to him the novelty of the achievement seemed all the greater because it was made by "an American lady" he did not even name. Yet to the Bonds Miss Mitchell's discovery was neither a surprise nor an oddity because of the discoverer's sex. They looked upon her as a fellow astronomer, a keen observer, a serious student of the science whom they welcomed at the Observatory and were happy to assist in every way. It is fair to say that she owed to her close association with the Bonds not only the beginning of her career, but its later distinction.

The hospitality shown Miss Mitchell, the professional, they extended also to the perceptive amateur. One of the earliest and sprightliest volunteer observers was Miss Eliza Quincy,[4] the irrepressible daughter of the founder of the Observatory and throughout her life a devoted friend of it. By no means a negligible observer, she let no important astronomical event escape her notice, and, in letters always spiced with humor and self-deprecation, she reported every "celestial visitor" she saw from her parlor window in Quincy.

Although during the Winlock administration one or two women applied to the Observatory as student assistants, none seem to have appeared, and it was not until 1875 that the Observatory embarked on its policy, unique at the time, of admitting women to its staff. The first three, Mrs. R. T. Rogers, Miss R. G. Saunders, and Miss Anna Winlock, were engaged that year as computers to assist William A. Rogers in the zone work assigned to Harvard for the revision of the *Durchmusterung*. President Eliot himself had actually promoted the employment of women by asking Arthur Searle, then acting director, to engage Miss Saunders for a year at a salary of $600. Miss Saunders remained at the Observatory thirteen years.

Anna Winlock[5] needed no special introduction. Elder daughter of the third director, she was born in 1857, and at the age of twelve, gifted in mathematics and deeply interested in her father's work, she had accompanied him on the solar eclipse expedition of 1869. Immediately after graduation from high school, she came to the Observatory to help support the family after Winlock's unexpected death. She began as a computer without experience, but by the completion of the Harvard zone program she had developed into a full-fledged mathematical astronomer. Rogers, with whom she collaborated on a number of papers, invariably credited her with the major portion of the basic research. One of her most important contributions was an exhaustive study of stars close to the north and south poles, which were needed for the photographic work of the Observatory. Published in volume 18 of the *Annals,* parts 9 and 10, they constituted at the time the most complete catalogue of stars near the poles ever attempted. For part 9 she closely studied sixty-five early catalogues and examined over a hundred others. For part 10, stars near the south pole, she used altogether forty-two catalogues, some dating from 1750. She also made a long series of calculations of the orbit of Eros (shown on three series of Harvard plates, 1893–1896), which led to the prediction of its future path. Much of her other work or that done under her supervision appears in volumes 37 and 38 of the *Annals.* At the time of her death on January 3, 1904, after twenty-eight years at the Observatory, she was preparing a supplementary catalogue of stars near the south pole, photographed on Arequipa plates, and was supervising the reduction of some earlier investigations by Rogers on star positions. By her own development as a scientist and her lasting contributions to the stellar program of the Observatory she demonstrated convincingly that astronomy would do well to call on the hitherto untapped skills of women.

Like Miss Winlock, the fourth member of the women's corps belonged to the Observatory family. In 1877 William Bond's daughter Selina, having been defrauded of her small capital by a rascally trustee, applied to Pickering for work. She was not altogether inexperienced, since during her father's and brother's administrations she had assisted them in some of their computations. For a year and a half she desperately importuned Pickering for employment, as copyist, computer, anything that would allow her to earn enough to live on. Though troubled by her plight and mindful of the debt owed the Bonds, Pickering could make no addition to the staff until the successful outcome of his campaign for funds. In February 1879 he engaged Miss Bond as a computer, but allowed her to work at home, chiefly on the zone work under Rogers's direction. When she had to be absent from Cambridge, Pickering sent her the papers and books she needed, but always left her free to work at her own pace. In 1888, learning of her illness, he offered to advance her wages out of his own pocket, and a few years later, when she was in desperate need of rest but fearful of losing her position, he assured her that it was safe and that she must take as much time off as necessary. On this kindly basis, though living meagerly and proudly refusing all aid from her none-too-prosperous nieces, she maintained her independence for twenty-seven years as a useful staff computer. In 1906, as she approached her seventy-fifth birthday, she still eked out a livelihood of about $275 a year plus a small legacy that yielded $75, but, as she was ineligible for any Harvard pension, Pickering with Elizabeth and Catherine Bond (George Bond's daughters) entered into a benevolent conspiracy to retire her with dignity. With $2500 somehow provided by the two nieces (they had refused Pickering's offer to contribute about a fourth of the sum) and supplemented by unrestricted funds of the Observatory, Pickering submitted a plan to the Corporation that would give Miss Bond a "salary" of $500 for life.

In presenting the proposal to President Eliot, he offered two suggestions—that the President and Fellows of Harvard act as trustees of the annuity and that Miss Bond be named Emeritus (hastily corrected to Emerita!) "in consideration of the distinguished and long-continued services to astronomy of her father, her brother, and herself." He hoped that the move would establish a precedent, he wrote in his report, for "such appointments, in the case of women, have been rare." Miss Bond died in 1920, proud of her unique title and happily ignorant of the compassionate ruse that not only secured her future but demonstrated Pickering's respect for the name of Bond, his generous personal response to need, and his persistent efforts to gain recognition for women's work in science.

To the decade of the 1880's belongs the true beginning of Pickering's program for the recruitment of women. By the close of his administration in 1919 he had opened the field of astronomy to nearly forty such pioneers.[6] For a few years limited funds kept the number of women assistants small—in 1881 there were but six—and their work consisted chiefly of simple copying, computing, or tabulating. In 1886, however, with the expansion of Pickering's program in celestial photography and substantial additions to the Observatory's resources, he could both add to the staff and give it a more important role. A large collection of plates, the results of his photographic experiments with the 8-inch Bache telescope, required examination, measurement, and reduction. For a decade a half dozen women, starting as complete novices, had shown steadiness, adaptability, and acuteness of observation, and there was every reason to believe that others could do as well. Acting on this conviction, between 1886 and 1889 alone Pickering augmented the staff by eleven new members, and as word of the Observatory's policy spread, often via glowing newspaper articles, his correspondence bulged with applications. Many of the candidates were touching in their naïve expectation that the hospitable doors of the Observatory would swing wide at their approach. Eager young girls just out of high school or college declared their intention of devoting their lives to astronomy, and donors and friends of the Observatory applied on behalf of protégées, some of them in genuine need of work. One desperate aspirant, writing from Malmö, Sweden, described herself as a middle-aged devotee of astronomy, but except in Paris where a "marvel" had happened (she referred to the exalted position of Dorothea Klumpke[7]) she had been unable to persuade anybody in Europe that a woman astronomer was anything but a monstrosity. She had tried London, but saw no stars, nothing but rain or fog. She had worked in Jamaica on the zodiacal light. Could Pickering use her services?

He took immense pride in his liberal policy, answered applications immediately, and suggested possible independent investigations. In fact, as early as 1882, he had prepared a pamphlet, "A Plan for Securing Observations of the Variable Stars," expressly to interest women volunteers in astronomy. "The criticism is often made by the opponents of higher education for women," he had written, "that, while they are capable of following others as far as men can, they originate almost nothing, so that human knowledge is not advanced by their work. This reproach would be well answered could we point to a long series of such observations."[8] But he could offer little hope to applicants who needed to work for money. Never were even his increased funds sufficient to provide all the assistance he would have welcomed, for as hundreds of photographs of the sky accumulated, the burden on the staff increased. While their enthusiasm and devotion to science were most gratifying, he commented, obviously their health would be jeopardized if this pressure continued. In report after report he urged the Visiting Committee to find funds for more assistants to mine the rich vein of stellar information the photographs yielded. No such collection existed in any other observatory, he declared, nor any such possibilities for learning the past history of known celestial objects and for investigating the new as fast as they were discovered. He likened the collection to a library, each book of which was unique, fragile, and priceless in what it would show if adequately studied. As things stood, it was as if a vast library had no more than a dozen readers.

The women received 25 cents an hour, sometimes increased to 30 or 35 cents. They usually worked six days a week, averaging seven hours daily, which they could divide by spending five at the Observatory and two at home in afternoon or evening. On full time they averaged $10.50 a week, hardly an amount that encouraged high living, but, in a period of restricted opportunities, not to be scorned. The emergence of the typist, the secretary, and the business or professional woman belonged to the future. The low pay troubled Pickering, who protested against the injustice of using skilled personnel year after year on the same wage scale. "It is very hard for men or women," he wrote, "who begin when young to put their best energy into a work, to find after a few years, when it is too late to change their occupation, that as their expenses increase, there is not a corresponding increase in salary. This is especially the case in a subject like astronomy, in which the number of similar positions is limited." He regretted that Harvard's new salary increases to the regular faculty did not apply to his staff.

Yet if the pay was meager, the pleasant atmosphere of the Observatory and the amicable personal relations, thanks primarily to Pickering's own spirit and influence, were additional compensations. More than one of his assistants has praised his manner toward them. "He treated them as equals in the astronomical world," wrote Annie Cannon, who was in an excellent position to judge, "and his attitude toward them was as full of courtesy as if he were meeting them at a social gathering." Margaret Harwood recalls his gallantry—"a true Victorian gentleman in his attitude towards women and to everyone, men and women alike. One felt at ease with him—it was easy to talk to him." He saw to it that the rooms in which the corps worked were pleasant, even home-like, with flowered wallpaper fashionable in the period, mahogany writing tables, and conveniently placed shelves for notebooks, reports, and catalogues. Star maps and portraits of famous astronomers decorated the walls. A contemporary photograph (Fig. 49) shows a group of eight women seated informally about the room, two or three busily recording, some scrutinizing plates through magnifying glasses, still others apparently consulting reference books. To the modern eye, familiar with rows of typists and clerks at rigidly fixed steel desks, it is a rather charming

Fig. 49. A group of women computers, directed by Mrs. Williamina Fleming (standing). The seated figure in the front is Miss Evelyn Leland; at left rear, Miss Antonia Maury.

scene of quiet concentration and serenity. To Mrs. Draper, however, the picture of these intent workers, soberly clad in black, hardly seemed to do them justice—"a forlorn looking lot," she complained, "not at all up to the American standard of attractiveness, while in reality some of them are very nice looking."

After 1886 two workrooms seldom held fewer than fifteen occupants, and the numbers increased as the great stream of Bruce and other plates of the Draper and Boyden photographic programs flowed into the Observatory. The incredible labors of this devoted band are permanently recorded in many reports and volumes of the *Annals,* but, because the majority were overshadowed in their own time by the few who achieved wider renown, they have perhaps never received sufficient recognition outside the Observatory. Only a few can be named even here—Louise Winlock, who joined her sister in 1886 to carry on until 1915 an honored

The Harvard College Observatory

Observatory name; Louisa Wells, Edith and Mabel Gill, Evelyn Leland, Florence Cushman, Ida Woods, Lillian Hodgdon, Arville Walker—all of whom in various capacities not only served during a large part of the Pickering period, but brought their valuable experience to his successor.

The four stars of the first magnitude in this constellation of women, however, were Williamina Paton Fleming, Antonia Caetana Maury, Henrietta Swan Leavitt, and Annie Jump Cannon. As detailed accounts of their scientific work appear elsewhere in this book, only brief biographies and a general statement of their contributions are given here.

Williamina Paton Stevens (Fig. 50) was born in Dundee, Scotland, in 1857.[9] In her early teens, following a common practice there, she became a pupil-teacher. With her husband, James Orr Fleming, she arrived in this country in December 1878 and settled in Boston. In need of immediate employment, she became a second maid in the Pickering household. In 1879, struck by her obviously superior education and intelligence, Pickering offered her part-time work in the Observatory. In 1881, as a permanent member of the staff, she began the remarkable career that spanned three decades, during which she held one of the most important posts in the Observatory and became one of the best-known women astronomers in the world. Her arrival on the scene at the time of Pickering's first experiments in stellar photography gave her the kind of experience that by 1886 fitted her to take a major part in the organization of the Henry Draper Memorial program and to assume the management of its many mechanical details. To her Pickering entrusted the enormous responsibility of examination, physical care, classification, and indexing of the growing number of photographic plates in Harvard's unique collection. The results of some of her earliest research on these plates appear in volume 18 of the *Annals:* "Detection of New Nebulae by Photography" and "A Photographic Determination of the Brightness of Stars." Her first major achievements, however, occupy volumes 26 and 27, published in 1890: "Description and Discussion of the Draper Catalogue" and "The Draper Catalogue of Stellar Spectra," which embody the results of the initial investigations under the Henry Draper Memorial.

In addition to such original work and care of the photographic collection, Mrs. Fleming gave considerable attention to the preparation of the *Annals* and other publications and saw them through the press.

Fig. 50. Mrs. Williamina Paton Fleming, curator of astronomical photographs, 1881–1911.

She also supervised the women's work and recruited and interviewed new applicants. Having been convinced by her own unique experience and by the devoted application of the staff that women were especially apt in astronomical work, she tirelessly pressed their case. One of her best opportunities to do so came in 1893 when she was invited to participate in the Congress of Astronomy and Astrophysics at Chicago's Colum-

393

The Harvard College Observatory

bian Exposition. In an impressive paper, "A Field for Women's Work in Astronomy,"[10] she not only reviewed the achievements of women at this Observatory but emphasized the greatly enhanced opportunities that astronomical photography had lately opened up to them elsewhere. At Harvard alone, she said, seventeen women were now engaged in studying photographs of the sky made at Cambridge and Arequipa. They had measured thousands of stars, examined existing catalogues for comparison, computed or tabulated innumerable observations, searched intensively for new objects, and made some significant discoveries of their own. Moreover, the gift of one woman had made possible the Henry Draper Memorial, which the gift of another, the $50,000 Bruce telescope, would immensely enrich.

In a second essay Mrs. Fleming pointed to still more promising vistas. Women could now take excellent courses in astronomy not previously available. For example, Mrs. John Whitin had recently established a student's observatory at Wellesley, unmatched in equipment by any other of its kind, and at Smith, Mount Holyoke, and Vassar similar instruction was being offered and eagerly received. In fact, Mrs. Fleming concluded, the experience of the past decade had conclusively shown not only that women had a natural aptitude for astronomical work but that they had already made positive contributions to our knowledge of the universe.

In 1899 Mrs. Fleming received the official title of Curator of Astronomical Photographs, the first such Corporation appointment of a woman at Harvard, and in 1906 the Royal Astronomical Society elected her an honorary member, a distinction shared by only three other women—Lady Huggins, Mary Somerville, and Agnes Clerke. Wellesley made her an Honorary Fellow in Astronomy (Professor Sarah Whiting had been unable to persuade her trustees to confer an honorary degree instead), and although Pickering was unsuccessful in winning the Bruce medal for her, she did receive shortly before her death the gold medal of the Astronomical Society of Mexico. During the last years of her life her health failed, but in her strong personal attachment and loyalty to Pickering and her deep sense of responsibility for the uninterrupted work of the Observatory, she kept to a schedule of long hours, refused all invitations to attend meetings, denied herself vacations, and for thirty years hardly missed a day. When she died on May 21, 1911, she left unfinished enough material to fill several quarto volumes of

Fɪɢ. 51. Miss Antonia Caetano Maury, research associate, 1888–1933.

the *Annals*. Messages of sympathy that poured in from all parts of the world unanimously lamented the great loss to astronomy and to Pickering personally, who for so many years had relied on her sturdy devotion to him and to the welfare of the Observatory.

Antonia Maury (Fig. 51), the most original as well as the most elusive personality among the women astronomers at Harvard, joined the staff

The Harvard College Observatory

in June 1888.[11] A granddaughter of John W. Draper and niece of Henry Draper—her mother was his sister Virginia—she was born in Cold Spring, New York, on March 21, 1866, and was graduated from Vassar in 1887. Her introduction to Pickering actually came through her father, the Reverend Mytton Maury, a cousin of Matthew Fontaine Maury. Contemplating a move from Goshen, New York, to the vicinity of Boston, in March 1888 he called at the Observatory ostensibly to inquire about the Draper Memorial program but actually to find out if Pickering would employ his daughter as a computer. Although the appropriateness of the idea appealed to Pickering, he demurred at first, reluctant to offer a Vassar graduate such low pay as 25 cents an hour. He wrote to Mrs. Draper to ask about the girl's qualifications and her possible interest in such work. Mrs. Draper, whose attitude toward her niece by marriage remained somewhat ambivalent throughout, had no illusions about her brother-in-law's tendency to run his daughter's life. She doubted that Antonia had been consulted at all or that, having been prepared to teach chemistry and physics, she would have the slightest interest in such dull routine as computing.

In fact, however, Antonia accepted her father's suggestion at once. She wrote Pickering that she would welcome a chance to join a project begun in her uncle's name but would have to postpone doing so for some weeks. After waiting for a reasonable time without further word, Pickering concluded that Miss Maury did not want to commit herself to a long indeterminate period of work and wrote Mrs. Draper that he would have to look elsewhere. He was impatient to begin a close investigation of the spectra of the brighter northern stars. On this occasion Mrs. Draper defended her niece: "The girl has been very busy, for when she arrived at Waltham her father had taken no steps toward getting a house, so she quietly took the matter in hand herself, and found a house, and settled the family in it."

Although, as her subsequent experience at the Observatory showed, Miss Maury was by temperament not entirely suited to the day-by-day tedium that was accepted as a matter of course by the less creative members of the staff, her scientific training enabled her to plunge at once into the complex problems of stellar classification. Pickering assigned to her the examination and analysis of plates of all the bright stars north of declination $-30°$. Within a few weeks of her arrival he reported to Mrs. Draper that he was highly satisfied by Antonia's work, and in August he wrote that her catalogue of the spectra of bright stars was growing daily and needed only a few remaining touches to complete it. He next assigned her to work on the spectrum of Sirius. She found the astonishing number of over 500 faint lines, Pickering wrote Mrs. Draper, and some months later he was pleased to report Miss Maury's "discovery of no little importance." She had aided in confirming his own discovery that ζ Ursae Majoris was a double star—the first spectroscopic binary. In that same year also she detected a second spectroscopic binary, β Aurigae, a discovery that led to the overstated comment of one authority, "If β Aurigae does not constitute a satisfactory memorial, I am at a loss to conceive the kind of tombstone which the relatives of a man of science would prefer."

Throughout 1890 Miss Maury continued her investigation of the spectra of bright northern stars, which Pickering intended to publish in a forthcoming volume of the *Annals*. He did not anticipate, however, that seven years would elapse before it appeared. In 1891 she left the Observatory for a teaching post, and thereafter the chronology of her association with the Observatory becomes difficult to follow. She had apparently been restless and unhappy with her status even during 1889, for in October of that year her father asked Pickering to recommend her for the curatorship of the Natural History Museum of Nova Scotia. In March 1892, disturbed by receiving some intimation that her unfinished work would be assigned to someone else, she wrote Pickering:

I have had in mind for some time to explain to you how I feel in regard to the closing up of my work at the Observatory. I am willing and anxious to leave it in satisfactory condition, both for my own credit and in honor of my uncle. I do not think it is fair to myself that I should pass the work into other hands until it is in such shape that it can stand as work done by me. I do not mean that I need necessarily complete all the details of the classification, but that I should make a full statement of all the important results of the investigation. I worked out the theory at the cost of much thought and elaborate comparison and I think that I should have full credit for my theory of the relations of the star spectra, and also for my theories in regard to β Lyrae. Would it not be fair that I should, at whatever time the results are published, receive credit for whatever I leave in writing in regard to these matters?

Her anxiety to protect her theories perhaps arose from the statement in Pickering's report for 1891 that Mrs. Fleming had made "the interesting discovery . . . that the bright lines in the spectrum of β Lyrae change their position in a manner somewhat like the doubling of the dark lines in β Aurigae." However that may be, Miss Maury informed him that she intended resigning from her present position and returning to Cambridge to complete her work.

Pickering was invariably scrupulous in acknowledging the work of his assistants, and in Miss Maury's case had been especially careful to do so. He valued her skill highly, but by leaving an important piece of investigation in midair yet being unwilling to relinquish it to others she had placed him in a quandary. He felt a deep obligation to Mrs. Draper to show results as rapidly as possible. Moreover, he was accustomed to the regular schedule of the other women on the staff. Every day, one of them recalled, he came to their workrooms to find out what progress they had made—if measuring stars on photographic plates, how many and at what rate: "He timed us, he really seemed to keep tabs on people." To Miss Maury, by her own admission naturally unsystematic, such supervision could only have been irksome.

Nevertheless, Miss Maury returned to the Observatory in the spring of 1893. Mrs. Draper acquiesced in the arrangement, but having been annoyed by her niece's "dilatory habits," and doubtless not sufficiently impressed by her fundamental contribution to the spectral work, urged Pickering to treat her as if she were a stranger, on a strictly business basis. "Between ourselves," the aunt wrote, "if Antonia would write up the work she has done,—it seems as if for the future we might supply her place more satisfactorily—she is not a valuable member of the corps . . . I shall be happy when you are rid of the annoyance." Perhaps now that she was back "there will be some hope of her piece of work being finished eventually and we will bid her goodbye without regret."

For a year and a half Miss Maury stuck valiantly to her task, but toward the end of 1894, evidently suffering from fatigue and imagined slights[12] at Pickering's hands, she once again left the Observatory. In a harsh letter unjustly accusing Pickering of mistreating his daughter, her father insisted that she must now be relieved of all further responsibility and announced that he was sending her abroad with her brother. At the same time, he did not mind asking Pickering for letters of introduction to foreign astronomers stressing her relationship to her grandfather, John Draper, and her cousin, Matthew Fontaine Maury. Pickering complied, duly identifying her family, but took care to point out a distinction more to the point—her outstanding work on stellar spectra and her discovery of the duality of β Aurigae. Before sailing in January 1895, she sent Pickering two farewell letters in which she expressed her embarrassment at her father's complaint, explained her own difficulties, and apologized for her delays. She hoped that, though their association was at an end, she could still keep his friendly regard.

As it turned out, her connection did not end then or for many years thereafter. Immediately upon her return in December 1895 she wrote to ask about the state of her manuscript and offered to fill in any remaining gaps and assist with the proof. It is a mark of her astonishing precision and deep involvement in the subtleties of classification that she had retained in her memory annotations she still wished to supply. Her catalogue finally appeared in 1897, in volume 28, part 1, of the *Annals*, "Spectra of Bright Stars Photographed with the 11-inch Draper Telescope as Part of the Henry Draper Memorial."

For some years after the publication of this volume, Miss Maury, apparently either preferring freedom from rigorous schedules or unable to find the kind of appointment she wanted, lectured on astronomy to both popular and professional audiences. She gave four lectures at Cornell in 1899 and others in Brooklyn, New York City, and elsewhere. During intervals between lectures she accepted private pupils and an occasional teaching position. In 1908 Pickering recommended her highly for an adjunct professorship in physics and astronomy (at an unnamed institution) "as a painstaking investigator" and an important contributor to Observatory research, but Miss Maury chose to resume her investigation of spectroscopic binaries, a subject of increasing interest to her. In December of that year she returned to the Observatory as a research associate, and, although she kept up her outside activities, her chief interest remained there. After Pickering's death, Professor Solon Bailey, unable to afford even the minor expense of her irregular hourly pay, tried to place her at the Victoria Observatory, British Columbia. The director, J. S. Plaskett, to whom her qualifications were well known, would have welcomed her assistance in his research on radial velocities, but she was now fifty-three years old, and he feared that the physical inconveniences of the location and the high cost of living there would make such a radical move impossible for her. In the end she remained at or near Cambridge. In 1919–20 she received the Pickering Fellowship for women[13] and worked on spectroscopic binaries, with special attention to β Lyrae. Sometime during the 1920's she had taken charge of the Henry Draper Memorial Museum at Hastings-on-Hudson, but each year until her official retirement in 1948 she returned to the Observatory to study photographs of β Lyrae for comparison with her earlier predictions of its spectral changes. In 1933 the results of her later work on this star and on the spectroscopic binaries μ' Scorpii and V Puppis appeared in the *Annals,* volume 84, parts 6 and 8. Miss Maury died January 8, 1952, at the age of eighty-five. She has long been recognized here and abroad as an important contributor to sidereal astronomy and as one of the most distinguished women astronomers this country has produced.

Like Antonia Maury, Henrietta Leavitt (Fig. 52) not only made significant discoveries[14] but was the cause of literally far-reaching discoveries by others (see Chapter X). She was born in Lancaster, Massachusetts, July, 4, 1868, and attended Radcliffe College, then still known as the Society for the Collegiate Instruction of Women. After graduation in 1892 she divided her time between travel and school teaching, but in 1895 she returned to Cambridge and became a volunteer research assistant and advanced student of astronomy at the Observatory. Like the other women on the staff, she began with very humble tasks, but soon showed unusual aptitude for the work. Her name appears for the first time in Pickering's report for 1896: "An interesting investigation has been made by Miss H. S. Leavitt on the photographic brightness of circumpolar stars." In 1900 she had completed the first draft of a careful study on the photometric measurements of variable stars when she was called to Wisconsin by a family crisis. She remained there for two years, but in 1902, ready to take up her investigation again, she wrote Pickering, "I am more sorry than I can tell you that the work I undertook with such delight, and carried to a certain point, with such pleasure, should be left uncompleted." Her interest in astronomy had not waned. Could she carry on from such a distance?

Pickering, who had been impressed from the first by Miss Leavitt's exceptional ability, urged her to return to Cambridge at his expense, proposed a slight increase in pay, and promised that if she had to leave again she could take with her all the materials she would need to carry on her work. This offer she accepted at once, and in August 1902 joined the staff permanently. As a steady stream of photographs made in

Fig. 52. Miss Henrietta Swan Leavitt, staff member, 1902–1921.

Arequipa with the Bruce telescope was arriving at the Observatory, Pickering assigned to Miss Leavitt the study of the brightnesses of variable stars shown on them. In the course of her investigation of these plates she discovered hundreds of variables, the major portion of them in the Large Magellanic Cloud. It was her assiduous measurement of twenty-

401

five stars in the Small Magellanic Cloud, all Cepheid variables, that led eventually to her most important achievement, the discovery of the relation between their magnitudes and their periods. Known as the "period-luminosity law"—the brighter the star, the longer the period—it was the basis on which Harlow Shapley made the first sound estimates of the scale of the universe (Chapter X).[15]

In 1908 Miss Leavitt, after her intensive study of the variables, fell ill and was forced to return to her family home in Wisconsin. During a prolonged convalescence she nevertheless kept up her work on the photographic material Pickering supplied from Cambridge. Upon her return she continued her study of the photographic magnitudes of the North Polar Sequence, so that by February 1912 Pickering issued Harvard Observatory Circular 170, "Adopted Photographic Magnitudes of 96 Polar Stars," which he believed represented a true scale. Her extensive monograph, "The North Polar Sequence," published in the *Annals,* volume 71, in 1917, gave even more nearly definitive results. This volume also contains her list of sequences for the forty-eight "Harvard Standard Regions" into which, for convenience of investigation, Pickering had earlier divided the sky.

In 1913 Pickering accepted the chairmanship of a subcommittee of the International Committee on Photographic Stellar Magnitudes and turned over to Miss Leavitt the portion of the work assigned to Harvard. Her results on the northern zones, "Standards of Magnitude for the Astrographic Catalogue," were published in the *Annals,* volume 85, No. 1, but the similar work she had begun on the southern zones she did not live to complete. While engaged in the investigations for the Astrographic Catalogue, with the assistance of others on the staff, she had also given considerable attention to the problem of standard photographic magnitudes for the "Kapteyn Selected Areas."[16] These included 206 small regions distributed over the sky, plus 46 others, for which Pickering had supplied Kapteyn with photographs taken at Arequipa and Cambridge. The first of three volumes of this "Durchmusterung of Selected Areas" appeared in volume 101 of the *Annals* (1918), but the subsequent volumes, 102 and 103, were not published until 1923 and 1924 under the supervision of Harlow Shapley. Miss Leavitt, who had made the Observatory work virtually her whole life, died in Cambridge on December 12, 1921, at the age of 53.

The contributions of Mrs. Fleming, Miss Maury, and Miss Leavitt are now perhaps best appreciated by the professional astronomer. Recognition of the fourth member of this notable quartette has extended far beyond observatory walls. Annie Jump Cannon (Fig. 53), author of nine monumental volumes of the Henry Draper Catalogue, joined the staff in 1896.[17] She was born in Dover, Delaware, December 11, 1863, and at the age of 17 entered Wellesley College, just five years after its opening. Already attracted to science, she had the good fortune to come under the immediate influence of Sarah Frances Whiting,[18] the dynamic professor of physics and astronomy, who through her association with Pickering became Miss Cannon's direct line to the Observatory. Like other members of the Wellesley faculty, Miss Whiting was the personal choice of the founder of the college, Henry Fowle Durant, whose basic intention was to establish an institution administered by and for women. "Women can do the work," Miss Whiting recalled his saying; "I give them the chance."

The particular chance he gave them, immediately after their appointment, was time to study the curricula and methods of other institutions and as long a period as they needed to acquire the additional training not otherwise readily available to women. Under the influence of his close friend and Wellesley's other great benefactor, Eben Horsford, professor of chemistry at Harvard, Durant gave high priority to science, for which he planned in great detail. Instead of traditional textbook instruction, he insisted on experimental laboratory science, even though it required costly apparatus that in many instances would have to be made to order. For special work in physics he sent Miss Whiting to the first students' physical laboratory established in this country, founded and directed by E. C. Pickering at MIT.

That institution itself had no women students, but as early as 1867 one of its professors had written to President Rogers:

Application has come from one young woman, a rather remarkable teacher, who desires to avail herself of this Institute. I was sorry to have to reply that nothing was open to her . . . . There is a large and increasing class of young women who are seeking for something more systematic in the way of a higher education. If we continue a special technical school, ours will not be the place for them; but if we should expand into a modern university, and I am confident there is room for one, by taking the bold step of opening our doors freely to both sexes, I believe we should distance all competitors. It is a step sure to be taken somewhere.[19]

Fig. 53. Miss Annie Jump Cannon, shown in her Oxford robes, 1925; staff member, 1896–1932, curator of astronomical photographs, 1911–1932.

A Field for Women

Although the Institute postponed that step for a long time, Pickering himself took it. He invited Miss Whiting to attend his lectures as his special guest. Four times a week she journeyed to Boston to hear him and to study the apparatus he had devised and ordered from local workmen. In addition to that aid, she traveled to Philadelphia to look over the exhibits at the Centennial Exposition, where, curiously enough, another link with Harvard was created. Her guide was Henry Draper, not the man to find it strange that a woman should be interested in science.

With such assistance from Pickering and consultations with other professors (one of whom was almost shocked to find himself talking about the Wheatstone bridge with a lady), Miss Whiting felt ready to establish her own physics course. In the converted organ loft on the fifth floor of College Hall, the elaborate all-purpose building that housed chapel, library, gymnasium, infirmary, parlors, dining room, classrooms, dormitory for teachers and students, kitchen, and pantries, she set up her new apparatus for experiments in light, heat, and electricity. It was the second such students' science laboratory in the United States and the first for women. Here Miss Cannon studied physics, then required, and took her first course in astronomy, given originally as applied physics. For astronomy, until the establishment of the Whitin Observatory in 1900, there was very little equipment beyond a globe and a portable 4-inch Browning telescope, but Miss Whiting at Pickering's invitation kept up with astronomical developments at Harvard and saw to it that her class missed no current phenomenon. Miss Cannon has recalled vividly Miss Whiting's marshaling her young astronomers day after day to view the Great Comet of 1882 as long as it remained visible.

After graduation in 1884 Miss Cannon returned to her home in Delaware, where she remained for almost a decade. Sometime during that period she fell ill with a severe case of scarlet fever that impaired her hearing. In 1892 she viewed a solar eclipse in Spain, and in 1894 returned to Wellesley to assist Miss Whiting in a physics course that now included experiments with x-rays. In the spring of 1895, however, she decided to take up astronomy in earnest and arranged, with Pickering's assistance, to enroll in the course at Radcliffe labeled "Practical Research." As he had suggested supplementing it with actual observation, she began in February 1896 to spend her free time at the Observatory, where she studied photographic plates in the daytime and in the evening

405

The Harvard College Observatory

observed variable stars. He saw to it that Miss Leavitt furnished information about the plates and that a telescope or two became available for work on the variables.

At the end of that academic year Miss Cannon gave up her Wellesley appointment and became a permanent member of the Observatory corps, at the same time working for her Radcliffe degree, for which in June 1898 Pickering unreservedly recommended her:

Miss A. J. Cannon has devoted about five hours a day during the past year to astronomical research at this Observatory. On clear evenings she has also spent from one to three hours observing with the west equatorial, and has made 365 observations of variable stars in 63 evenings. During the day-time she has studied the spectra of the southern stars, in continuation of the work done by Miss Maury and published in the Annals, Vol. XXVIII, Part 1. She has examined 1400 photographs of about 400 stars, classified them according to their spectra. This work is of excellent quality and will be published in our Annals. I trust, therefore, that proper weight may be given to it in considering Miss Cannon's application for a degree.

That summary of one year's work was the forerunner of similar productivity which marked her brilliant career at the Observatory. Although she gave considerable attention to both visual and photographic study of variable stars, that interest did not hamper her activity in other areas. Assigned to the study of bright stars south of declination −30°, by 1900 she had completed her classification of the spectra of 1122 southern stars, which was published in 1901 in the *Annals,* volume 28, part 2.

In 1903 Miss Cannon produced her "Provisional Catalogue" of 1227 variables and supplemented it shortly afterward by two additions. She continued meanwhile classifying stars, many fainter than those given in volume 28. Her bibliography of variables, which she had been compiling for some time, was nearly ready for publication when, in order to avoid competing with a similar plan of the Astronomische Gesellschaft, she decided to abridge it, though she continued indexing information about every star observed in the Harvard photographic library. Her second catalogue of variable stars, the most complete of its kind in print at the time, appeared in volume 55, part I, in 1906. Part II contained "Maxima and Minima of Variable Stars of Long Period." When Pickering drew up his plan for a revised Draper Catalogue to take in all parts of the sky, including the spectra of some 50,000 stars of the eighth magnitude and brighter, he turned the responsibility over to Miss Cannon.

She began by classifying 5000 stars a month, and each year, as her expertness grew, the number increased. Pickering's appreciation of her skill and great productivity did not stop with tributes in his reports. On October 5, 1911, he wrote to President Lowell:

> If it meets with your approval, will you ask the Corporation to appoint Miss Annie Jump Cannon Curator of Astronomical Photographs, in the place of Mrs. Fleming? During the past few months she has performed a portion of the duties of this position in a very satisfactory manner . . . Miss Cannon is the leading authority on the classification of stellar spectra, and perhaps on variable stars.

The president replied two days later, "I am not sure that it would be wise to give Miss Cannon a Corporation appointment," but offered to appoint her himself or authorize Pickering to do so. Pickering acquiesced, chiefly because it gave him a certain freedom about salary increases and vacations without troubling the Corporation, but he wanted to know if this arrangement would allow her name to appear in the University catalogue and give her the usual social advantages of a Corporation appointment. To this inquiry Lowell answered October 11, 1911:

> As I understand Miss Cannon's name has not hitherto appeared in the catalogue, and as it does not seem to me a matter of very great importance whether it does or not, it ought to be treated like that of any other department for work of similar character. I always felt that Mrs. Fleming's position was somewhat anomalous, and that it would be better not to make a regular practice of treating her successors in the same way. I think it would be rather better that Miss Cannon's name should not appear in the catalogue.

Pickering appointed Miss Cannon Curator of Astronomical Photographs in October 1911, with a salary for the first year of $1200. Like other salaried personnel, she was expected to work seven hours a day, eleven months of the year. Her duties included, under the director's supervision, responsibility for the care and use of photographic plates and such astronomical investigations as he stipulated. Between 1911 and 1915 Miss Cannon devoted virtually her entire time to the Henry

Draper Catalogue and had become a leading authority on stellar spectra. The chairman of the Visiting Committee wrote of her:

> At the present time she is the one person in the world who can do this work quickly and accurately. Through familiarity with it she has acquired such a perfect mental picture not only of the general types, but of their minute subdivisions, that she is able to classify the stars from a spectrum plate instantly upon inspection without any comparison with photographs of the typical stars.
>
> This gives her great speed in classification, amounting to no less than 300 stars an hour . . . At the same time her great speed in no way limits the accuracy of her estimates. From an investigation of her probable error upon parts of the sky where she has made independent duplicate estimates, it is found that her average deviation amounts to only $\frac{1}{10}$ of a unit.

The report went further. It reminded the administration that despite the prestige she had brought to the Observatory she still had not received the official recognition she merited:

> It is an anomaly that though she is recognized the world over as the greatest living expert in this area of investigation, and her services to the Observatory are so important, yet she holds no official position in the University. At present, as her appointment is by the Observatory, and not by the University, her name does not appear in the Catalogue, or any other official publication of the University. It is the unanimous opinion of the Visiting Committee that the University would be honoring itself and doing a simple act of justice to confer upon her an official position which would be a recognition of her scientific attainments.

Miss Cannon did receive an official appointment—in 1938!

The first volume of the revised Henry Draper Catalogue (*Annals*, 91) appeared in 1918. Miss Cannon's assistants were Grace Brooks, Alta Carpenter, Florence Cushman, and Edith and Mabel Gill. For the salary paid to two of these aides through the nine volumes the Observatory was indebted to Mr. George Agassiz.[20] Volumes 92 and 93 followed closely, but, although nearly all the material for the remaining six was ready, Pickering lived to see only the first three completed. The next four followed in 1919–1922, prefaced by Solon Bailey, and the last two (1923–1924) appeared under Harlow Shapley's direction. It is worth noting that the cost of publication was borne by a generous woman donor, Mrs. James R. Jewett, a member of the Visiting Committee of the Observatory, who also endowed a fellowship.[21]

When completed, the Henry Draper Catalogue listed 225,300 stars classified according to their spectra. The Extension, volume 100 of the *Annals,* also carried out by Miss Cannon between 1925 and 1936 with the able assistance of Mrs. Margaret Mayall, provided information on the spectra of still fainter stars.

In 1936, at the age of seventy-three, Miss Cannon remained as active as ever, undertaking in that year, at the special request of the Royal Observatory of the Cape of Good Hope, the study of 10,000 very faint stars. In honors she held a number of firsts: she was the first woman awarded an honorary degree by Oxford[22] (this was in 1925, after which her name appeared in the Harvard catalogue!), the first woman officer of the American Astronomical Society, the first woman to win the Henry Draper Medal. She was also one of the first women to be elected an honorary member of the Royal Astronomical Society. Honorary degrees were awarded her by the University of Delaware, Groningen University,[23] and Wellesley College. In 1932, having won the final Ellen Richards research prize of $2000, given at various times to women distinguished in science, Miss Cannon generously turned the sum over to the American Astronomical Society to establish the Annie Jump Cannon Prize for women astronomers. The first to receive it was Cecilia Payne (Gaposchkin). In 1938 Miss Cannon was named the William Cranch Bond Astronomer at Harvard. Besides producing her magnificent catalogues, she discovered 300 variables, 5 new stars, and 1 spectroscopic binary. She died on April 13, 1941, eulogized as "the world's most notable woman astronomer and a person beloved by all, both inside and outside the scientific realm."

To the four most prominent members of the women's corps should properly be added a fifth, Margaret Harwood (Fig. 54). Although she remained on the regular Observatory staff only five years, she began her long astronomical career there, undertook several special investigations chiefly at Pickering's behest, spent almost half of every year at the Observatory as a research assistant even while directing her own observatory, at Nantucket, aided in the photographic work at Arequipa during 1923, and still (1969) busies herself with current astronomical projects. She was born in Littleton, Massachusetts, March 19, 1885, and in 1903 entered Radcliffe College, where she studied physics, chemistry, and mathematics. Discouraged at first from registering for astronomy because of its reputed difficulty, she nevertheless determined to tackle the subject, inspired chiefly by Professor Arthur Searle, in whose

Fig. 54. Miss Margaret Harwood, staff member, 1907–1912, director of the Maria Mitchell Observatory at Nantucket, 1915–1951. Miss Harwood is the only woman to hold this position in an independent observatory.

household she boarded throughout her college years. Under his tutelage, besides regular class lectures, she read in theoretical astronomy, practiced computing orbits of comets, and enjoyed "field trips" to the Observatory, where she was permitted to use the 15-inch refractor and learned from Miss Cannon and Miss Leavitt their methods of discovering, identifying, and measuring stars.

Through Professor Searle, whom she calls her "father in astronomy," Miss Harwood became a member of the staff immediately after graduation from Radcliffe in 1907. During her first year she assisted him in computing observations of the zones assigned to Harvard for the revision of the Southern Durchmusterung.[24] She also read proof for a volume of the *Annals* and helped prepare for publication the remaining photometric measures made with the meridian photometer. In her second year Pickering suggested that she give a portion of her time to new study of the so-called Bond Zones of faint stars 1° north of the equator, which had been observed by George Bond and his father between 1852 and 1859. This work, "Bond Zones of Faint Equatorial Stars," was published in the *Annals,* volume 75, part 1, in 1912.

Between 1907 and 1912, among other assignments Miss Harwood assisted Miss Leavitt and Miss Cannon in computation, and, instructed by Miss Cannon, observed variable stars and measured long-period variables. In 1912 she received the first $1000 Astronomical Fellowship of the Nantucket Maria Mitchell Association,[25] which was founded in 1902 by relatives of Miss Mitchell and various alumnae of Vassar College. Under the terms of the fellowship one half-year must be spent at Nantucket, the other at one of the larger observatories for research and study. Except for one year at the Lick Observatory and the University of California, during which she received an advanced degree in astronomy, her free half-years were usually spent at Harvard. In 1914 she began investigations on the light curves of Eros, using both Harvard plates and a number of her own excellent photographs made at Nantucket. In 1916 she became permanent director of the Maria Mitchell Observatory, a position she held with distinction until her retirement in 1957. Since that time she has published her studies on 65 known and 354 newly discovered variables in or near the Scutum Cloud,[26] based on plates of the Maria Mitchell, Harvard, and Leiden collections. She also completed a bibliography of all American articles on astronomy published between 1881 and 1899—a task she undertook at the special request of Donald Menzel, former director of the Harvard Observatory, on behalf of the International Astronomical Union. It was intended to fill the gap in the *Astronomische Jahresberichte* for that eighteen-year period of American contributions.

As a result of Miss Harwood's successful performance during her fellowship and the relation she had established with the Harvard Observatory, the officers of the Maria Mitchell Association made a determined drive for funds to establish a fellowship for the support of a second woman astronomer. To be known as the Harvard Astronomical Fellowship for Women,[27] it would offer a stipend of $500 a year toward study at the Observatory, with special emphasis on stellar research. Pickering had contributed to this fund since 1914. By November 1916 the Committee appointed to complete the drive announced that over a hundred donors from fifteen states had contributed $12,263. At a surprise ceremony on November 16 the Association presented this fund to Pickering in honor of his fortieth year as director of the Observatory and in grateful recognition of his interest in the Association and continuous support of its work during the past ten years. "It is our wish," the Committee declared, "that you will accept it with full power as to its use, and that in the future it will be managed with the same broad-minded fairness to women which has characterized your administration."

President Emeritus Charles W. Eliot, who was present, recalled with some amusement the consternation aroused in various quarters by his "peculiar" appointment in 1876 of a physics professor to direct the astronomical observatory. The enormous development of physical astronomy, he observed, and Pickering's unusual administrative skill had long ago justified the choice. Pickering, surprised and touched by the personal tribute, in thanking the donors said the gift would be a welcome aid to astronomy and the work of women in it. It would "keep green the memory of Maria Mitchell, and will enable one woman after another to advance the science" for which the founders of the Association had done so much.

An unexpected complication arose, however, over the name of the fellowship. On December 13 President Lowell wrote the donors that, while the Corporation appreciated their generosity, since the fund had been placed in Professor Pickering's hands and would not be administered by the governing board of the University, it was considered preferable to omit the name of Harvard from the title. As a result, the

Association duly voted to call it the Edward C. Pickering Astronomical Fellowship for Women. Pickering's first announcement of the fellowship sent to the women's colleges resulted in an astonishing number of applicants. The first fellow was Miss Mary H. Vann, a graduate of Cornell and graduate student at California. The second was Miss Dorothy Block.[28] In 1919–20 it was held by Antonia Maury. Between the year of its establishment and 1952 twenty-eight young women were fellows at Harvard, in some cases for more than a year.[29] Unquestionably the most distinguished recipient was Cecilia Payne (Gaposchkin),[30] a fellow in 1923 and 1924, who came to the Observatory from Newnham College, Cambridge, where she was a student of Eddington's, and whose brilliant career at Harvard has brought her renown not merely as a woman astronomer but as one of the leading contemporary astronomers of the world. The Pickering Fellowship, though not awarded every year, remains the one such prize designated exclusively for women students at the Observatory.

II

No account of the relation of the Observatory to the work of women in astronomy would be complete without brief mention of the encouragement and assistance Pickering gave to those at other institutions. Knowing all too well the obstacles his own staff faced, whether in such petty matters as the right to tickets for special Harvard events or the question of inclusion in the official University catalogue, he responded all the more readily to the needs of women who were struggling to promote astronomy in their own colleges. His part in the training of Sarah Frances Whiting has already been touched upon, but his interest in her subsequent career never wavered. He advised her on the size, quality, and arrangement of instruments at her physics laboratory, and when she began to teach astronomy he regularly welcomed her classes at the Observatory. Finally given an observatory of her own, she depended on him to help choose a suitable site and persuaded him to inaugurate the building at the dedication exercises. Later, he drew on his Advancement of Science Fund (to which he contributed anonymously) to aid her study of variable stars in the hope that his example would influence other donors. "I am not surprised at your cordial response to my requests," she wrote him gratefully; "you have been uniformly helpful to my work at Wellesley."

413

The Harvard College Observatory

Like Miss Whiting at Wellesley, Mary Emma Byrd[31] at Smith College received great help from Pickering. A graduate of the University of Michigan, she arrived at the Observatory in 1882 for a year of special study under his direction. Between 1883 and 1887 she taught mathematics and astronomy at Carleton College and became the first assistant at its new observatory, where, among other duties, she twice daily sent out time signals over 10,000 miles of railway telegraph lines. In 1887 she became the director of the observatory at Smith College. Although her method of teaching astronomy as a laboratory science resembled Miss Whiting's, in other respects her situation compared unfavorably. She had no academic rank; in fact, during her second year she was compelled to correct Pickering's reference to her as "professor": "It is, I understand, contrary to the traditions of this institution to give to any women the title and pay of professor." He must just speak of her as "teacher."

From beginning to end of her work at Smith she sought Pickering's advice. Scarcely a month passed during her first year there without some request she confidently knew he would respond to. When with the assistance of Professor Mary Whitney at Vassar she undertook the determination of the longitude of Smith College, Pickering cooperated by placing observatory instruments at her disposal and published the results in volume 29 of the *Annals*. With scant means for buying sky maps, charts, and other astronomical material, she turned to Pickering, who gave her Harvard publications and obtained from Europe those not customarily sent to obscure observatories. During the summer she used instruments in Cambridge. In preparing her own articles and her textbook, *A Laboratory Manual in Astronomy,* she benefited by his criticism, and when she wanted to submit a paper to *Astronomische Nachrichten* Pickering paved the way by recommending her to the editor. After nineteen years at Smith Miss Byrd resigned in protest against the acceptance by the College of funds from a source she thought "tainted," but it is fair to say that without the support Pickering had given her during her administration her efforts to establish astronomy firmly in the Smith College curriculum would have been far less successful.

With Mary Watson Whitney,[32] professor of astronomy and director of the Vassar Observatory from 1889 to 1912, Pickering, as a trusted counselor, maintained a long correspondence, dating from 1885, when she first visited the Observatory. Her ties with Harvard actually began much earlier, although they did not benefit her professionally at the time. After graduation from Vassar, where she spent three happy years as Maria Mitchell's favorite pupil, she determined to pursue her deep interest in mathematics and astronomy, but in the absence of graduate courses for women she had to take a roundabout way. Benjamin Peirce, at Maria Mitchell's suggestion, invited her to attend his lectures on quaternions and celestial mechanics, along with two fellow students, William Byerly and James Mills Peirce,[33] whose academic vistas she could hardly share. She journeyed to Chicago, where Truman Safford, George Bond's former assistant, welcomed her at the Dearborn Observatory, but despite all her preparation here and (later) abroad, she could find no position worthy of her. At Vassar Miss Mitchell held the only professorship, Smith as yet had no department of astronomy, and Wellesley would not have welcomed a Unitarian. In 1881 she returned to Vassar as Miss Mitchell's assistant, but with only the status of a graduate student and without opportunity for independent research.

In 1887 she accepted Pickering's invitation to spend a year at Harvard, where she was assigned work under Arthur Searle's direction, but remained only a few months. Summoned to Vassar to succeed Miss Mitchell, she nevertheless throughout her directorship kept in close contact with the Observatory at Harvard. In 1901 she became one of Pickering's most consistent observers of variable stars.

Through Miss Whitney her most notable protégée, Caroline Ellen Furness,[34] joined the company of those women whose careers Pickering did much to further. After graduation from Vassar in 1887 and a few years of teaching mathematics, she returned to the college in 1894 as Miss Whitney's special assistant. In 1901, when they began their long series of observations of variable stars, they were guided by Pickering, and in 1903, in connection with a project undertaken for Professor Harold Jacoby at Columbia,[35] Miss Whitney sent her assistant to the Harvard Observatory for reference material nowhere else obtainable. In 1908, on the conclusion of that work, she arranged for Miss Furness to assist Kapteyn at Groningen on his "Plan of Selected Areas," the great star project which Pickering had helped initiate and promote. In 1909 Miss Furness began her independent study of variable stars, which led not only to her editing the results of all the previous observations prepared at Vassar during Miss Whitney's administration, but to the publication in 1915 of her own *Introduction to the Study of Variable Stars*. For this work and for a series of articles on astronomy in *Popular Science Monthly* Pickering gave her full use of all Harvard material she wanted for illustration. He also reviewed the outline of her proposed book and recommended its publication. In the summer of 1911 he welcomed her to the Observatory, where she used instruments not available at Vassar. Having followed her career closely, he recommended her without reservation as Miss Whitney's successor. In 1913 she became director of the Vassar Observatory and in 1915 Maria Mitchell Professor of Astronomy.

Two somewhat peripheral names belong in the roster of women Pickering assisted in their work: Anne Young, niece of the noted astronomer Charles Augustus Young, and Mary Proctor, daughter of Richard A. Proctor,[36] best remembered today perhaps for his provocative book, *Other Worlds Than Ours*. Miss Young, professor of astronomy and director of the John Williston Payne Observatory at Mount Holyoke, was an ardent observer of variable stars, about which she reported to Pickering with great regularity. He in turn sent her numerous Harvard plates, provided her with letters of introduction to European astronomers, supported her nomination for membership in the American Association of University Professors, conferred with her on candidates for the Maria Mitchell and Pickering Fellowships, and at various times even personally undertook repair of her observatory instruments. A long and pleasant correspondence reveals her reliance on his interest in her work.

Miss Proctor,[37] on the other hand, held no academic post. Daughter of an unrivaled popularizer of science, she determined to carry on his work by lecturing, editing a department of *Popular Astronomy* ("Evenings with the Stars"), and sending to newspapers and magazines an almost endless stream of articles on astronomy. For her material she unabashedly turned to the leading astronomers in America, who, in deference to her father and out of genuine respect for her knowledge and her ability to interpret astronomy to the public, readily responded to her requests. In 1896, when first undertaking her columns for *Popular Astronomy*, she wrote to Pickering "in the interest of science" for "something about celestial photography." He sent her a pamphlet and some kind words, on the strength of which she shortly afterward solicited his endorsement for her new lecture circular. Others giving it were Professors Rees, Young, Newcomb, Burnham, and Barnard. Thereupon followed for a number of years a series of appeals, prefaced by her ingenuous hope that with such aid she would "make astronomy popular some day." For her lectures and other work she wanted nothing less than slides of the station at Arequipa, the Observatory at Cambridge, portraits of Henry Draper, Pickering, and Mrs. Fleming, sky charts, star-spectrum plates, the Bruce telescope, the nebulae in Orion, Andromeda, and Argus, the large Magellanic Cloud, and Harvard Circulars on every conceivable subject. She also asked for information about William Pickering's tenth satellite of Saturn, forthcoming eclipse expeditions, illustrative material for her own book (*Giant Sun and His Family*), exact references to sources of various memoirs, names of publishers, and the place of meeting of the American Association for the Advancement of Science. Busy as he was, Pickering not once in all his correspondence with her showed the slightest impatience, questioned her right to any Harvard material or information he could give, or treated her with condescension. It would not have occurred to him that in promoting astronomy she might at the same time be promoting herself. On the contrary, in a curious way, she served to publicize Pickering's conviction that women, given adequate support and recognition, could move into the world of science and by their special qualities of endurance and keenness of observation contribute substantially to its enlargement.

XII

BEYOND THE OBSERVATORY

If a man's influence on his own and a later time can be judged by the nature of his personal achievement and the extent to which he affected and fostered the work of others, no one can question Pickering's impact on astronomical development in the United States and throughout the world. He early staked out the areas in which he proposed to concentrate. He devised and obtained the necessary equipment for them, and by superb administration of resources and staff produced the results that have made more than half the history recorded here. Yet success in a single establishment was not enough. Compelling as was his determination to equal the standard of his predecessors and to move beyond them into the full stream of the "new astronomy," Pickering could content himself with nothing less than the promotion of all science. At the outset of his directorship he began his crusade, waged incessantly until his final illness; even in delirium, as Solon Bailey[1] has written, he was preoccupied with plans for the future. He had two main convictions about science—that it must be well supported if national life were to benefit and that only collaboration among the men and institutions engaged in it could render that support effective. His efforts to organize such cooperation often met with frustration, but he never abandoned them, and no account of his administration would be complete without a review of them.

One of Pickering's first public pronouncements on the subject, made in 1877, just six months after he came to Harvard, appears in his vice-presidential address read at the Nashville meeting of the American Association for the Advancement of Science. Entitled "The Endowment of Research," it dealt, interestingly enough, not with astronomy but with the organization of a new kind of institution for research in mathematics, physics, and chemistry. His description of it in some respects anticipates for the United States the kind of research institute already familiar in Europe, now an accepted feature of American scientific activity, but then a novel concept in this country. Among other advantages, he argued, a centralized institution would permit study of problems too difficult for one man alone to investigate. The institution would draw together a number of men to solve such problems. Science would thereby constantly extend its boundaries and, with proper management, such an institute could become the fountainhead of scientific experiment in this country. "Whoever will supply this want," he wrote, "will leave the name of one who extended human knowledge and benefited his fellow men."

In 1886 Pickering published his "Plan for the Extension of Astronomical Research," the first of several pamphlets on the subject, and in a real sense the key to his whole subsequent program. Each emphasized the need for new funds. Each insisted on the collaboration of astronomers and observatories both at home and abroad. As matters stood, although lavish gifts of expensive telescopes had come to astronomy, the lack of money to pay for observers often kept the telescopes idle. Too many observations remained unpublished for lack of money to print them. Too many able astronomers were deflected from independent research by heavy teaching schedules. It was all a question of finding money, but not, Pickering believed, from government, which had as yet indicated little disposition to support research unrelated to its own utilitarian purposes. Moreover, the Naval Observatory—long a source of deep dissatisfaction to American astronomers—had not demonstrated that the government would always spend appropriations wisely.[2] Pickering could scarcely have foreseen the tremendous impetus that two world wars would give to nationally funded research. Although the Observatory had not neglected the study of the moon, any notion that the government half a century after his death would spend $25 billion for man to land on it Pickering would have regarded as one of the wildest fantasies of Jules Verne. Private support had always been his mainstay. It would in his view continue to be. No vast sums were required to supply the needs of individual astronomers, but if a single large endowment proved impossible, he recommended the plan he had successfully adopted in 1878, an annual subscription by various donors of $5000 for five years. His campaign had produced such striking astronomical results that two great bequests had followed—the Paine legacy of over $300,000 and the Boyden Fund of more than $200,000.

Once the money was obtained, the next essential would be to put it in the hands of skilled managers of endowments, and for this he could suggest none better than the Harvard Corporation. If this choice appeared to be selfish concern for his own Observatory, he hastened to say, he meant to use any undesignated funds to assist other astronomers, and the grants would be made independently of persons or places and would carry no restrictions.

In 1890, thanks to Miss Bruce's experimental gift of $6000 (Chapter VII), Pickering had the great satisfaction of proving that even small sums substantially helped fifteen astronomers—eight in Europe, one in Asia, one in Africa, and five in the United States—an excellent illustration of international cooperation that could well serve as a model. At the same time, having become painfully aware of the lot of the numerous unsuccessful applicants, he was all the more resolved to do everything possible for astronomers everywhere.

In a paper of 1903, "The Endowment of Astronomical Research,"[3] Pickering no longer had to argue theoretically. He proposed setting up an institution affiliated with the Observatory, but one that would be administered by an independent board of leaders in the major fields of astronomy. They would try to find out which areas were most neglected, prevent duplication of research, furnish the necessary instruments, select the best person for a given project, and provide him with what he required to carry it out. In this way, contrary to the practice of trustees of other institutions, who wait to receive applications, the new board would actually initiate research. Neither nationality nor personal favor should weigh. Funds would be safely invested to produce the largest possible income, and for this purpose he repeated his trust in the Corporation of Harvard College. Meanwhile, as a beginning, he invited astronomers to report to him personally what they needed, so that he could refer their requests to potential private donors and to the trustees of the funds he helped administer—the Rumford, the Elizabeth Thompson, the Henry Draper, administered by the National Academy of Sciences, and his own Advancement of Astronomical Science Fund.[4]

In 1904 a supplement to this pamphlet suggested several practical methods of aiding astronomy: fellowships to students; expeditions (limited to carefully chosen and achievable goals); publication of valuable observations and memoirs; grants to astronomers for the employment of assistants; supplements to the income of existing observatories for salaries, buildings, and so forth; and finally—to him the most important item of all—international cooperation. Aside from such specific suggestions, the pamphlet also reveals the kind of resistance Pickering met. It not only reviewed his campaign to date, but also gave the results of a questionnaire he had sent to hundreds of astronomers, here and abroad, for information that might guide further planning: How much money could be spent on astronomy advantageously at this time? What kind of plan seemed most likely to appeal to potential donors? What help would be of greatest benefit to the man questioned? What additional items should be considered?

About a hundred persons replied. From Europe several suggestions proved helpful, but to Pickering's dismay some Americans were highly critical—there was too much Harvard in the plan! Two men flatly refused even to serve on an informal advisory committee to discuss any aspect of the scheme, not, they insisted, because they were jealous of Harvard or doubted Pickering's ability to succeed, but because they objected to domination by any one observatory and to its control over money. In view of such opposition, Pickering saw no recourse but to proceed alone. "As other astronomers have not expressed a desire to aid astronomers elsewhere," he wrote, "there seems to be no objection to making it a part of the policy at Harvard." In fact, he reminded his critics, it had always been so. The statutes of the Observatory, adopted during the administration of the first director, had expressly declared its purposes: "In general, to promote the progress of knowledge of Astronomy and the kindred sciences." Many annual reports and volumes of the *Annals* testified to the steady pursuit of that objective.

One should read Pickering's papers of 1903 and 1904 in the light of the new situation created by the establishment in 1902 of the Carnegie Institution of Washington.[5] In his deed of trust, Andrew Carnegie had stated explicitly that his aims were the promotion of "original research, paying great attention thereto as one of the most important of all departments," and the discovery of "the exceptional man in every department of study, whenever and wherever found, inside or outside of schools." As a first step in the organization of the enterprise, the executive committee, one of whose most powerful members was Dr. John Shaw Billings,[6] appointed eighteen subcommittees to advise the Institution on plans, procedures, and recommendations. Pickering, who had promptly submitted an outline for the endowment of research along the lines of his 1886 pamphlet, was invited to become chairman of the committee on astronomy. Before accepting, he inquired whether the Institution wanted merely advisors, or active investigators who would request grants for their own and others' work. When Daniel Coit Gilman, the first president, answered in favor of the latter, Pickering chose to serve along with him a strong group that included Lewis Boss, George Hale, Samuel Langley, and Simon Newcomb. In March he supplemented his first communication by a plan for an independent astronomical foundation under Carnegie trusteeship.

On October 20, 1902, the committee as a whole made its first report, to which each member attached his own specific statement.[7] On certain matters they all agreed: grants for assistants and computers, establishment of an observatory somewhere in the Southern Hemisphere, collaboration in research, extensive investigations of the stellar and solar systems, publications, and a cautious examination of new instruments. In their separate reports each man emphasized his own special interest. Langley pressed for a solar observatory in the Southern Hemisphere, while Boss outlined at great length more than a dozen areas of study. Hale, differing from Pickering's view that the full use of present instruments was preferable to acquiring new ones, called for radical development of apparatus to meet the growing demands of astrophysics. Pickering stressed celestial photometry and photography, and Newcomb introduced the question of the "international unification of and cooperation in astronomical work."

The differences among the astronomers were mainly those of emphasis, but both Boss and Hale opposed Pickering's idea of any central direction of astronomical research. Boss wrote that "the massing of miscellaneous astronomical investigations under a single head in a great institution does not commend itself to my judgment." Hale was even more explicit:

In seeking to further the interests of astronomical and astrophysical research through cooperative effort, it seems to me that great care must be taken not to defeat what is perhaps the highest aim of the Carnegie

Institution—the encouragement and development of individual genius. Cases will doubtless arise in which a guidance of many workers by some well-known authority may be desirable and even necessary to the accomplishment of far-reaching results; but for the most part such cases will involve only the application of routine methods, sufficient for the collection of a mass of data needed in certain investigations. My argument is directed mainly against the centralization of authority and its concentration in a few individuals, however able they may be. It seems probable that the greatest progress will come rather through the advent of new minds, each competent to select its own point of view and to plan researches unhampered by the regulations of established systems. For this reason I believe that the best returns will be realized from assistance given to individual workers, cooperating, it may well be, with others, but constantly encouraged to advance science through the development of ideas and methods of their own.[8]

Hale's view prevailed, initially at least. The Institution made its first grants to various departments of science, the largest amount, $21,000, going to astronomy (Pickering had earlier suggested $30,000). Several astronomers who had requested $5000 received $4000. Pickering's application, first for $4000, then for $3500, was granted but was cut to $2500—a curious act of discrimination that drew from President Eliot the sardonic comment, "Great cry & little wool, as the devil said when he sheared his pig." Still, as Eliot rightly pointed out, it was better than nothing. With that addition to the payroll Pickering could engage new assistants to study his rapidly growing collection of photographic plates.

In his letter of thanks to Carnegie, Pickering described his new corps as "readers who will extract from this storehouse facts heretofore unknown, and which, except for this collection, could never have been learned, since it contains the only record of them upon the Earth . . . You have given us the key by which new facts regarding unknown worlds are daily revealed to us."

"Go ahead," Carnegie replied encouragingly from his castle in Scotland, "you are on the right lines. I hope to see you upon my return. I thought probably that I beat the record by selling three pounds of steel for two cents, but the whole constellation of Orion for a cent knocks me out. Take the cake!"

Unfortunately, Carnegie's agents failed to share his enthusiasm. When

The Harvard College Observatory

Pickering applied for a renewal in October 1903, he met with the result best described in his letter (marked personal) to President Gilman, October 16:

> To place you in possession of all the facts, I wish to say that I met Dr. Billings at Mrs. Draper's last Sunday. He is violently opposed to my plan for international aid to astronomy, and stated that if I continued to advocate it the continuation of the appropriation made by the Carnegie Institution might be refused. All I ask is that my application may be judged on its scientific merits. He supposed that we had ten thousand dollars which could now be used either for this Observatory or for others, and argued that if I gave it to others the Carnegie Institution ought to give nothing to me. In fact it is only the interest of this sum that can be used. I told him it was hard that the Harvard Observatory should be punished, or fined several thousand dollars in order to carry out the wishes of the donor . . . My object is simply to secure aid for those astronomers who can use it to best advantage, and whether the Harvard Observatory or I have a share in accomplishing this result is, it seems to me, a matter of little importance.
>
> I sincerely hope that I shall not be obliged to abandon the work described in my Carnegie report, and discharge our assistants.

Gilman was sympathetic. He had visited the Observatory during the year and had nothing but praise for the work: "This is an admirable beginning. Not every one who knows his way through the myriad hosts of heaven can find his way through the maze of earth—but you are equal to both tasks." Now, in his reply to the report of Billings' threat, though regretting Pickering's annoyance by "the positive utterances of a very positive man," Gilman assured him that one voice alone could not make decisions. "I am not always in accordance with the action taken," he admitted, "and have often been disappointed when projects in which I was very much interested did not meet with their approval; but it is not an easy matter to adjust the differences of men who have all the qualities of 'human nature.' " Nevertheless, Billings carried the rest of the executive committee with him. Ten days before the expiration of the original grant, a telegram announced the adverse vote. Without the subsidy Pickering was compelled to dismiss nine of the ten assistants he had carefully trained for a year. "Clearly," President Eliot wrote when informed of the blow, "the methods of the Carnegie institution are tentative and provisional . . . Apparently, too, President Gilman has no firm hold on the work done." This was true—though the arrange-

424

ment was afterward altered, policy had been placed originally in the hands of the executive committee, with little power given to the president.

To Professor Turner of Oxford, Pickering wrote that his plan for international research was hardly taking a path strewn with rose leaves, but that he did not propose giving it up. In fact, at the very moment he was preparing another pamphlet "that will probably further excite the wrath of those who believe that a man's entire efforts should be devoted to his own advancement." And to Sir David Gill he observed ruefully, "You see that astronomical research in this country is affected by other considerations than those of science."

Whether or not Pickering's initial defiance of Billings colored his further relations with the Institution, the fact remains, as will appear, that every subsequent appeal for support of the "exceptional man whenever and wherever found" met with rebuff, not always tactfully delivered. After the establishment of the solar observatory at Mt. Wilson in 1904 (annual appropriations reaching $220,000, plus hundreds of thousands for large telescopes), and of the department of meridian astronomy at the Dudley Observatory (annual appropriation $25,000), the Institution concentrated on these and other enterprises of its own creation, and grants to individuals, except in special cases, were no longer a factor in its policy.

Not all was lost, however. Besides the valuable information gained from the year's routine computations, Pickering managed to retain Miss Leavitt, whose research led to the profound discovery described earlier.

With the General Education Board, to which Pickering also turned, he fared little better than with the Carnegie Institution. In 1906 he proposed establishing an International Southern Telescope[9] to be mounted at a favorable site, preferably South Africa, at a total cost of $500,000 for instruments and administration. "My plan is not for the benefit of Harvard or any local institution," he wrote Frederick Gates, chairman of the board and advisor to John D. Rockefeller, "but its fundamental basis is that it should be international and benefit actually all the astronomers in the world, competent to do good work." Although this appeal failed, the courteous tone of the refusal led Pickering in July 1909 to suggest an even more modest program, an annual appropriation of $25,000 from which a carefully selected committee of experts could grant $500 to $1000 to advanced students and to teachers anxious to undertake original investigations. He cited several excellent projects awaiting such support: E. B. Frost at the Yerkes Observatory had material now ready for a catalogue of the spectral lines of sunspots; Frank Schlesinger of the Allegheny Observatory needed an assistant to measure photographs for a determination of stellar parallaxes; the Reverend Joel Metcalf (later one of Pickering's most valuable associates and a member of the Observatory Visiting Committee) could keep at work his own magnificently designed photographic doublet, the largest of its kind in this country and one of the largest in the world; Solon Bailey needed money to continue his photographs of the Milky Way; Miss Leavitt wished to prepare a catalogue of photographic magnitudes of some 10,000 stars.

Gates willingly submitted the application, but he warned that the Board had avoided entering a field already "pre-empted" by the Carnegie Institution and was not likely to change. Although Pickering tried to point out that the Institution had by then abandoned the making of most individual grants, the Board, as President Eliot regretfully informed Pickering, had given barely a moment to discussing the matter, which they still regarded as in the Carnegie domain. In his answer to Eliot the undaunted Pickering declared, "I have been urging this plan [of aid to science] since my first pamphlet in 1886 and am confident that I shall sometime meet the man who will give a fair trial to unrestricted international research and whose only object will be to obtain the greatest scientific return for the expenditure."

He did not find that man, he never again met another cosmopolitan-minded Miss Bruce, and he was unable during his lifetime to convert the foundations to his point of view, but he cannot be charged with either lack of persistence or timidity in tackling every possible source of support. Pickering had a very practical attitude toward money, but he was not afraid of it or of the men who possessed it. He had a clear idea of what he could do with it, and in one case at least had proved that he could get it without directly asking for it. In 1902 at a "Captains of Industry" breakfast given by several magnates to honor Prince Henry of Prussia, then visiting this country, Pickering had sat next to one of the hosts, whom he apparently charmed with his talk of the Observatory and his ideas on the organization of research. The wholly unexpected response was a check for $20,000, sent with the sole proviso that the donor[10] remain anonymous. Encouraged by this unsolicited bounty, Pickering offered several other "captains" a similar opportunity to invest in a scientific program that should appeal to them by its good business management. He was attempting, he explained, to introduce into astronomy the administrative methods by which they had brought the United States to its present industrial supremacy.

It was a gallant effort, but hardly successful, and Pickering, who, as he once remarked to President Eliot, was not a "patient waiter" in such matters, continued to set his own pace. Postponement of grants seemed to him an irretrievable loss to science, and since all of his colleagues were well aware of his concern, they did not hesitate to appeal to it. Indeed, from the vast network of correspondence carried on with astronomers all over the world one might readily compile a Who's Who in astronomy during the half century between 1877 and 1919, virtually every man mentioned being a beneficiary in one way or another of Pickering's interest and assistance. Most of the great Europeans with whom Pickering cooperated have already appeared in this book—the Herschels, Turner, Gill, Lockyer, Huggins, the Struves, Kapteyn, Hertzsprung, Schwarzschild, and many more. For the still unwritten history of American astronomy, as well, one could find no more rewarding source than the exchange of letters between Pickering and such eminent contemporaries as Young, Newcomb, Holden, Campbell, Barnard, Frost, Stebbins, Schlesinger, Russell, Hale, Keeler, and others. Aside from the evidence of constant professional collaboration and cordial personal relations, one may also trace there the beginning and development of several distinguished careers that owed an enormous debt to Pickering's insight and counsel. Two or three at least are worth recording in some detail.

James Edward Keeler[11] visited the Harvard Observatory first in 1877 to consult Pickering about a future in astronomy, but more immediately to ask for an introduction to the Clarks. Having built a telescope of his own, he now wanted to acquire more expert knowledge and hoped he might find a temporary apprenticeship in their workshops. The Clarks were cordial but had no orders at the moment. Keeler then entered the Johns Hopkins University, where he concentrated in science and, to help pay his way, assisted in the laboratories, but he was determined to be an astronomer. His opportunity came sooner than he might have hoped. In December 1880 Langley wrote to Pickering:

I am getting into a very interesting line of work, and I want some additional assistance, perhaps temporarily, possibly for some time. Do you know of any quite young man with a turn for physical research . . . who could assist me & Very[12] and can be had cheap?

The requirements, he added facetiously, were quite simple:

Let him have a touch of Faraday,—a good deal of Melloni, a strong flavor of both Herschels, the best points of Humboldt (I may take him on a scientific journey) and feel these modest gifts overpaid with 5 or 600 a year! It is a hard case if I can't have at least as much as that for my money!

The "quite young man" proved to be James Keeler, whom Langley took on his famous journey to Mount Whitney[13] in 1881, where he measured the radiation of the sun's heat on the earth (the "solar constant") and gathered unexpected data on the solar spectrum beyond the infrared region. In 1891 Keeler succeeded Langley as director of the Allegheny Observatory, where, among other original investigations, he confirmed Clerk Maxwell's theory of the structure of the rings of Saturn. Hale, attracted by Keeler's work, hoped to bring him to Chicago by securing, with Pickering's help, an endowment of the post from Miss Bruce, but while negotiations were pending, Keeler accepted the directorship of the Lick Observatory. Before leaving for California, he wrote to Pickering:

I have learned that you were good enough to recommend me for the directorship of the Lick Observatory, and it is evident that to this letter and a few others of the same kind I owe my appointment. Of course I know that personal considerations would have no weight with you in such a matter, but I am much obliged to you for your high opinion of my abilities, and I will endeavor to justify it by my future exertion.

Justify it he did, as all familiar with his short but brilliant tenure there will recall. In 1899 by unanimous vote he received the Rumford medals of the American Academy of Arts and Sciences, which Pickering accepted on his behalf. In forwarding them, Pickering wrote that in twenty years of membership on the committee no choice had given him greater satisfaction. Keeler also received the Draper medal in 1899. When he died in 1900, Pickering wrote, "Astronomy suffers a loss that cannot be overestimated, while I lost a much valued personal friend. There was no one who seemed to me to have a more brilliant future than Professor

Keeler or on whom we could better depend for important advances in work of the highest good."

Like Keeler, George Ellery Hale came to Pickering seeking a career in astronomy. Between 1886 and 1890, while attending the Massachusetts Institute of Technology, he spent many Saturday afternoons at the Observatory to acquaint himself with its apparatus and its work. When he returned to Chicago after graduation to outfit his private Kenwood Observatory, the gift of his father, he depended heavily on Pickering's help in the design of the instruments, especially an advanced type of spectroscope. Pickering sent exact specifications, which Hale then passed on to Brashear, the maker. Occasionally also, while awaiting equipment, he borrowed apparatus from Harvard, "emboldened to do so," he wrote Pickering, "by your indulgent liberality in the past." For his first trip to Europe Pickering furnished letters of introduction to the great European astronomers—Huggins, Christie, Vogel, Tacchini, Mouchez, and Struve—thanks to which Hale not only had a highly successful journey but established connections of lasting importance to his later career. In 1895, confronted with a deficit in financing his newly established *Astrophysical Journal,* Hale confidently turned to his mentor. Within a few days a check for $1000 from Miss Bruce took care of the matter.

In 1897 the dedication of the Yerkes Observatory brought to Chicago what Pickering regarded as the most impressive gathering of astronomers ever held in America. Pickering attended as Hale's special guest and as the representative of Harvard University. ("I do not know," Eliot wrote him before his departure, "whether you will find Chicago society divided about this ceremony. Mr. Yerkes is considered by a good many respectable citizens of Chicago the embodiment and representative of corruption in municipal affairs. These opinions may, however, be temporarily suppressed on the 21st of October.") Pickering spoke at the official banquet, using the occasion not only to describe the work of his own older observatory, but to convey to Mr. Yerkes as delicately as possible the importance of an endowment worthy of the superb establishment he had created. Shortly afterward, Hale wrote that he had additional reason for gratitude to Pickering: the Rumford Committee had just voted to appropriate $400 for his new design of a spectroheliograph, the result, he was certain, of one single influence. "I feel that no one has shown a greater interest than yourself in this Observatory and its work and feel very grateful to you for so much kind assistance." This instrument, which took monochromatic photographs of the sun, was destined to revolutionize solar astronomy.

Even after Hale became an undisputed master in his own right, he remained in close communication with Pickering. They shared a deep concern for the promotion of American science and for broad scientific cooperation through the Solar Union, the American Astronomical Society, the National Academy of Sciences, and the wartime National Research Council. Both men also took an active part in the postwar restoration of international relations that, after Pickering's death, culminated in the organization of the International Astronomical Union. In a letter to Solon Bailey (Fig. 55), February 4, 1919, Hale acknowledged his great personal debt to Pickering:

I remember with gratitude his friendly welcome the first time I met him at the Harvard Observatory in 1886, when I was a freshman at the Institute. It was the day of the celebration of the 250th anniversary of the founding of the University, if I am not mistaken, and in the exhibit of photographs at the Observatory I found, to my delight, some of the original photographs of stellar spectra made by Henry Draper. Soon afterwards I was permitted to work Saturday afternoons and evenings as a volunteer—the only time I could spare, in view of my work at the Institute. Beginning thus, and developing in subsequent years, my acquaintance with Professor Pickering has been a source of much pleasure and value to me. I warmly admired his great ability, his originality of view, his power of organization, and his unwearied initiative. I also appreciated how much he did, in so many ways, to stimulate research and to help astronomers everywhere. The great development of the Observatory under his direction, and its immense contribution to the progress of astronomy, mark an epoch, universally recognized, in the advancement of science.

What I shall remember with greatest pleasure, however, is his kindly interest in me as an unknown amateur when I first came to the Observatory. Many others have enjoyed this experience, for the circle of amateurs he touched and helped was a wide one.

Henry Norris Russell, too, bore witness to Pickering's gracious welcome to the beginner and to many succeeding acts of generosity. Shortly after his first interview with Pickering, in which he mentioned his initial astronomical work on stellar distances, he received a letter suggesting the usefulness of determining the magnitudes and spectral types of his stars, and offering to have the work done at Harvard. This research

FIG. 55. Solon Irving Bailey, acting director of the Observatory, 1919–1921.

involved "the photometric and spectroscopic observation of some three hundred stars (the photometric setting being made by Professor Pickering himself)," Russell explained, "and was offered as an unsolicited contribution to the work of a young and unknown instructor."[14] "Indeed," as he wrote Bailey "he really did more than any one else to give me a good start in my scientific career, and his friendship and counsel have ever since been among the things that I most prized. There are many

431

others who owe much to his ever-willing spirit of cooperation, but there can be few who owe more."[15]

Russell did not overestimate the extent of that debt, as the correspondence amply proves. To Pickering's strong recommendation he doubtless owed his appointment as director of the Princeton Observatory, grants from the Elizabeth Thompson Fund, and in 1915 an especially welcome boon to the observatory's income—a subscription of $1000 a year for five years, which Pickering persuaded Archibald Russell, a wealthy trustee of Princeton, to contribute. "This gift," the grateful recipient wrote Pickering,

we owe to your ability in formulating such an effective plan for the presentation of the needs of the various observatories, and your friendly aid in putting it into effect. It adds one more important item to the list of benefits which we here—and I personally—owe to your generous and cordial friendship, and I can only express once more, my hearty thanks, and my admiration, equally of the spirit and the effectiveness, of your way of doing things.

It is worth noting here that Russell was asked to succeed the man who had so consistently served his interests and that he declined the offer in favor of his most brilliant pupil, Harlow Shapley (Fig. 56), who while at Princeton began with borrowed Harvard plates the studies for which he became renowned.[16] Russell later became a member of the Visiting Committee of the Observatory.

The range and variety of Pickering's innumerable acts of assistance can only be suggested here. No appeal, whether for advice, materials, publications, recommendations, or money, went unheeded. He lent, gave, or otherwise procured instruments and distributed with a lavish hand the results of Harvard's research. When asked if copies of photographs from the Observatory collection could be used without restriction, Pickering invariably answered, "Of course. That is what they are for." Of this liberality Henry Norris Russell wrote, "Who among us has not gone to Harvard, and returned . . . loaded down with data for investigations, new or old, and inspired by his experience with new enthusiasm alike for the magnificent researches of the great observatory, and his own humbler work?"[17] Far from jealously hoarding the treasures there, Pickering shared them with all who came.

Because, like George Bond, he saw little use in expensive telescopes and elaborate buildings given by enthusiastic donors unless sustained

FIG. 56. Harlow Shapley, who became fifth director of the Observatory, 1921–1954.

financial support was also provided, he directed a large portion of his efforts toward making poorly endowed observatories as self-sustaining as possible. Among these was the Yerkes Observatory, which in the early years of Edwin Frost's direction operated almost entirely without income. Pickering offered to write or call on the founder of the observatory to get immediate help, but in the meantime he arranged for a grant from the Rumford Fund, and from his own Astronomical Fund sent modest sums to pay an assistant working on Frost's stellar spectrograms. In 1906 the Yerkes bequest relieved the major difficulties, but a year later Pickering found it necessary to contribute money for compiling an index to the *Astrophysical Journal,* which Frost had edited since 1902.

The Allegheny Observatory, always of special interest to Pickering because of his long and intimate association with Langley and afterward with Keeler, was another object of his concern during the directorship of Frank Schlesinger.[18] Before coming to Pittsburgh, Schlesinger had been in charge of the International Latitude Observatory[19] at Ukiah, California, and in 1903 had moved to the Yerkes Observatory, where he resumed the investigations in photographic astrometry he had first undertaken as a graduate student at Columbia. In 1905, when he became director of the Allegheny Observatory, his prospects for productive research looked dim indeed. Aid from the Thompson Fund, the loan of plates from the Draper collection, chemicals to develop his own photographic plates, and gifts of several small instruments—all provided by Pickering—soon made it possible for him to undertake the work in stellar astronomy for which he became widely known. In 1912, at the dedication of a new building at the Allegheny Observatory, long left unfinished for lack of funds, Pickering gave the principal address, and in succeeding years promoted Schlesinger's work and the welfare of the observatory with as much vigor as if they had been his own.

In 1914, when Pickering learned from Samuel A. Mitchell, the newly appointed director of the Leander McCormick Observatory in Virginia, that less than $1500 a year was available for maintainance, improvements, and assistance, he promptly secured modest grants and then spurred the McCormick family to renew its support. Pickering found funds also for the Sproul Observatory at Swarthmore, and it should be recalled that between 1890 and 1900 he endorsed liberal appropriations by Miss Bruce to a number of precariously financed observatories—for example, $35,000 to the Dudley, almost $25,000 to Columbia, more than $15,000 to Yerkes, small grants to Vassar and other observatories both in this country and abroad.

Nor did Pickering confine his concern to astronomers. He took an early interest in the brilliant joint experiments of A. A. Michelson and Edward Morley on the "ether drift," welcomed them at the Observatory, encouraged them to apply for grants certain to be given, and in 1903 recommended Michelson for the Nobel Prize. (It was awarded him, again with Pickering's endorsement, in 1907; he was the first American to receive it.) With the young Henry Rowland, later the distinguished Johns Hopkins physicist, Pickering while still at MIT had planned a joint experiment on the mechanical equivalent of heat. Pickering's appointment to Harvard ended his share in the project, which Rowland eventually carried out alone, but it was doubtless gratifying to learn from him that there were more good ideas to be found in Pickering's laboratory than in all those he had seen abroad. In 1893, acknowledging a small grant from the Rumford Fund (the first of several), Rowland wrote Pickering, "I have not forgotten that you were the first in this country to address to me a kind word of encouragement after my first paper." In 1890 Rowland was one of the fifteen men Pickering chose to aid from Miss Bruce's fund.

One of Pickering's most interesting efforts was on behalf of Percy Bridgman, as a graduate student at Harvard, at the outset of his experiments in thermodynamics. The correspondence on the matter furnishes not only useful facts about Bridgman's early struggles, but also valuable documents for a case study of the initially narrow policy of a great foundation. In September 1907 Bridgman came to Pickering with a letter from Professor John Trowbridge,[20] chairman of the Physics Department, a former classmate of Pickering's at the Lawrence Scientific School, his associate at MIT, and probably his most intimate friend on the Harvard faculty:

Mr. Bridgman, the bearer of this note, wishes to explain his work to you. I have advised him to apply for an additional $400 from the Rumford Committee. He is in the position to make good, for he has entered upon a great field hitherto unexplored in the thermodynamics of matter. His apparatus is in condition to obtain many new results; but he needs additional mechanical work & assistance.

I think he can convince you that the Rumford Committee can aid one who is well equipped, both from the mathematical & the experimental sides.

The Harvard College Observatory

In Pickering's view an ideal case, Bridgman's request needed little argument. Bridgman received a number of grants and, as his research attained increasingly impressive results, Trowbridge confidently applied to the Carnegie Institution for $1000 toward the work of the "exceptional man" Carnegie had declared it his express purpose to discover and foster. This request and one made by Pickering at the same time for another graduate student received similar replies: "Our practice has now substantially crystallized into a rule to the effect that we will not give aid to students in colleges and universities who are pursuing work for the higher degrees." To the refusal was appended the plaintive statement that the president of the Institution was swamped with such appeals. "He is there to be swamped," observed Trowbridge acidly.

In 1910 Bridgman, who had now become an instructor in the Physics Department, went on to magnificent achievements, determining the effect of high pressures on the physical nature of various substances, thus completely refuting the Institution's dogmatic view that "men who teach cannot conduct systematic investigations."[21] In 1913 Trowbridge, who had retired but was still anxious to ensure the continuation of Bridgman's brilliant experiments, again turned to Pickering. After having been so roundly snubbed, he said, he did not wish to "kow tow" again. Would Pickering see what could be done? Armed with Bridgman's statement, "Brief Summary of Work in the Jefferson Physical Laboratory with High Hydrostatic Pressures," and, as a physicist himself, aware of the significance of the research, Pickering began at once to present the case to all who might be interested, both in this country and abroad. One immediate result was an invitation to Bridgman from Dr. W. W. Keen, president of the American Philosophical Society, to present his paper at the annual meeting in April 1914.

In this same year Pickering's attempts to assist men in science were given official reinforcement by the American Association for the Advancement of Science, which had authorized the establishment of the Committee of 100 on Research.[22] Its membership represented the scientific and academic elite of the country—university and college presidents, distinguished professors in every branch of science, administrators of several government bureaus, directors of laboratories of the Rockefeller Institute, officers of the Carnegie Institution and its affiliates, the secretary of the Smithsonian Institution, and research directors of two great industrial companies. Besides Pickering, who was made chairman of

436

the executive committee, the astronomers included Hale, Leuschner, Moulton, Schlesinger, and Campbell.

Since some of the important functions of the committee were to examine the use of existing research funds, to find new resources, and to make them available to competent persons, Pickering launched at once into a vigorous attack on the tasks assigned to him. In these he had the unflagging support of J. McKeen Cattell,[23] secretary of the committee, the editor of *Science* and the owner of the journal until April 1925 when, by agreement, it became the property of the American Association for the Advancement of Science, to serve as its official organ. Shortly after the meeting of the committee on April 20, 1914, Pickering submitted to the Carnegie Institution four research projects in dire need of funds. Three were for astronomers, in amounts varying from $200 to $500. The first on the list, however, was for Bridgman, with a request for $1000 to cover labor and materials. The reply he received, dated May 14, is worth quoting at length. It epitomizes the then prevailing concept of the Institution's role:

I have watched with warm interest the admirable researches of Professor Bridgman, and was close to the point at one time a few years ago of recommending his connection with the Institution as a Research Associate. It did not appear, however, at the time that we could clearly differentiate his work of research at Harvard from his work of instruction in that institution. Our experience forces us to draw the line sharply between men who are primarily devoted to research and those who are primarily devoted to work of instruction. Any attempt on our part to play the role of paternalism for the benefit of educational institutions would clearly result unfavorably for all concerned . . . We are not disposed, therefore, to appoint any man connected with an educational institution to a Research Associateship unless he is able to give all or nearly all of his time directly to research. Moreover, we cannot afford to have our annual budgets and their construction conditioned by or in any wise tangled up with those of educational institutions. We must be permitted to proceed as independently of them as they do of our establishments.

The letter went on to say that for the same reasons the Institution would not support observatories, laboratories, or other academic institutions for research. "I am coming to consider, in fact, that there is no more reason why colleges and universities and other establishments should ask us for aid than there is why we should apply to them for support of our operations." The letter complained also of the "stupid and wide-spread notion" that all of Carnegie's and Rockefeller's fortunes were at the disposal of the foundations, and of the harm "caused by our own friends and associates who appear to have lost all capacity to deal with numbers having more than three figures." Not until "men in our own camps recover their sanity" will we be able to "proceed under more favorable auspices than those with which your humble servant has been obliged to contend during the past decade."

This document was followed by another to Pickering about a week later:

Nearly all of our best men of affairs, educators, publicists, and other highly esteemed colleagues are very apt to confound research with education, with invention, with patent rights, copyrights, and the necromancies of "geniuses" and "male witches." It has been appalling to me to observe during the past decade how many otherwise well educated men have held that the chief object of a research institution should be to rake over the piles of intellectual rubbish which come to such establishments, with the expectation of finding in them new roads of knowledge and discovery.

One of the best ways to correct these false views will be, I think, to get a considerable number of institutions and hence individuals committed to the business of research, and consequently responsible to themselves and to society for the application of funds designed not to give poor boys and inventors special advantages but to attack obvious problems in research with continuity of effort and intelligent appreciation of well known effective methods. When such at present very much needed views come to be taken of research, inside as well as outside of colleges and universities, we shall get on much better than we are able to at present.

Precisely why it could have been supposed that Pickering merited such a diatribe is not clear, but in any case he saw that there was no hope in that direction. He regretted it all the more when in the fall of 1914 Bridgman wrote to him, "Acting on the advice of Professor Lyman and emboldened by the interest in my work on high pressures which you showed last spring in Philadelphia, I am venturing to ask advice." Funds at the laboratory were low, a number of corporations he had counted on had passed dividends, and it was impossible at the moment to pay for mechanical assistance. The work was of great value. Did Pickering know of any source for funds, at least $1000? Bridgman enclosed reprints of two papers recently published and offered to send some new unpublished material, relating not only to previous experiments but also to more recent investigations. Pickering replied at once, promising to do everything possible to get some appropriations. His first move was to gather together representatives of the various trust funds for concerted action on Bridgman's application.

In 1917 and 1918, asked to nominate a candidate for the Nobel Prize, Pickering named Bridgman. In 1936 the historian of the Physics Department wrote of him, "He is easily the world's foremost authority on the laboratory applications of high pressures and the behavior of solid and liquid materials under such pressures . . . Sometime the general public may discover him."[24] In 1946, twenty-nine years later than Pickering, the Nobel Prize Committee did so.

Between 1914, when the Committee on Research began to act, and 1918, when Pickering resigned[25] from it, he devoted a large share of his time to its work, for he meant to show that it could succeed. The work involved him in endless correspondence with astronomers and possible benefactors, with efforts to publicize the needs of science, with meetings, and especially with a series of new appeals to the Rockfeller Foundation, which, in view of the Carnegie Institution's position, remained his chief hope. Ten or twenty thousand dollars a year, he urged, not only would prove of immense benefit to individual investigators, but would go far to remedy what he considered the discreditable position of research in this country. The war in Europe made the matter all the more urgent, for it threw upon the United States the major responsibility for advancing science. This was especially true in astronomy. Nationally supported European observatories could not fail to suffer serious interruption from reduced or even halted appropriations and from the loss of young men to the armies. Pickering suggested a scheme that would serve American investigations and at the same time keep certain great establishments in Europe actively at work. If small sums could be made available to the Committee on Research, he argued, it could assign certain portions of projects initiated here or in process before the war to some of the most eminent astronomers in Europe, at least in England if nowhere else. He had already shown that such a plan was feasible. A minimal grant from the Elizabeth Thompson Fund to Professor Turner of Oxford had prevented the interruption of a piece of research in which Pickering had long been interested. "Now that is a really kind act," Turner had written. "It is an immense relief to feel that the safety of the work is assured." For Pickering the significance was even greater. It promised the unbroken international cooperation which he had advocated and practiced for decades, and which, when peace came, would again include all those who had once participated in it.

Deeply as Pickering cared about his astronomical friends in England, he originally had grave reservations about American participation in the war. "For a long time," he wrote a correspondent there when the United States finally entered it, "I felt we could maintain our position as neutrals and might help more than as combatants. My feeling has been that nothing should disturb our relations to science." In this respect he of course differed markedly from Hale,[26] an ardent interventionist from the first and a tremendously effective mobilizer of scientific talent for American preparedness. Even so, as early as 1915, shortly after the Navy set up its Consulting Board,[27] Pickering wrote to the chairman, Thomas A. Edison, to offer the services of the Committee on Research and to remind him that the National Academy of Sciences, under the law enacted during another grave crisis, had been officially empowered to advise the government. His more immediate objective, however, was to suggest a plan for combatting the submarine menace to American ships. Once the United States had joined the fight, Pickering outlined a series of measures the government should take at once to organize its efforts. His various "Memoranda for Preparedness" discussed such problems as submarine detection, automatic SOS signals, range finders, and statistical computations for speedy mobilization of manpower. He also listed the names of experts on his various committees who were equally ready to serve as scientists. A device he himself designed, the Pickering Polaris range finder, was among the few practical inventions adopted and put into production by the War Department.

Never regarding war as the main business of science, however, Pickering continued to accept applications for research in astronomy and other sciences and to press as vigorously as ever for funds. The hysteria engendered by war propaganda disturbed him greatly, not only because it represented a loss of perspective but because it directly affected the work of the Committee on Research when Cattell, abruptly dismissed by Columbia University because of his opposition to the war, resigned from it as secretary and editor of *Science* to avoid prejudicing its work. Pickering, who stood by him staunchly throughout, wrote Cattell, "I sincerely hope that you will withdraw your resignation. I have discussed the subject with various persons, including a majority of the Committee

440

on Policy, and have not found anyone who did not agree with me, that you were within your rights in your letters [opposing the war] to members of Congress."[28] In January 1918 Pickering wrote again: "I hope . . . you will resume your duties as Secretary. It seems to me that, in view of the expressions of opinion regarding your work, there is every reason why you should again take up all your duties to the American Association and to other scientific societies."

At no time, in fact, did Pickering relinquish his abiding goal—the restoration of all scientific relations. A few months before the war ended he made his position clear. Early in August 1918, for example, in answer to a long communication from Hale outlining his plan for an Interallied Research Council, Pickering wrote:

The problem of International Research interests me very much . . . It seems to me that there are two questions. I am strongly in favor of an inter-allied research council which should secure from scientific men every aid possible for our successful prosecution of the war, and the continuation of necessary routine observations.

The advancement of pure science internationally is a very different question. This is especially the case with modern astronomy, which stands on a different basis from almost every other Science. It is purely impersonal, seeking after truth, independently of individuals or nations. Every consideration should give way to the fulfilment of this object. No ordinary punishment is adequate for those responsible for barbarities contrary to the laws of nations and humanity, yet we ought not to ignore the work of those who, laboring quietly in their observatories, have done their best to extend our knowledge in these terrible times. I hope, therefore, that in Astronomy no definite action in this matter will be taken during war times, until we know the attitude of those whom we once admired and respected.

A month later, alarmed by a suggestion in *Science* (September 6), that the Royal Society was considering the advisability of expelling its members who belonged to enemy nations, Pickering protested to Arthur Schuster, the Secretary, "It seems to me that it would be unfortunate to take such action at this time. After the war, we can determine the attitude of individual members . . . Common justice allows every accused person to be heard." Writing to Elis Strömgren at Copenhagen on January 7, 1919, he was even more emphatic:

The announcement of an Armistice was welcomed here by such a celebration as we have never had before. I am looking forward with much anxiety

441

to the conditions of peace. As soon as may be, I want to reestablish relations with such of my good friends among the German astronomers as were not in any way responsible for what we consider deeds contrary to the laws of war. When the proper time comes, I have thought of writing to them . . . Of course, at present, there is no mail communication.

I do not approve of the plan of establishing international societies in which scientific man of the central and neutral nations are not included, at least in the case of such subjects as have no relation to war. Astronomers, for example, ought to be able to carry out international work by correspondence, even if it is not desirable at present to hold meetings. I am anxious to know how far European observatories have suffered and what is likely to be their condition when the peace treaties are actually signed. All of this seems a little premature, but large delays are likely to occur. I believe that many astronomers agree with me that we should make every effort for the advancement of our science, regardless of personal or national considerations.

This view he restated eloquently in a letter to an Italian correspondent who had proposed a "Scientific Quadruple Entente":

I hope that the astronomers of the world can again cooperate in a friendly way, as they did so successfully only five years ago. This has been the policy of the Harvard Observatory for many years, and it should lead not only to a great increase in human knowledge, but to the establishment of friendly relations which should exist between men in all parts of the world, in so ennobling a science as the study of the worlds among which the Earth is but an atom.

Pickering had looked forward to a revival of the Solar Union. He would have rejoiced infinitely more in the knowledge that July 1919 saw the birth of the International Astronomical Union, in Brussels, and that fifty years later (June 1969), the sixth director of the Harvard Observatory would add to Pickering's words his own fervent plea for internationalism and cooperation in science:

We must be able to meet one another, to exchange data and theories in the early stages of researches. And that is one of the reasons why these annual conferences at Liége have played, and I hope, will continue to play for a long time to come, an important role in the development of astronomical science.[29]

Pickering has had no full-length biography, but in a genuine sense the history of his administration is the story of his life. Throughout his more than four decades as director, he lived for astronomy. As Miss

Cannon remarked, "His joy in taking part in what he called the greatest problem ever presented to the mind of man, the study of the stellar universe, never left him."[30] Over the years he contributed liberally to the funds of the Observatory for research there and elsewhere, the sum amounting to considerably more than he had received in salary while director. He had always regretted that he was not a wealthier man, so that he might have given more. By his will, he left the bulk of his estate to the Observatory.

A notable host, he enjoyed the company of friends, and he and Mrs. Pickering for years "made a temporary but delightful home for many visitors from all parts of the world, and especially for astronomers."[31] His other recreations—the theater, opera, music,[32] bicycling, mountain climbing, working jigsaw puzzles (with the pieces kept picture side down so that they must be fitted from their shapes alone), chess—were always subordinated to his major passion, accumulating facts about the celestial universe. During his lifetime he received almost all the honors that could be bestowed on an astronomer—medals, honorary degrees, election to the academies of science in most of the countries of Europe.

Pickering had always believed, however, that the principal reward received by a scientist was "to find that his work is appreciated by those competent to judge."[33] This reward he had in abundance. As he entered his seventy-third year, still active and hard-working, he must have felt that the course he had chosen for the Observatory was fully justified by letters such as that of Harlow Shapley, who wrote (February 18, 1918), in regard to a projected study of stellar parallaxes:

I have gone far enough with the work to be sure of the suitability of the Harvard material for the determination of accurate parallaxes of at least a thousand bright stars, and to be impressed again with the wealth of astrophysical knowledge contained in the Harvard photographic library. I share with Russell, Chant, Charlier and a few others (too few others, unfortunately) the conviction that the answers to most of the astrophysical problems of the present time were already recorded on Harvard plates some 15 or 20 years ago. Of course I may be prejudiced because my own work has been on variable stars, general photometry, objective prism spectra, and star clusters—lines of work in which the Harvard Observatory has for long had first place.

Edward Pickering died on February 3, 1919, after a brief illness. Letters of grief came to the Observatory from astronomers in all parts of the world. They were acknowledged by Solon Bailey, who had become acting director. All expressed their sense of personal loss, as well as of the loss to astronomical science. The many obituaries that appeared in this country and abroad all paid tribute to Pickering's energy, his generosity, the magnitude of his plans, and the unique accomplishments effected under his direction. In one of the most perceptive of these notices, his old friend Professor Turner commented that "it is safe to say that if he had been free to choose any career the world has to offer, he would have chosen to be what he was."[34]

What were the contributions the Observatory made to astronomical science during the 42 years of Pickering's directorship?

Aided by the generosity of patrons such as Robert Treat Paine, Uriah Boyden, Mrs. Henry Draper, and Miss Catherine Bruce, he had transformed a small, poorly endowed research center engaged chiefly in the visual observation of celestial objects into a large, well-endowed, well-equipped institution that fully exploited the newest methods of spectroscopy and photography. He had established an observing station in the Southern Hemisphere so that all parts of the sky might be studied, and had devised new instrumentation, such as the meridian photometer and the photographic doublet.

Pickering, with his talented and devoted assistants—Professor Bailey, Professor King, Professor Searle, Mrs. Fleming, Miss Maury, Miss Leavitt, Miss Cannon, and the many others mentioned in the pages of this book—had established a uniform system for measuring the brightnesses of stars, and produced fundamental catalogues of stellar magnitudes. They amassed a photographic library containing a history of the stellar universe over a continuous period of more than 30 years. They discovered many hundreds of variable stars, in globular clusters and elsewhere, and made the first thorough study of the Magellanic Clouds. They devised a system of spectral classification that, with some modification, is still in use today, and classified the spectra of more than a quarter of a million stars. With the exception of Miss Maury and Miss Leavitt, they produced few notable advances in astronomical theory but they did provide basic data on which the theories of others could be built.

And, equally important, Pickering encouraged and helped the younger generation of astronomers, such as Hale, Douglass, Upton, Keeler, Barnard, Schlesinger, Russell, and Shapley, who in their turn helped to shape the astronomy of the twentieth century.

NOTES

CHAPTER I

1. This description of the American observatory is quoted from W. C. Rufus, "Astronomical Observatories in the United States Prior to 1848," *Scientific Monthly,* 19:120–139 (August 1924). As will be obvious, for much of the information about early instruction at Harvard this chapter leans on Samuel Eliot Morison's *Harvard College in the Seventeenth Century* (2 vols., Cambridge: Harvard University Press, 1936); I. Bernard Cohen, *Some Early Tools of American Science* (Cambridge: Harvard University Press, 1950; reissued by Russell & Russell, New York, 1967); Theodore Hornberger, *Scientific Thought in the American Colleges* (Austin: University of Texas Press, 1945); Solon I. Bailey, *The History and Work of Harvard Observatory, 1839 to 1927* (Harvard Observatory Monographs, No. 4; New York: McGraw-Hill, 1931), esp. chap. 1. For the more general scientific interests of the country a number of works have been helpful, among them Dirk J. Struik, *Yankee Science in the Making* (Boston: Little Brown, 1948); A. Hunter Dupree, *Science in the Federal Government* (Cambridge: Harvard University Press, 1957); Brooke Hindle, *The Pursuit of Science in Revolutionary America, 1735–1789* (Chapel Hill: University of North Carolina Press, 1956). On textbooks at Harvard, see Frederick G. Kilgour, "The First Century of Scientific Books in the Harvard College Library," *Harvard Library Notes,* 3:No. 29 (Mar. 1939), pp. 217–225; Arthur O. Norton, "Harvard Text-books and Reference Books of the 17th Century," *Publications of the Colonial Society of Massachusetts,* 28:361–438 (1935); an edition of *Compendia Physicae,* Charles Morton's influential text (Theodore Hornberger, ed.) *Publications of the Colonial Society of Massachusetts,* 33 (1940), reviewed by I. B. Cohen, *Isis,* 33:657–671 (1942).

2. Morison, *Harvard in the Seventeenth Century,* I, 216. See also Donald Fleming, "The Judgment upon Copernicus in Puritan New England," in *Mélanges Alexandre Koyré* (Paris: Hermann, 1964), vol. 2, pp. 160–176.

3. See Leo Goldberg in Gerard P. Kuiper, ed., *The Sun* (Chicago: University of Chicago Press, 1955), p. 10: "The appearances of the vortices suggested to Hale the rotary motion of electrically charged particles and the attendant generation of magnetic fields in sunspots."

4. *New England Quarterly,* 7:3–24 (1934).

5. Cohen, *Some Early Tools,* pp. 8, 124, 130.

6. John W. Streeter, "John Winthrop, Junior, and the Fifth Satellite of Jupiter," *Isis,* 39:159–163 (1948); T. F. Waters, *A Sketch of the Life of John Winthrop the Younger* (Ipswich, 1899); *Dictionary of American Biography,* 20:411–413.

7. *Proceedings of the Massachusetts Historical Society,* 2nd ser., 4:264–267 (1887–1889). This letter was first called to the Society's attention in January 1889 by Robert Winthrop, Jr., whose father had long served the Society as president and the Harvard Observatory (succeeding Josiah Quincy) as chairman of the Visiting Committee. "It cannot fail to be regarded as an interesting circumstance," he observed, "that at that early period, when Isaac Newton, then a young professor at English Cambridge, was engaged in the discoveries which made him famous, three Tutors and Resident-Fellows of a poverty-stricken New England college should have attempted similar researches." This comment incited the journalist D. W. Baker, author of a series of excellent articles in the Boston *Traveller* ("History of the Harvard College Observatory During the Period 1840–1890," Boston, 1890) to suggest to E. C. Pickering (1894) that Harvard might even have anticipated Newton!

8. Morison, *Harvard in the Seventeenth Century,* I, 221; Bailey, *History,* p. 5.

9. Frederick G. Kilgour, "Thomas Robie (1689–1729), Colonial Scientist and Physician," *Isis,* 30:473–490 (1939). The letter on his observations is quoted in Rufus, "Astronomical Observatories," p. 124.

10. For the establishment of the Hollis Professorship and the career of Isaac Greenwood, see Josiah Quincy, *History of Harvard University* (2 vols., Cambridge, 1840), II, 11–22; Cohen, *Some Early Tools,* chap. 2, n. 17; Hornberger, *Scientific Thought,* pp. 26, 27, 44–48, 57, 65, 74, 83; Samuel Eliot Morison, *Three Centuries of Harvard* (Cambridge: Harvard University Press, 1936), pp. 63, 79, 89; *DAB,* 7:591–592. Greenwood was appointed in 1727, previous to which the Harvard Corporation at its meeting of November 14, 1726, as a result of reports from Hollis, stated "the opinion that they would not at present proceed to elect him . . . Yet ye Corporation expressed to Mr. Greenwood their hopes & desires, yt matters might be made up with Mr. Hollis, as yt upon writing him and receiving Answers, he might be fixed on his Foundation." *Publications of the Colonial Society of Massachusetts,* vol. 16 (*Harvard College Records,* 2:545). On November 25, 1737, "notwithstanding repeated Warnings and Admonitions," Greenwood was reported guilty of "various Acts of gross Intemperance by excessive drinking," and warned that if not reformed he would be removed in five months, but on December 7, since he "hath twice relapsed into the said Crime," the Corporation decided not to wait. *Ibid.,* pp. 670–671.

11. *Scientific Thought,* pp. 44–45.

12. From the voluminous literature on Winthrop useful articles include: Frederick E. Brasch, "The Royal Society of London and Its Influence upon Scientific Thought in the American Colonies," *Scientific Monthly,* 33:336–355, 448–469 (1931); "Newton's First Critical Disciple in the American Colonies—John Winthrop," in *Sir Isaac Newton, 1727–1927: A Bicentenary Evaluation of His Work*

(Baltimore: Williams and Wilkins, 1928), pp. 301–338; "John Winthrop (1714–1779), America's First Astronomer and the Science of His Period," *Publications of the Astronomical Society of the Pacific,* 28:153–170 (Aug.–Oct. 1916). See also John C. Greene, "Some Aspects of American Astronomy, 1750–1815," *Isis,* 44:339–358 (1954); numerous references in *DAB,* 20:414–416.

13. Winthrop's MS, "The Summary of a Course of Experimental Philosophical Lectures," in Harvard Archives, is analyzed by Cohen, *Some Early Tools,* pp. 41–44.

14. *Bibliographical Contributions,* No. 32 (Justin Winsor, ed.), compiled by Henry C. Badger (Harvard Library, 1888).

15. See Jacob Bigelow, "The Life and Writings of Count Rumford," *Memoirs of the American Academy of Arts and Sciences,* 4:i–xxiii: "At the age of about 17, Mr. Thompson obtained permission to attend the lectures of Professor Winthrop on Natural Philosophy in the University at Cambridge, in company with his friend the late Col. Baldwin of Woburn. A rich field of information was thus opened to his inquisitive mind . . . he seems to have acquired here the rudiments of that philosophical knowledge which afterwards ripened into the source of so much honour to himself and benefit to the community."

16. Anon., *The American Magazine and Historical Chronicle* (Feb. 1744), printed in Hornberger, *Scientific Thought,* pp. 68–69; also Cohen, *Some Early Tools,* facing p. 3.

17. For accounts of the fire see Quincy, *History,* II, 111–116; Benjamin Peirce, *History of Harvard University* (Cambridge, 1833), pp. 288–299; Morison, *Three Centuries,* pp. 95–96. Mrs. Mascarene's letter is reprinted entirely in Henry F. Waters, "The College Fire in 1764—A Contemporary Account," *Harvard Register,* 3:294–297 (aside from the disaster, a delightful report); and quoted in part in F. Apthorp Foster, "The Burning of Harvard Hall, 1764, and Its Consequences," *Publications of the Colonial Society of Massachusetts,* 14:2–43 (1911–1913), esp. pp. 3–4. (This account gives an illuminating inventory of the students' losses of books and belongings.) The broadside is quoted in full in Quincy, *History,* II, 480–483. Cohen, *Some Early Tools,* chap. 2, describes Harvard's scientific instruments before the fire.

18. Foster, "The Burning of Harvard Hall," p. 9.

19. *Corporation Records,* 2:141.

20. See Harry Woolf, *The Transits of Venus: A Study of Eighteenth Century Science* (Princeton: Princeton University Press, 1959). The Winthrop expedition and observations are discussed on pp. 93–95, 97–98, 133, 170–174.

21. Quincy, *History,* II, 221.

22. *Corporation Records,* 2:142.

23. *Relation of a Voyage from Boston to Newfoundland, for the Observation of the Transit of Venus, June 6, 1761* (Boston, 1761).

24. Hindle, *Pursuit of Science,* p. 100.

25. Winthrop's results appeared in *Philosophical Transactions,* 54:279–283 (1764). The most recent such measurements give a parallax of 8.7945 seconds of arc, slightly larger than the figures obtained in the transits of 1761 and 1769.

The best modern determinations, however, are based on radar measurements of the distance of Venus. They give the value 92,955,600 miles as the distance between the sun and the earth. See Donald H. Menzel, "Venus Past, and the Distance of the Sun," *Proceedings of the American Philosophical Society,* 113:197–202 (1969).

26. Note that incorrect initials have been used in some of the following letters.

Eskdale

Fortis Green

No 2

London

July. 25. 1918

Professor Pickering.

Dear Sir

I think you might be interested to hear that a friend of mine W. T. Court who has a fine collection of old microscopes & telescopes has among them a very fine example of a Gregorian telescope by the celebrated mathematical instrument maker & writer, Benjamin Martin of Fleet Street London (1704–1782) which has the following inscription in contemporary engraving on the body of the instrument.

"The gift of the Honorable Thomas Hancock of Boston to Harvard College Cambridge."

The telescope is mounted on a strong brass tripod with folding claw feet each with a levelling screw. It has slow motions in altitude & azimuth which can be thrown in & out of gear, & is fitted with two levels. As Thos. Hancock died in 1764, & Martin came to London in 1750 this fixes the date very closely.

If you can give me any information as to its history, & especially how it has found its way back to England, I should be much obliged.

I am studying the early history of these instruments of of [*sic*] early microscopes with a view to publication, & if you can give me any help I would of course give you full acknowledgment.

Yours very faithfully

(Dr) A. S. Clay

August 16, 1918.

Dear Mr. Pickering:

It may be doubted whether the telescope referred to in Dr. Clay's letter, which I return, is still the property of Harvard College. At any rate, it would be impossible to prove it. If it was carried away without being sold, it was probably done by some officer who fled at the time of the Revolution.

I will try and see if there is anything on our records about it, but I think it highly improbable.

Very truly yours,

A. Lawrence Lowell

August 31, 1918.

Dr. A. S. Clay,
 Eskdale, Fortis Green, N. E.,
  London, England.

My dear Sir:
Your letter of July 25 is received. I enclose extracts from the records of the Corporation of Harvard College regarding the telescope to which you refer. It could not have been sold by the College, and President Lowell suggests that during the Revolutionary War the telescope was probably carried to England by some one whose sympathy was not with the Americans. Obviously, any legal claim we might have in the instrument was outlawed long ago, but I should be glad to know if Mr. Court would name a price for which it might be purchased and restored to its former owner.

Yours very truly,
E. C. Pickering

31 Harewood Avenue
London, N. W. 1
12 October 1918

Edward C. Pickering
Harvard College
Dear Sir:
 Dr. R. S. Clay has shown me your letter, re the old telescope formerly belonging to Harvard College & I promised him that I would write you about it. With this letter I am sending you a photo of it, & I should be glad if you accept it so that it may be kept at Harvard till (I hope at no very distant date), you get the original instrument itself back.

 I am now & have been for over 20 years a collector of these old optical instruments & I have a very large number selected to show the gradual development both mechanical, & optical, of all kinds of Microscopes, Telescopes, Surveying, Navigation Engineering & Mathematical Instruments & all kinds of Instruments for the determination of time & position.

 About 11 years ago I purchased the Telescope, belonging to Harvard, in the Town of Stafford, here in England & although since then it has been together with a number of others, on loan to the Astronomical Section at the Science Museum South Kensington London W. & I have had several tempting offers to purchase during that time, I have always refused to part with it. First, I have always felt, that although the telescope was mine by purchase, as a matter of justice, its final destination should be Harvard. 2nd It is an extremely fine example of Martin's early work, & up to the present I have never been able to get anything to take its place in my collection. 3rd I am together with a friend, writing a small book on the early opticians, & I want it to illustrate this book. Benj Martin who was born in 1704, died in 1782, (he educated himself) he was originally a farm labourer, afterwards an assistant to Dr. Desaguliers F.R.S. in his Lecture on Natu-

449

ral Philosophy, then a Schoolmaster in Chichester. about 1750 he came to London & till his death he carried on the Businesss of a Mathematical Instrument Maker at 171 Fleet St. he was a clever mathemathician & the author of over 70 books on Science.

The Telescope is a Gregorian, 3-inch aperture in very good condition & in original case. the photo shows only 1 eyepiece but I feel sure that there are two. The photo was taken by the Authorities at South Kensington, they gave me two copies so I can spare one. Finally I do not want to sell this Telescope just now but of this you may be assured that I will never think of selling it without telling Harvard first & giving them the first refusal.

I may mention that in the event of my death the Telescope would be returned to Harvard College & I left directions to that effect in my Will when disposing of my Collection and I have also told the Curator Mr. David Baxandall that I intended it to go back to Harvard.

If you are in London at any time & care to call at the Science Museum Western Galleries, South Kensington, & ask for Mr. Baxandall he will be glad to show it to you & let you examine it in any way you think fit.

I am sorry that I cannot part with it just at present, but when the war is over & the book is published, I will then write you again.

With Compliments
I remain
Yours faithfully
Thomas H. Court

P.S. Although not very plain in this photo the inscription can be made out on the body tube at the end of the Telescope.   THC

November 8, 1918.

Mr. Thomas H. Court,
   31 Harewood Avenue,
     London, N.W. 1, England.
My dear Mr. Court:—

Your letter of October 12 and the photograph have just reached me. Your plan of providing in your will for the return of the telescope to Harvard is a very graceful and generous one. As long as you keep your collection of astronomical instruments, this telescope should remain in your care. The photograph is an excellent one and I am having it mounted and framed.

I am sending you a photograph of a small reflecting telescope, apparently made in the eighteenth century.

With high appreciation of your friendly attitude,

I remain
Yours very sincerely,
E. C. Pickering

27. For donations after the fire see Quincy, *History,* II (Appendix xi), 484–496; Cohen, *Some Early Tools,* pp. 45–49, and Appendix II, pp. 145–151. See also

David Wheatland's beautifully illustrated work, *The Apparatus of Science at Harvard 1765–1800* (Collection of Historical Scientific Instruments at Harvard; Cambridge: Harvard University Press, 1968).

28. *Two Lectures on the Parallax and Distance of the Sun, as Deducible from the Transit of Venus. Read in Holden-Chapel at Harvard-College in New-England in March 1769* (Boston, 1769). See Woolf, *Transits of Venus,* p. 173.

29. *Philosophical Transactions,* 59:351–358 (1769).

30. A. E. Lownes, "The 1769 Transit of Venus and Its Relation to Early American Astronomy," *Sky and Telescope,* 2:3–5 (April 1943).

31. Quoted by Samuel E. Mitchell, "Astronomy During the Early Years of the American Philosophical Society," *Proceedings,* 86:13–21 (1943), esp. p. 16.

32. The extracts quoted on negotiations and other matters connected with Williams's career are drawn from *Corporation Records,* vol. 3 (May 1778–Aug. 1795). See also Bailey, *History,* pp. 8–9; Cohen, *Some Early Tools,* pp. 16, 45, 51–56; Quincy, *History,* II, 263–264; Hindle, *Pursuit of Science,* pp. 335–336; references in Hornberger, *Scientific Thought,* p. 98, n. 40. Williams's own account of his expedition appears in *Memoirs of the American Academy,* 1:86–102 (1785). See also Struik, *Yankee Science,* p. 49. The rediscovery of the site of Williams's observation was reported in the *Christian Science Monitor,* October 17, 1940.

33. *College Papers,* 3:23. On this same page, but not in Williams's own handwriting, there is a slightly different version of his letter of resignation from that in *Corporation Records,* 3:312. See C. K. Shipton, *Sibley's Harvard Graduates* (Boston: Massachusetts Historical Society, 1970), 15:141.

34. *Corporation Records,* 3:314.

35. Cohen, *Some Early Tools,* pp. 57–59.

36. Quincy, *History,* II, 264, 286–287, 298; Morison, *Three Centuries,* pp. 190–191, 195, 211. See "Eulogy at the interment of the Rev. Dr. Webber, by Henry Ware, D. D., pronounced 20th July, 1810" (Harvard Archives).

37. John Lowell (A. B. 1786) served as Fellow in 1810–1822 and Overseer 1823–1827. He was a great-grandfather of President A. Lawrence Lowell. See "The Lowell Dynasty" in Morison, *Three Centuries,* p. 159; Ferris Greenslet, *The Lowells and Their Seven Worlds* (Boston: Houghton Mifflin, 1946).

38. *Corporation Records,* 4 (pt. 2): 79–80 (1806); 5:192 (1815).

39. *Annals of the Astronomical Observatory of Harvard College,* 1:ii (1856). This volume, pp. i–cxc, provides an admirable background for the early history of the Observatory, written by the first director, William Cranch Bond. The only substantial account of the Bonds is by their kinsman, the astronomer Edward S. Holden, *Memorials of William Cranch Bond and of His Son George Phillips Bond* (San Francisco and New York, 1897). See also *DAB,* 2:430–431, 434–435.

40. Farrar, "On the Comet of 1811," *Memoirs of the American Academy of Arts and Sciences,* 3:308; Bowditch on the elements of the comet, *ibid.,* p. 313.

41. Elizabeth Bond, George Phillips Bond's daughter, "Family Sketches Compiled from Old Letters and Diaries and From Descriptions of Persons Given by Old Family Friends" (MS, Harvard Archives), provides invaluable details about the Cranches, Bonds, and other family connections. The Cranches, originally from

Devonshire and "rigid Calvinists" of the narrowest sort, after coming to the New World found their "temper sweetened as they ranged themselves on the Liberal side of Religion and Politics." She tells a delicious anecdote about the Smiths' welcome of Richard Cranch (1726–1811) as a husband for their daughter Mary and their reluctance to accept the country lawyer John Adams for Abigail. (For numerous references to the Cranches, see Page Smith, *John Adams* [2 vols., Garden City: Doubleday, 1962].) Miss Bond paints a somewhat romantic picture of her family's early affluence in Cornwall, of her great-grandfather's success as a silversmith in Plymouth, England, but of repeated failures and mishaps in this country that resulted in the early privations of William Cranch Bond. His tour of inspection in 1815 apparently fired his astronomical ambitions even more, for Miss Bond writes that after his return he was "always speaking of scientists: 'I mean in time to be one of those great men'." Miss Bond died in 1943 at the age of 89.

42. *Annals,* 1:iii. For Farrar see *DAB*, 6:292–293, and other works on early science at Harvard previously cited.

43. One of these letters President Kirkland had written to be delivered by Bond to John Quincy Adams, then United States Minister at the Court of St. James's, to which Adams replied, November 30, 1815:

"Mr. W. C. Bond some time in the month of September delivered to me your obliging favour of the 23rd of June immediately after which I accompanied him to Greenwich with the purpose of introducing him in person to the Astronomer Royal, Mr. Pond; it happened, however, unfortunately that this gentleman when we called at the Observatory was not at home. I was obliged to return myself the same evening to my own house [Ealing, near London] and Mr. Bond remained at Greenwich with the intention of calling on the Astronomer Royal the same evening or the next morning to deliver to him your letter and that of Professor Farrar.

"I have not since that time had the pleasure of seeing or of hearing from Mr. Bond, but I have no doubt that he obtained from Mr. Pond all the information concerning the object of his visit which he could desire. I should have been happy to have given him every other assistance in my power not only because he came furnished with your recommendation but because I felt high satisfaction at the purpose which you have now undertaken of erecting an observatory at Cambridge. If in this or in any other object connected with the venerable Institution over which you preside I can during my residence in this Country render you any service whatever, I flatter myself that you will not only freely require it, but that I shall receive every command from you to that effect as a favour." *College Papers,* 8:5. The rest of the letter is a highly informative view of the conditions in Europe.

44. In the entry of the Bond diary for December 24, 1848, Bond wrote: "Dined yesterday with President Quincy. Miss [Eliza] Quincy read to me some Biographical notices of Miss Herschel; she was 96 years old when she died—retained her faculties, and cheerful disposition to the last. I recollect passing an afternoon at Slow [Slough], Sir William being absent, she kindly undertook the task of. Cicerone and told and shewed me all about the Great Telescope and what she called the more convenient ones of 20 feet, she described to me the modes of observing, with both

kinds—this was 33 years ago." Bessie Z. Jones, ed., "Diary of the Two Bonds, 1846–1849," *Harvard Library Bulletin,* 16:188 (April 1968). For two previous portions of the Diary see *ibid.,* 15:368–386 (October 1967); and 16:49–71 (January 1968).

45. *Corporation Records,* 6:64.

46. The Perkins bequest was officially noted on August 28, 1822, to be applied when payable "for the purpose of founding such a professorship as the said President & Fellows with the concurrence of the overseers shall judge will be most useful to the University and the public." The funds, however, did not come to the University until nearly twenty years later.

47. Adams's letters to Davis and to George Adams, his son (Adams Ms. Trust, on reels 147 and 148, Documents Division, Widener Library, Harvard), quoted by permission of the Massachusetts Historical Society.

48. *Annals,* 1:lxx–lxxii, esp. item 4.

49. *College Papers,* 10:72. On the loss of the state subsidy see Morison, *Three Centuries of Harvard,* pp. 218–221.

50. By the will of James Smithson, an Englishman, the sum of over half a million dollars was left to this country for the "increase and diffusion of knowledge among men." The debate about its use lasted for ten years. John Quincy Adams's determined fight for application of the fund to an observatory failed, but he prevented its dissipation for purposes he considered unsuitable, and in 1846 the struggle resulted in the establishment of the Smithsonian Institution. See William H. Rhees, "Extracts from the Memoirs of John Quincy Adams," *Miscellaneous Collections,* Smithsonian Institution, 17:763–801; John Quincy Adams, *The Great Design: Two Lectures on the Smithson Bequest by John Quincy Adams* (Wilcomb E. Washburn, ed.; Washington: Smithsonian Institution, 1965); Bessie Z. Jones, *Lighthouse of the Skies: The Smithsonian Astrophysical Observatory, Background and History* (Washington: Smithsonian Institution, 1965), chap. 1 and pp. 292, 308, n. 12.

51. *An Oration Delivered before the Cincinnati Astronomical Society on the Occasion of Laying the Cornerstone of an Astronomical Observatory the 10th of November* (Cincinnati, 1843); *The Centenary of the Cincinnati Observatory, 1843–1943* (Cincinnati: Publications of the Historical and Philosophical Society of Ohio, 1944). In a letter of October 3, 1843, to Ormsby MacKnight Mitchel, Adams wrote that he was leaving home on the 25th and allowing himself 13 days "to arrive at Cincinnati by the way of Buffalo, Ashtabula and Cleveland . . . If a spark of enthusiasm for the cause of Science, and the honour of our Country, burns in my bosom, it shall live until the cornerstone of your observatory shall have been laid, nor shall it be delayed an hour by any neglect, indolence, or indifference of mine."

52. See *Annals,* vol. 1, Appendix, for various reports during Adams's chairmanship of the Visiting Committee, and esp. pp. cxxvi–cxxviii for Quincy's tribute to him; also Quincy's *Memoir of the Life of John Quincy Adams* (Boston, 1859), *passim.*

53. Elias Loomis, *The Recent Progress of Astronomy, Especially in the United States* (New York, 1850, 1856). See also W. C. Rufus, "Proposed Periods in the

History of Astronomy in America," *Popular Astronomy,* 29:393–404, 468–475; and his "Astronomical Observatories," *Scientific Monthly,* 19: no. 2, 120–139; David F. Musto, "A Survey of the American Observatory Movement, 1800–1850," *Vistas in Astronomy* (Arthur Beer, ed., London and New York: Pergamon Press, 1967), 9:87–92.

54. G. B. Airy, *Report of the First and Second Meetings of the British Association for the Advancement of Science, at York in 1831, and at Oxford in 1832* (London, 1833), pp. 125–189 (May 2, 1832).

55. James Love, *The Nation,* Aug. 16, 1888; F. P. Venable, *Sigma Xi Quarterly,* Sept. 1920.

56. David F. Musto, "Yale Astronomy in the Nineteenth Century," *Ventures* (magazine of the Yale Graduate School), vol. 8, Spring, 1968.

57. The development of the firm of Alvan Clark & Sons as the leading American makers of telescopes is one of the remarkable achievements of the latter half of the nineteenth century. On April 24, 1846, William Bond wrote in his diary, "recd a note from Mr. Alvan Clark inviting George & myself to try his new Reflecting Telescope made by himself he says he has now done his utmost to perfect it. I have been three times already when it was said to be in good order, but its performance was quite inferior—giving nothing but a round disc to a star. It is the same old story, I think; dozens of these small things have been made heretofore and the same story told about them—thus far there is no proof of his having accomplished any thing uncommon, and yet one is blamed and thought hardly of, for not extolling it as a wonderful affair"; *Harvard Library Bulletin,* 15:372–373 (October 1867). Within a short time the Clarks' services to the Harvard Observatory became indispensable. Among the important telescopes they built were the 36-inch for the Lick, the 30-inch for Pulkovo, and the 18½-inch (Dearborn Observatory) instrument with which Alvan Clark discovered the companion star of Sirius. See Chapter II, part v, and note 93 below; *DAB,* 4:119–120; Deborah Jean Warner, *Alvan Clark & Sons, Artists in Optics* (Washington: Smithsonian Institution Press, 1968).

58. Quincy, *History,* II, 656.

59. *Ibid.,* p. 661.

CHAPTER II

1. Josiah Quincy, *History of Harvard University* (2 vols., Cambridge, 1840), II, 391–392; Appendix, pp. 566–568.

2. *Proceedings of the American Academy of Arts and Sciences,* 4:167–169 (February 8, 1859).

3. The Expedition discovered the Antarctic Continent and explored islands of the Pacific and the Continental Northwest. *The Narrative of the United States Exploring Expedition* was published in 5 volumes (1844), and many of the collections made formed a large part of the Smithsonian Museum.

4. *Annals of the Astronomical Observatory of Harvard College,* 1:lxxxv (1856).

5. Bond Correspondence, Harvard Archives. The Dorpat telescope was later

surpassed by the 15-inch instrument at Pulkovo, which provided the model for Harvard's.

6. Mitchell (1791–1869) was an able amateur astronomer, member of the Visiting Committee for many years, and father of the noted astronomer Maria Mitchell.

7. *A Popular History of Astronomy During the Nineteenth Century* (2nd ed., Edinburgh, 1887), pp. 129–132, esp. p. 130.

8. See his delightful chapter, "The End of the World," in *Errand into the Wilderness* (Cambridge: Harvard University Press, 1956), pp. 217–239.

9. François Arago, *Tract on Comets* (John Farrar, tr., Boston, 1832), p. 58.

10. Followers of William Miller (1782–1849), who predicted the Second Coming in 1843. When it failed to happen he moved the date to 1844.

11. Peirce's lecture, contemporary accounts of the comet, and details of the meeting to plan for a telescope are found in the Observatory Scrapbook, 1843–1850 (Harvard Archives).

12. (1787–1871), a Harvard graduate of 1810. His exchange of letters with Quincy is in the Bond Correspondence.

13. The list of donors to the Observatory, 1839–1855, is given in *Annals,* 1, lxiii–lxvi.

14. F. G. W. von Struve (1793–1864) was the first director of the Russian Imperial Observatory at Pulkovo and founder of a distinguished family of astronomers (see Chapter V, note 15). The cordial relation between the Harvard and Russian astronomers, especially with the Struves, is a striking feature of Harvard Observatory history.

15. E. S. Holden, *Memorials of William Cranch Bond and of His Son George Phillips Bond* (San Francisco and New York, 1897), p. 22.

16. On Tuesday, April 28, 1846, Bond records in his diary, "Mr. Eliot has decided that the Dome shall be placed on spherical balls of eight inches diameter."

17. Holden, *Memorials,* pp. 68–69.

18. *Annals,* 1:lxxx. Quincy's entire letter, pp. lxxii–lxxxii, provides an excellent statement of the steps taken in the establishment of the Observatory.

19. See Chapter I, note 44.

20. S. E. Morison, *Three Centuries of Harvard* (Cambridge: Harvard University Press, 1936), p. 276.

21. O. M. Mitchel published Mädler's papers in *Sidereal Messenger,* vol. I (September and October 1846); the quotation is from vol. 1, no. 3.

22. Mädler had used Bradley's star catalogues for an important part of his analysis.

23. *Outlines of Astronomy* (London, 1851), p. 589.

24. *The Centenary of the Cincinnati Observatory, 1843–1943* (Cincinnati: Publications of the Historical and Philosophical Society of Ohio, 1944), pp. 47–56.

25. Johan Gottfried Galle (1812–1910) found the planet Neptune, September 23, 1846, on the basis of calculations by the French astronomer-mathematician Urban J. J. Leverrier (1811–1877). Elias Loomis, *The Recent Progress of Astronomy* (New York, 1850, 1856), chap. I, discusses it in clear outline just four years after its discovery. An excellent recent history of the discovery, with extensive bibliography, of the persons and circumstances connected with the famous astronomical

event is by Morton Grosser, *The Discovery of Neptune* (Cambridge: Harvard University Press, 1962). The Bonds' observations on Neptune were reported in the Royal Astronomical Society's *Monthly Notices,* 7:273; *Proceedings of the American Academy of Arts and Sciences,* 1:50–51 (1848); and *Astronomische Nachrichten,* 25:231–234, 301–302 (1847).

26. An amusing sidelight on Challis's failure to recognize the planet appears in a query addressed to E. C. Pickering by Edward Emerson, January 23, 1912, when editing his father's journal. An entry for January 1847 cryptically said, "A Scholar brooks no interruption. For want of posting books at Greenwich the star was lost by Adams and England." Pickering doubted that England lost the "star" (Neptune) because of failure to post the books. In fact, Galle had a better star map.

27. The version of the letter in *Annals,* 1:cxxi–cxxii, has undergone some editing, but the substance is the same.

28. In 1882 Holden prepared an elaborate monograph, *The Central Parts of the Nebula of Orion* (Washington: U. S. Navy), in which he carefully reviewed the work of the Bonds and others on this nebula.

29. Benjamin Apthorp Gould (1824–1896), though at first not particularly unfriendly to the Bonds, later disparaged their work (see Holden, *Memorials,* pp. 37–39) and was one of the organizers instrumental in keeping George Bond out of the National Academy of Sciences (see Chapter II, part v). For the Dudley controversy see *Dudley Observatory: Inauguration and Other Papers* (Albany, 1856–1859).

30. John Adams Whipple (1822–1891), for whose daguerreotype and photographic work the best sources are Robert Taft, *Photography and the American Scene* (New York: Dover Publications, Inc., 1964), pp. 69, 74, 76, 114, 120, and notes; M. A. Root, *The Camera and the Pencil* (Philadelphia and New York, 1864), esp. pp. 364–365, 371.

31. *Comptes Rendus,* 8:4–6 (1839). See also Taft, *Photography and the American Scene;* Helmut and Alison Gernsheim, *L. J. M. Daguerre: The World's First Photographer and Inventor of the Daguerreotype* (Cleveland and New York: World Publishing Co., 1956).

32. Taft, *Photography,* pp. 9, 29, 31, 104, 105, 112, 113, and notes.

33. William Henry Fox Talbot (1800–1877), pioneer English photographer. See Taft, *ibid.,* pp. 102ff, 416, 436, and notes. He considers Herschel and Talbot founders of photography. See also Root, *Camera and Pencil,* pp. 340, 400, 415.

34. *Comptes Rendus,* 9:250ff. (1840), esp. p. 263. This account is said to be the first in any scientific medium.

35. See Taft, *Photography, passim,* and Root, *Camera and Pencil,* pp. 339–343.

36. Donald H. Fleming, *John W. Draper and the Religion of Science* (Philadelphia; University of Pennsylvania Press, 1950), pp. 38–39. See Daniel Norman, "John W. Draper's Contributions to Astronomy," *Telescope,* 5 (no. 1): 11–16 (1938), and "The Development of Astronomical Photography," *Osiris,* 5:560–594 (1938). Norman states, "There seems to be no question that Draper and Becquerel discovered, simultaneously and independently, the ultra-violet lines; but Draper undisputedly was the discoverer of the new infrared lines" (p. 561, n. 8).

37. See Chapter V below.

38. Root (*Camera and Pencil,* pp. 371, 380–381) praises Black's skill and energy as the first successful "heliographer" to attempt images taken from a balloon. On September 15, 1860, George Bond wrote Black, "Learning that you are proposing to take photographic views from a balloon, I should be glad to have you try to ascertain in how much less *time* the upper surface of a dense cloud in full sunshine will make a strong negative than the landscape view of the earth requires." (Holden, *Memorials,* p. 189.)

39. *Annals,* 1: cxxviii.

40. John White Webster (1793–1850), professor of chemistry at Harvard, was executed for the murder of George Parkman, to whom he was in debt. There are numerous accounts of this famous case, but see *DAB,* 14:592–593. It is reported that a reexamination of the medical and legal questions involved in the case is being undertaken by Dr. Warren R. Guild, of the Harvard Medical School, and Justice Robert Sullivan, of the Superior Court, Suffolk County, Massachusetts, in a projected book, *Twine and Noose (Harvard Crimson,* May 13, 1969).

41. *Annals,* I: cxlix. It is interesting to note that Whipple himself did not recall the year of this experiment, for in 1888, suggesting "the neighborhood of 1850 or 1851," he asked Pickering to consult the records, and a few months later, January 25, 1889, wrote again to say of Pickering's stellar photography, "It is wonderful the results you have attained in that line. I am thankful for it."

42. Observatory Scrapbook. Experiments at the Observatory and items in the notebook are traced in detail in the excellent pamphlet by Dorrit Hoffleit, "Some Firsts in Astronomical Photography" (Harvard Observatory Monograph, 1950), esp. pp. 24–27. See also Daniel Norman, "The Development of Astronomical Photography," *Osiris,* 5:560–595 (1938), which has numerous references to George Bond's significant work in celestial photography; earlier articles in *Photographic Art Journal,* 5:334–341 (1853), esp. pp. 337–338 for the Bonds; and Edward S. Holden, "Photography the Servant of Astronomy," *Overland Monthly* (November 1886).

43. *Annals,* 1:clvii.

44. Quoted in *American Journal of Photography,* n.s., 3:180 (1859–1860).

45. *Memoirs,* n.s., 8:221–286 (1860). See also, "On the Relative Brightness of the Sun and Moon," *ibid.,* pp. 287–298. For modern recognition ot the "Bond albedo," derived from Bond's photometric studies, see Gerard P. Kuiper and Barbara M. Middlehurst, eds., *Planets and Satellites (The Solar System,* vol. 3; Chicago: University of Chicago Press, 1961), pp. 306ff.

46. *Proceedings,* 3:386–389 (1857).

47. Elizabeth Bond, "Early Recollections of Harvard College Observatory," MS, Harvard Archives.

48. *Monthly Notices* 17:18 (1858); quoted by Dorrit Hoffleit, "Some Firsts in Astronomical Photography," pp. 31–32.

49. Pp. 77–84. A previous article in the *Almanac* for 1858 (pp. 80–82) had discussed the photographing of Vega (α Lyrae).

50. "Stellar Photography," *Astronomische Nachrichten,* 47:1–6 (1858); 48:1–14 (1858); 49:81–100 (1859).

51. Quoted in Holden, *Memorials,* pp. 154–159. A portion of this letter is quoted also by Harlow Shapley and Helen E. Howarth (eds.), *A Source Book in Astronomy* (New York: McGraw-Hill, 1929), pp. 267–269.

52. Quoted in Holden, *Memorials,* pp. 211–212. See Draper's answer, *ibid.,* pp. 212–213.

53. Letter to J. Ingersoll Bowditch, March 31, 1860.

54. Daniel Norman, "The Development of Astronomical Photography," *Osiris,* 5:573–578 (1938).

55. The priority of Bond's invention of the clock became a sharp issue between him and John Locke (1792–1856). In *Annals,* I:xxi–xxviii, Bond quotes at length from the account of Sears Walker on the question, drawn from the Coast Survey records. Clippings of a brisk correspondence between Locke and Bond may be found in the Observatory Scrapbook, 1843–1850.

56. Gould began his *Astronomical Journal* in 1849, directed to the professional astronomer. He transferred it to Albany during his brief unhappy connection with the Dudley Observatory, suspended it in 1861, and resumed it in 1886. Articles by the Bonds appeared in some of the earlier issues, but they were not altogether happy in their relation with Gould. Bond wrote C. H. F. Peters, August 20, 1858, "I agree with you in thinking that Dr. Gould's publications have not been ·impartial, of this I have had ample experience and he is aware of it." George Bond wrote this same correspondent in 1859 that "there should be some concert among our astronomers in regard to publication. The *Astronomical Journal,* aside from the personal relations of the editor, is late in appearance, expensive and of little use." The need was for something cheap and prompt.

57. *Memoirs,* n.s., 4:189–208 (1849). Encke's letter to Bond is quoted in Holden, *Memorials,* pp. 153–154. Encke reviewed Bond's paper in *Astronomische Nachrichten,* 34:349–360 (1852).

58. Communicated to the American Academy April 15, 1856, in *Memoirs,* n.s., 6:179–212 (1859).

59. See Samuel E. Morison, ed., *The Development of Harvard University* (Cambridge: Harvard University Press, 1930), pp. 413–427; *The Autobiography of Nathaniel Southgate Shaler* (Boston: Houghton Mifflin, 1909); William B. Rogers, *Life and Letters* (2 vols., Boston, 1896), I, 272–274, 279–280, 309–310; II, 30.

60. Edward Lurie, *Louis Agassiz* (Chicago: University of Chicago Press, 1960), pp. 136–141, 163–165, 321–322, 325–327.

61. In 1887 President Eliot wrote Pickering: "A serious demand has arisen for a good course in observational astronomy. The demand comes from five good students who want to learn all the astronomical processes which are used in geodetic work. The application is backed by several of our best professors, and it seems to me for the interest of the University as a whole that a course be provided . . . The Scientific School as such has no means of paying for such a course and unless the Observatory can give it, I fear that it cannot be given." In his answer Pickering outlined a program for instruction in astronomy but insisted that, since the Observatory was organized strictly for research and had neither staff nor money to under-

take teaching, he could not offer a course or spare any of his assistants. To this letter Eliot replied (November 3, 1887) that, though bowing to Pickering's decision, he had serious doubts about excluding teaching from the functions of the Observatory: "An institution which trains nobody in astronomy is sure to be dependent upon outsiders for its astronomical staff. This inevitable result is, I confess, somewhat unwelcome to me. The more the Observatory gets organized as an observing and computing machine, the more will this be true. For persons employed in it on what may be called the machine plan, will never be trained by that employment for the highest functions." The appointment of Robert W. Willson (A.B. 1873) as instructor in astronomy in 1891 brought new life to the teaching of the subject in the College. At first a part of the Engineering Department, in 1897 it became a special section of the Department of Physics. In 1903, appointed professor of astronomy, Willson introduced laboratory methods, and his course in Navigation and Nautical Astronomy, a major interest of his from the beginning, proved of great value to the government in World War I and afterward. Willson was succeeded by Harlan T. Stetson. Under the Shapley administration a closer association of the Observatory and the University was established. See Solon Bailey, "Astronomy, 1877–1927," in S. E. Morison, ed., *The Development of Harvard University*, chap. 18, pp. 292–306.

62. *Annals,* 1:clxxxv.

63. *Ibid.,* cxxv. Actually, Lassell's discovery came two days later (September 18). See Chapter X, note 7.

64. William Bond, "Description of the Nebula about the Star θ Orionis," *Memoirs of the American Academy,* n.s., 3:87–96 (1848); George Bond, "An Account of the Nebula in Andromeda," *ibid.,* pp. 75–86.

65. *Annals,* 1:lix–lxiii.

66. Bond's observations, drawings, and analysis and the reports of Peirce's lecture are found in newspaper clippings, Harvard Archives.

67. Reported in the Boston *Traveller,* February 14, 1852.

68. In 1863 George Bond visited Clerk Maxwell in London and discussed the rings with him. See Holden, *Memorials,* pp. 129, 203–206, 258; Clerk Maxwell's prize essay (1856), "On the Stability of the Motion of Saturn's Rings," is given in Shapley and Howarth, *Source Book,* pp. 274–276; Keeler, "A Spectroscopic Proof of the Meteoric Constitution of Saturn's Rings," *ibid.,* pp. 394–396.

69. For extracts from his diary of the journey see Holden, *Memorials,* pp. 87–122.

70. Peirce maintained that Neptune was not the planet "to which geometrical analysis has directed the telescope," that the orbit "was not contained within the limits of space which have been explored by geometers searching for the source of the disturbance of Uranus," and that its discovery by Galle "must be regarded as a happy accident." His communications to the American Academy on March 14, May 4 (1847), and April 4 (1848) were published in *Proceedings,* 1:57–68, 144–149, 331–342. See *Sidereal Messenger,* 2:68–71 (1848), for an exchange of letters between Leverrier and Peirce.

71. Paine (1803–1885) was a grandson of the signer of the Declaration of Independence, a devoted friend of the Bonds and the Observatory, to which he be-

queathed his entire fortune amounting to over $325,000. The will was unsuccessfully contested. A member of the Visiting Committee for many years, he was an able observer and computer. The Paine Professorship of Practical Astronomy, usually held by the director, is named for him. See Solon Bailey, *History,* pp. 280–281.

72. Charles Wesley Tuttle (1825?–1881), was a staff member 1850–1854; see Bailey, *History,* p. 253. For Safford see *DAB,* 16:287–88; "Diary of the Two Bonds," *Harvard Library Bulletin,* 15:385 (1967); 16:182 (1968); *Popular Astronomy,* 21:473–479 (1913); Bailey, *History,* pp. 257–258; Chapter III below.

73. The marked catalogue of items purchased and prices paid (Bond Correspondence, Harvard Archives), reveals Bond's discernment about works worth acquiring. He sold off duplicates and bought other books with the proceeds. See Dorrit Hoffleit, "The Library of Harvard College Observatory," *Harvard Library Bulletin,* 5:102–111 (1951). An interesting list compiled by the Bonds of their own possessions and Phillips Library books includes many indispensable works, a number in other languages.

74. On April 28, 1855, at a meeting of the Corporation Josiah Quincy announced that he wished to connect his father's name with the College by devoting to the "permanent interests of science" a fund left by him. Quincy stated that during his own lifetime he would contribute $600 annually for the publication of the astronomical observations of the principal observer, and would bequeath the sum of $10,000 to be known as the Quincy Fund. In accepting the gift the Corporation noted it as "a mark of his interest in the University and especially in that important branch of it, the Observatory, which has been so munificently endowed mainly through his exertions and influence."

75. See Marcus Baker, "Survey of the Northwest Boundary of the United States, 1857–1861," *Bulletin of the United States Geological Survey, No. 174* (Washington, 1900); Charles H. Carey, *A General History of Oregon* (2 vols., Portland: Metropolitan Press, 1936), II, 459–461. An item in the *New York Times,* May 5, 1968 (p. 44), points out the inaccuracy of the survey and the loss to Canada of 822 feet of coastline.

76. See Eufrosina Dvoichenko-Markov, "The Pulkovo Observatory and Some American Astronomers in the Mid-Nineteenth Century," *Isis,* 43:243–246 (1952). In the siege of Leningrad, when German artillery razed the Observatory, the 15-inch refractor was destroyed. The article concludes, "Its companion at Cambridge [is] the only remembrance and symbol of the cultural collaboration which united the world of science of both countries but which is so difficult to reconstruct in our atomic age." See Frances Montgomery Cowan, "An Account of the Founding of the Central Observatory of Poulkova, Russia," *Popular Astronomy,* 20:415–423 (1912).

77. Everett's articles in this series were originally published in the *New York Ledger.* In Number 5 he described a visit to the Observatory (Houghton Library).

78. Peirce Papers, Houghton Library. Quoted by permission of the Harvard College Library.

79. Holden, *Memorials,* p. 36.

80. *Ibid.* p. 164.

81. *Ibid.,* p. 162; letter to Peirce, p. 163. Bache was head of the Coast Survey; he had been professor of physics and chemistry at the University of Pennsylvania.

82. Peirce Papers, Houghton Library. Quoted by permission of the Harvard College Library.

83. Lurie, *Agassiz,* p. 182 and references in index, p. 443. See also Nathan Reingold, *Science in Nineteenth Century America* (New York: Hill and Wang, 1964), index under Lazzaroni, p. 334.

84. James Melville Gilliss (1811–1865) led his expedition to Chile, 1849–1852, for a determination of the sun's distance (see *DAB,* 7:292–293). Peirce evidently followed Gould's line in what William Bond called "Dr. Gould's misrepresentations" about the Harvard Observatory contributions to Gilliss's work. In a sharp letter to William Mitchell, chairman of the Visiting Committee, September 22, 1858, William Bond had written that either "through gross ignorance, or evil intent," Gould "speaks in a very disparaging manner of our observations," calls them useless, though "previous to penning these remarks Dr. Gould had been informed of the series of differential observations . . . made at this Observatory on the planet Mars E and W of the meridian, near the opposition in 1849–50, which have since been reduced and have given consistent and satisfying results and proved that this method is the best, all things considered, that has ever been practised . . . I have been told that Dr. Gould has been formerly quite merry on the subject of the Harvard College Observatory right ascension observations . . . of late however he begins to see that 'there are more things in Heaven and Earth than are dreamt of in his Philosophy'."

85. See Chapter XI below.

86. Christian Henry Frederick Peters (1813–1890), whose letters to George Bond are among the most interesting in the entire collection, and who maintained his friendly association with Bond's successors, Winlock and Pickering. See *DAB,* 14:502–504; see also the *Dudley Observatory: Inauguration and Other Papers,* for his role in the Gould controversy.

87. Bond's daughters sent recollections and copies of letters to Holden for his *Memorials.* For touching reminiscences of their father, see pp. 48–63.

88. Quoted in Holden, *Memorials,* pp. 167–168, esp. note, p. 168.

89. See Hall, "My Connection with the Harvard Observatory and the Bonds, 1857–62," in Holden, *Memorials,* 77–80; Bailey, *History,* pp. 254–255; H. S. Pritchett, "Asaph Hall," *Popular Astronomy,* 16:67–70 (February 1908); *DAB,* 8:117–118. Horace Tuttle, a brother of Charles, was a staff member 1858–1862. Aside from Bond's memorial pamphlet (in Archives), the most definite information about Coolidge and his romantic career is found in Shaler's *Autobiography.*

90. Lurie, *Agassiz,* p. 165.

91. Almost a hundred years after its appearance, Henry Norris Russell wrote, "Bond's beautiful drawings of Donati's comet of 1858 are, and probably will remain, the classic example of what can be accomplished by the combination of a keen eye and a skilful hand"; "America's Role in the Development of Astronomy," *Proceedings of the American Philosophical Society,* 91:1–16 (1946).

92. Interesting confirmation of Bond's doubts about the viability of new observa-

tories improperly supported appears in a letter Joseph Winlock wrote to Leverrier, March 4, 1874, in which, after describing the work at Harvard since he became director, he adds, "The Chicago Observatory is . . . not now doing anything on account of lack of means. Prof. Safford was Director of it but I believe that he is now in Washington employed upon the Government Surveys of our Western Territories."

93. George Bond contributed to the *American Journal of Sciences and Arts,* 33 (March 1862), his own observations of the companion of Sirius with the Harvard refractor, and T. H. Safford communicated to the American Academy, May 26, 1863, a paper "On the Observed Motions of the Companion of Sirius."

94. For extracts from his diary see Holden, *Memorials,* pp. 122–149.

95. In *Proceedings,* American Academy, 6:169–175 (1863).

96. See F. W. True, *The National Academy of Sciences: A History of the First Half Century, 1863–1913* (Washington: Smithsonian Institution, 1913), which contains a list of the first members, pp. 104–105. For a less orthodox account see A. Hunter Dupree, *Science and the Federal Government,* chaps. 6 and 7, esp. pp. 135–148; see also Dupree, *Asa Gray, 1810–1888* (Cambridge: Harvard University Press, 1959); Reingold, *Science in Nineteenth Century America,* pp. 200–225; Lurie, *Agassiz,* pp. 331–335. Unless otherwise noted, the letters quoted about the Academy are in Reingold, pp. 203, 204, 206, 209, 212–214, 221, 222.

97. *Life and Letters,* II, 154–155.

98. *Ibid.,* 161–162.

99. See Bailey, *History,* pp. 255–256.

100. Quoted in Holden, p. 214.

101. "Observations upon the Great Nebula of Orion," *Annals,* vol. 5 (1867), edited by Safford.

102. "Address on Presenting the Gold Medal of the Society to Professor G. P. Bond," *Monthly Notices,* 25:125–137 (1865).

103. William Rogers wrote his brother Henry, February 25, 1865, "George P. Bond, the most brilliant of American astronomers, was buried a few days since, having died of consumption at the early age of thirty-nine. His merits have earned him a high reputation abroad, and have brought to his tomb, but too late to cheer his heart, the highest testimonial of the Royal Astronomical Society of London" (*Life and Letters,* II, 226). For an eloquent obituary of George Bond, see *Proceedings of the American Academy of Arts and Sciences,* 6:499–500 (1862–1865).

CHAPTER III

1. Details associated with Winlock's election may be found in *Overseers Records,* vol. 10, under dates noted. For his biography see *DAB,* 20:389–390; F. W. True, *The National Academy of Sciences* (Washington: Smithsonian Institution, 1913), pp. 195–197; Joseph Lovering, *Biographical Memoirs of the National Academy,* I:329–343 (1877).

2. Walker and Mitchell had been associated for some time with the affairs of

the Observatory. Muzzey, appointed Overseer in 1860, was more closely concerned with the Divinity School.

3. *Reminiscences of an Astronomer* (Boston: Houghton Mifflin, 1903), p. 67.

4. Hilgard evidently had advance word of the confirmation, for he wrote Professor Benjamin Peirce on January 3 that he was glad to hear of it but wished Winlock's "present place could be made more acceptable to him by an increase of salary and that by his declining the Observatory Gould would have another chance." (Peirce Papers, Houghton Library.)

5. Solon I. Bailey, "The Cambridge Astronomical Society," Harvard Reprint No. 45 from *Popular Astronomy*, 36:226–229 (April 1928). MS. copy of Minutes (1854–1857), Phillips Library.

6. A. H. Dupree, *Science in the Federal Government* (Cambridge: Harvard University Press, 1957), p. 103.

7. For Bunsen's use of Draper's tithonometer, see Sir Henry Roscoe, "Bunsen Memorial Lecture," *The Golden Age of Science* (Bessie Z. Jones, ed., New York: Simon and Schuster, 1966), p. 378. This is a slightly condensed version of the full text reprinted in *Smithsonian Annual Report* (1899), pp. 605–644. For Draper's own articles on his tithonometer, see the bibliography in Donald H. Fleming, *John W. Draper and the Religion of Science* (Philadelphia: University of Pennsylvania Press, 1950), p. 188.

8. *Philosophical Magazine*, ser. 3, 3:343 (1847). See Daniel Norman, "John W. Draper's Contributions to Astronomy," *Telescope*, 5:11–16 (1938). Norman states that Kirchhoff mentioned Draper's work in a footnote in his first report in *Poggendorf's Annalen*, 9:275 (1860), but later removed it.

9. "On the Nature of Flame, and on the Condition of the Sun's surface," *Philosophical Magazine*, ser. 4, 15:90–93 (1858). See also "On the Measurement of Chemical Light," *ibid.*, 14:161–164 (1857); "On the Diffraction Spectrum," *ibid.*, 13:153–156 (1857).

10. *Golden Age of Science*, pp. 380–381.

11. Lectures on Spectrum Analysis, delivered at the Royal Institution of Great Britain, 1861; published in *Chemical News*, 4:118–122 (1861); 5:218–222, 261–265, 287–293 (1862), and *Spectrum Analysis*, six lectures, delivered in 1868, before the Society of Apothecaries (London, 1869). By 1888, when the Smithsonian Institution published a 423-page *Index to the Literature of the Spectroscope*, compiled by Alfred Tuckerman, it contained 3829 titles by 799 authors, and a continuation in 1902 ran to 373 pages. *Smithsonian Miscellaneous Collections*, vols. 32 (1888) and 41 (1902).

12. *Scientific Papers* (Publications of Sir William Huggins Observatory, 2 vols, London, 1909), II, 6.

13. "The Celebration of the Semi-Centennial of the Chicago Astronomical Society and the Dedication of a Tablet to the Memory of Truman Henry Safford," *Popular Astronomy*, 21:473–479 (1913). E. C. Pickering sent a paper for this occasion, in which he described Safford's life at Harvard.

14. There is no adequate biography of Langley. See *DAB*, 10:594–597; Bessie Z. Jones, *Lighthouse of the Skies* (Washington: Smithsonian Institution, 1965),

esp. chaps. 5–7; "Samuel Pierpont Langley: Memorial Meeting," *Smithsonian Miscellaneous Collections,* vol. 49 (with addresses by Andrew D. White, E. C. Pickering, and others); Paul H. Oehser, *Sons of Science: Story of the Smithsonian Institution and Its Leaders* (New York: Schuman, 1949), chap. 7.

15. William Allen Miller (1817–1868), a close collaborator of Huggins. See *Dictionary of National Biography,* 13:429.

16. Although Peirce joined the staff in 1868, during 1872 and 1875 he combined research for the Coast Survey and the Observatory. The available material by and about Peirce is now voluminous (*Collected Papers,* 8 vols.; Cambridge: Harvard University Press, 1931–1968), but sources most relevant to his Observatory work include correspondence in the Winlock and Pickering papers; Solon I. Bailey, *The History and Work of the Harvard Observatory* (New York: McGraw-Hill, 1931); Victor F. Lenzen, "Charles S. Peirce as Astronomer, in E. C. Moor and R. S. Robin (eds.), *Studies in the Philosophy of Charles Sanders Peirce,* Second Series (Amherst: University of Massachusetts Press, 1964); letters in Nathan Reingold, *Science in Nineteenth Century America* (New York: Hill and Wang, 1964), pp. 226–235; Introduction to Philip Weiner (ed.), *Values in a Universe of Chance* (Garden City: Doubleday, 1959); Carolyn Eisele, "Charles S. Peirce: Nineteenth Century Man of Science," *Scripta Mathematica,* 14:305–324 (1959); *DAB,* 14:389–403. Peirce's *Photometric Researches* appeared as vol. 9 of the *Annals* in 1878. For his part in the development of Harvard photometry, see Chapter IV below.

17. Austin was a staff member from 1869 to 1871, worked chiefly on investigations of nebulae, and assisted Rogers in preparation of material for *Annals,* vol. 10.

18. See Bailey, *History,* pp. 258–259, and numerous references throughout the volume; *DAB,* 16:534–535. Searle's unpublished manuscript, "Incidents of Eighty Years, 1837–1917," was consulted through the courtesy of Miss Margaret Harwood. George Searle (1839–1918), a Harvard graduate of 1857, served at the Dudley Observatory for 1 year (1858–1859), the Coast Survey, 1859–1862, the United States Naval Academy, 1862–1864, and the Harvard Observatory, 1866–1869. He then joined the Paulist Order, of which he was Superior General for many years. See Bailey, *History,* pp. 259–260.

19. The Club also included three members who became Presidents of Harvard—Quincy, Everett, and Walker. Pickering, commenting on the Club, noted that all its members were very young when elected, but that their longevity was remarkable.

20. *Annals,* 8:48ff.

21. The work on the zone allotted to Harvard, begun November 10, 1870, was completed on January 26, 1879. Final publication, however, came only in 1894 and 1896. An excellent account of the entire program is given in an unpublished MS in the Phillips Library, "The Harvard Studies in Stellar Astronomy, 1840–1890," by Brian Gee, University of London thesis, 1968. Pickering announced the completion of the observations in his report for 1878–1879.

22. Young, in an informative account of previous eclipses, "The Recent Solar Eclipse," *Princeton Review,* 2:865–888 (November 1878), esp. p. 867.

23. The interest in this hypothetical planet, which was called Vulcan, dates from 1859, when an amateur French astronomer, Dr. Lescarbault, announced having observed such a body. No one has seen it since.

24. See Bailey, *History,* pp. 239–240.

25. *Annals,* 8:59.

26. Young, "The Recent Solar Eclipse," *Princeton Review,* 2:868 (1878).

27. *The Sun and the Phenomena of its Atmosphere* (New Haven: University Series, No. 8, 1872), p. 197.

28. *Spectrum Analysis in its Application to Terrestrial Substances and the Physical Constitution of the Heavenly Bodies,* translated from the second enlarged and revised German edition by Jane and Caroline Lassell, ed., with notes by William Huggins (London and New York, 1872), pp. 334–335 and 341–342. In one of his notes (p. 125) Huggins describes Winlock's spectroscopes and his method of registering the different parts of the spectrum.

29. Winlock's few contributions to the literature of spectroscopy are briefly reported in "A Reliable Finder for a Spectro-Telescope," *Journal of the Franklin Institute,* ser. 3, 60:295 (1870); "Apparatus for Recording the Positions of Lines in the Spectrum, Especially Adapted to Solar Eclipses," *Proceedings of the American Academy of Arts and Sciences,* 8:299 (1868–1873); "Application of the Spectroscope to Observations of the Sun," *ibid.,* p. 330; "Lines in the Green near Σ (Corona Lines)," *Nature (London),* 7:182 (1872–73). Between 1867 and 1874 Winlock sent a number of contributions to *Astronomische Nachrichten* on observations of comets and asteroids which Pickering made use of in *Annals,* vol. 13 (1882).

30. "Researches on Solar Physics," *Philosophical Transactions of the Royal Society of London,* 159:1–110 (1870); 160:389–496 (1870).

31. See his "Memoir" of Bache, *Smithsonian Annual Report* (1870), p. 108.

32. By a compromise to which Henry agreed, the engravings were published in 1876 in *Annals,* vol. 8, part 2, as "Astronomical Engravings Illustrating Solar Phenomena Prepared at The Astronomical Observatory of Harvard College," with acknowledgement to the Bache Fund for the Advancement of Science for 12 of the 35 plates specifically identified by number. Explanatory notes, though not as originally planned by Winlock, were provided by Arthur Searle.

33. *Annals,* 8:7.

CHAPTER IV

1. Quoted by Nathan Reingold, *Science in Nineteenth Century America* (New York: Hill and Wang, 1964), p. 229.

2. Simon Newcomb, *Reminiscences of an Astronomer* (Boston: Houghton Mifflin, 1903), p. 67.

3. Benjamin Boss, *History of the Dudley Observatory: 1852–1956* (Albany, New York: Dudley Observatory, 1968).

4. C. A. Young, "American Astronomy—Its History, Present State, Needs and Prospects," vice-presidential address given at the meeting of the American Association for the Advancement of Science, August 1876, at Buffalo, New York.

5. See Solon Bailey, in S. E. Morison, ed., *The Development of Harvard University* (Cambridge: Harvard University Press, 1930), p. 294.

6. E. C. Pickering to Andrew F. West, April 14, 1917. When Charles J. Capen retired as headmaster of the Boston Latin School in 1909, Pickering was asked to serve on a committee to raise funds to provide an income for Mr. and Mrs. Capen. Pickering replied that he would serve with pleasure, provided he would not be examined for his knowledge of Latin grammar.

7. E. C. Pickering, "Progress of the Physical Department of the Mass. Institute of Technology, From 1867–1877," Report to President J. D. Runkle, January 31, 1877.

8. E. C. Pickering, *Elements of Physical Manipulation,* vol. I (Boston: A. A. Kingman, 1875); vol. 2 (Boston: Houghton, Osgood and Co., 1876).

9. Charles K. Wead to E. C. Pickering, January 1, 1877.

10. In 1874 Pickering married Lizzie Sparks, daughter of the historian Jared Sparks (1789–1866), who was president of Harvard from 1849 to 1853.

11. A man of genius, not fully appreciated during his lifetime, beset with domestic and financial troubles, Peirce (1839–1914) found work at the Coast Survey less congenial after the death of his father in 1880. Personal disagreements developed between him and his superiors and he finally was forced to resign in 1891. Thereafter he had great difficulty in finding work and was reduced to taking odd jobs, doing translations, and giving occasional lectures. In 1899 he applied to return to the Coast Survey as Inspector of Weights and Measures but was unsuccessful, in spite of the strong support of well-wishers, including Pickering. Living in great poverty, Peirce at one time was without food or the money to buy it. During his last years he existed chiefly on funds supplied by William James and other friends. No full biography has been published. For a brief account of Peirce's life, see Carolyn Eisele, "Charles S. Peirce, Nineteenth Century Man of Science," *Scripta Mathematica,* 24:305–324 (1959).

12. Derek Jones, *Norman Pogson and the Definition of Stellar Magnitude,* Astronomical Society of the Pacific, Leaflet No. 469 (July 1968), 8 pp.

13. *Annals,* 14:39 (1884). Although Polaris had always been supposed to be constant in its light, as early as 1882 at least one worker suspected it of having a slight variability. See *Sidereal Messenger,* 1:130 (1882).

14. The first Boston (and first American) performance of Gilbert and Sullivan's *H.M.S. Pinafore* took place on November 25, 1878.

15. Winslow Upton (1853–1914) graduated from Brown University in 1875 as class valedictorian and in 1877 obtained an M. A. in astronomy from the University of Cincinnati. The years 1877–1879 he spent at the Harvard Observatory, and then left to serve with various government agencies. He returned to Brown in 1884 as professor of astronomy and when the Ladd Observatory was built in 1891 became its first director.

16. For the year 1879, the Observatory expended a total of $11,235.14 for the salaries of Pickering and at least ten assistants, most of whom worked full time. The scale of pay was no higher at other observatories; see Langley's letter on the subject, Chapter XII.

17. Frank E. Seagrave (1860–1934) at the age of 14 had obtained his first telescope. Through the kindness of Professor Searle he was allowed to work at the Observatory from 1875 to 1877. He then left to establish his private observatory at his home in Providence, Rhode Island. Later he gave his private library to the Harvard Observatory.

18. S. C. Chandler and John Ritchie, Jr., "On the Telegraphic Transmission of Astronomical Data," *Science Observer*, 3:65–77 (1881).

19. An undated letter of this period from Chandler to Pickering says: "The amount of my voucher this month is so moderate that it won't be worth while turning it in, and I will let it lie over until next month."

20. During his student days at Vanderbilt University, Barnard was an ardent comet seeker. In a letter to Pickering (October 21, 1886) Barnard complained that on more than one occasion he had been deprived of deserved credit as the first discoverer of a comet. He added: "The discovery of a comet is not so slight a matter as one might think and the discoverer very seldom gets his just dues; the discovery of a single comet represents months and sometimes years of patient searching, at the risk of health from exposure to the biting cold of winter and the damp dewey nights of summer. I have many a cold bleak winter night searched ceaselessly from sunset until dawn, almost frozen and completely worn out and nothing to show for my labours and not being able to make up my lost sleep in the day time, before I came here I had to work every day, and since being here I have had to attend the classes in the day and get my lessons as any other student would."

21. For the pertinent correspondence among Ritchie, Baird, and Pickering, see *Smithsonian Annual Report for 1883*, Appendix, p. 87.

22. E. C. Pickering, "The Collection and Distribution of Astronomical Intelligence," *Science Observer*, 4:33–36 (1883).

23. Report in *The Observatory*, 6:202 (1883).

24. During the First World War the Herschel family wished to present these and other valuable manuscripts to Pickering, but because of the difficulties of transportation they did not reach Cambridge until 1919, after Pickering's death. This collection of Herschel papers, which includes a diagram of the "Georgian Planet" (Uranus) drawn by William Herschel on December 1, 1797, is now in the Houghton Library of Harvard University.

25. Cleveland Abbe wrote to Pickering, January 4, 1887, in connection with the photometric work, that "mention should also be made of the work done by George E. Searle, in 1861. I came to Cambridge to work for Dr. Gould on Coast Survey Longitude work, Oct. 1, 1860, and was almost daily with Searle until he went to Europe; the idea and the use of the Wedge Photometer occurred to us both independently and as we discussed its possibilities one day, he decided to order one made by Alvan Clark for his 4-inch telescope. He determined its constants in the Spring of 1861 but he was too expert in Argelander's step by step method to easily give that up. I recall that the varying sensitiveness of the different parts of the retina gave rise to trouble in using the wedge which however, was eliminated by proper inter-change of the direct and refracted images."

26. T. Lewis, "Harvard Photometry," *The Observatory*, 8:49–51 (1885).

27. Report in *The Observatory*, 8:232 (1885).

28. R. L. Waterfield, *A Hundred Years of Astronomy* (London: Duckworth, 1938), p. 93.

29. David Gill, "On Photographs of the Great Comet (b) 1882," *Monthly Notices of the Royal Astronomical Society*, 43:53–54 (1882).

30. Letter from George Bond to William Mitchell, July 6, 1857. Quoted by E. S. Holden, *Memorials of William Cranch Bond and George Phillips Bond* (San Francisco and New York, 1897), p. 158.

31. E. C. Pickering, "Stellar Photography," *Memoirs of the American Academy of Arts and Sciences*, 11:180–226 (1886).

32. E. P. Martz, Jr., "Professor William Henry Pickering: 1858–1938: An Appreciation," *Popular Astronomy*, 46:299–310 (1938).

33. W. H. Pickering, "Principles Involved in the Construction of Photographic Exposers," *Proceedings of the American Academy of Arts and Sciences*, 20:483–489 (1885).

34. E. C. Pickering, Report to the Bache Fund, December 21, 1885.

35. Ormond Stone, "Photographers Versus Old Fashioned Astronomers," *Sidereal Messenger*, 6:1–4 (1887).

36. A. G. Winterhalter, *The International Astrophotographic Congress* (Washington: Government Printing Office, 1889).

37. H. H. Turner, *The Great Star Map* (New York: Dutton, 1912).

CHAPTER V

1. See Henry Draper, "On the Coincidence of the Bright Lines of the Oxygen Spectrum with Bright Lines in the Solar Spectrum," *Monthly Notices of the Royal Astronomical Society*, 39:440–447 (1879).

2. "On the Changes of Blood Cells in the Spleen," *New York Journal of Medicine*, III, vol. 182 (September 1858).

3. Mrs. Draper to E. C. Pickering, April 10, 1887.

4. William Huggins, "On the Spectra of Some of the Fixed Stars," *Philosophical Transactions of the Royal Society of London*, 154:413–435 (1864), p. 428.

5. George F. Barker, "Henry Draper," *Biographical Memoirs, National Academy of Sciences*, 3:81–139 (1895), read before the Academy on April 18, 1888.

6. E. S. Holden, *Memorials of William Cranch Bond and George Phillips Bond* (San Francisco and New York: 1897), p. 212.

7. *Ibid.*, p. 213. Following Harvard's establishment of a station in Peru, a number of American and European universities also constructed observing facilities in the same general region, where the "seeing" is ideal. Among the most recent are the Interamerican Observatory (dedicated November 1967) on the summit of Cerro Tololo, operated by the Association of Universities for Research in Astronomy; and the European Southern Observatory (dedicated March 1969), at Cerro La Silla, established by a coalition of six European countries.

8. *Annals*, 8:60 (1877).

9. Boston *Globe,* July 29, 1878. The reference is of course to the inventor Thomas Alva Edison (1847–1931).

10. Annie J. Cannon, "Mrs. Henry Draper," *Science,* n.s., 41:380 (1915).

11. Reported in *The Observatory,* 3:73 (1879). Oxygen does of course exist in the sun but its presence is shown by dark absorption lines.

12. Annie J. Cannon; see note 10.

13. For a fuller transcription of the Draper-Pickering correspondence, see Lyle G. Boyd, "Mrs. Henry Draper and the Harvard College Observatory, 1883–1887," *Harvard Library Bulletin,* 17:70–97 (1969).

14. George F. Barker (1835–1910), chemist and physicist. He was head of the department of physiological chemistry and toxicology at Yale (1867–1873) and professor of physics at the University of Pennsylvania (1873–1900).

15. The remarkable Struve family (like the Herschels) produced several generations of astronomers. Friedrich Georg Wilhelm von Struve (1793–1864) was director of the Pulkovo Observatory from 1834 to 1862, and was followed by his son Otto Wilhelm (1819–1905), director from 1862 to 1889. Otto's sons Hermann (1854–1920) and Ludwig (1858–1920) became directors, respectively, of the Berlin and Kharkov Observatories. Ludwig's son Otto (1897–1963) migrated to the United States, where he became director of the Yerkes Observatory of the University of Chicago.

16. "Researches upon the Photography of Planetary and Stellar Spectra. By the Late Henry Draper, M. D., LL. D. With an Introduction by Professor C. A. Young, a List of the Photographic Plates in Mrs. Draper's Possession, and the Results of the Measurements of These Plates by Professor E. C. Pickering," *Proceedings of the American Academy of Arts and Sciences,* 19:231–261 (1884).

17. William Huggins, "On the Photographic Spectra of Stars," *Philosophical Transactions of the Royal Society,* 171, part ii, 669 (1880); reprinted in *The Scientific Papers of Sir William Huggins, K.C.B., O.M.* (London: William Wesley and Son, 1909), vol. 2, pp. 63–79.

18. Carried out by Professor Donald H. Menzel.

19. R. L. Waterfield, *A Hundred Years of Astronomy* (London: Duckworth, 1938), p. 66.

20. Antonia C. Maury, "The Spectral Changes in β Lyrae," *Annals,* 84, No. 8 (1933).

21. E. C. Pickering, "On the Spectrum of ζ Ursae Majoris," *Sidereal Messenger,* 9:80–82 (1890).

22. "The Spectrum of ζ Puppis," Harvard Circular No. 16, January 12, 1897.

CHAPTER VI

1. George Bond to William Mitchell, July 6, 1857, in E. S. Holden, *Memorials of William Cranch Bond and George Phillips Bond* (San Francisco and New York, 1897), p. 158.

2. George P. Bond, "Celestial Photography," *American Almanac,* 1859, p. 77.

3. *Corporation Records,* vol. 9, p. 417, July 16, 1856.

4. "Boyden Fund," Harvard College Observatory Circular, March 1, 1887.

5. A grandson of one of the signers of the Declaration of Independence, Paine was a Boston lawyer whose major interest was astronomy. He maintained an observatory at his house, and traveled to many places to see eclipses of the sun. By his will he left his entire fortune, more than a quarter of a million dollars, to the Observatory. The will was contested but the suit was amicably settled in favor of the Observatory.

6. Near Pikes Peak; it is now called Mount Almagre.

7. Pickering did help establish a department of astronomy at the college. Later an observatory was built near Colorado Springs by public subscription, chiefly through the efforts of Professor Loud, whom Pickering helped and encouraged. The observatory became the property of the Western Association for Stellar Photography, which elected Pickering its first president. Nearly forty years later Professor Loud, in turn, gave encouragement to Donald H. Menzel in his study of astronomy. In 1952 Dr. Menzel became the sixth director of the Harvard College Observatory.

8. Not far from Climax, Colorado, where in 1940 during the Shapley administration Dr. Menzel established Harvard's High Altitude Observatory for solar studies.

9. Edward S. Holden, *Handbook of the Lick Observatory* (San Francisco: Bancroft, 1888), p. 11.

10. Edward Skinner King (1861–1931) graduated from Hamilton College, New York, where he studied astronomy under Professor Peters. He came to the observatory in 1887. After returning to Cambridge from Wilson's Peak he superintended the photographic work at the Observatory. He held the Phillips Professorship from 1926 to 1931.

11. H. H. Turner, "President's Address," *Monthly Notices of the Royal Astronomical Society,* 64:388–401 (1904).

CHAPTER VII

1. See W. W. Payne, "The Late Catherine Wolfe Bruce," *Popular Astronomy,* 8:235–238 (1900). Besides giving the facts of her life, this article adds a complete list of her contributions. Father John Hagen, the recipient of considerable aid from Miss Bruce, supplied a brief obituary in *Astronomische Nachrichten,* 152:244 (1900). For George Bruce see *DAB,* 3:181.

2. In addition to the Draper and Boyden funds, the Observatory now had the munificent bequest of Robert Treat Paine. By 1886 half the legacy had become available for Pickering's use.

3. *The Observatory,* 12:308–311 (1889).

4. *Ibid.,* 6:202 (1883).

5. Communicated to the Academy March 10, 1886, and published under the title of "Stellar Photography," in *Memoirs of the American Academy of Arts and Sciences,* 11:180–226 (1886). Pickering's particular reference to his recommendation is on p. 207.

6. *The Observatory,* 12:375 (1889).

7. Mrs. Thompson was the daughter of William Thaw of Pittsburgh, who had been one of the chief benefactors of Langley at Allegheny. The Thompson fund was established in 1885 with a principal of about $26,000, the income of which was available to scientists of all countries. Pickering was one of the trustees. Grants usually averaged $300.

8. For the complete list and the projects supported, see E. C. Pickering, "The Endowment of Astronomical Research," *Science,* n.s., 17:721–729 (1903).

9. See Payne, *Popular Astronomy,* 8:237–238 (1900).

10. The statutes for the bestowal of the Bruce Medal were most carefully drawn. It was to be given "not oftener than once a year, for distinguished services to astronomy . . . international in character . . . awarded to a citizen of any country, and to a person of either sex." (No woman has ever received it, although Pickering twice unsuccessfully nominated Mrs. Fleming.) No one may receive the medal twice. The directors of six important observatories (Harvard, Lick, Yerkes, Observatorio Astronomico de la Nación Argentina, Greenwich, and Paris) each year would be asked to nominate no more than three astronomers, after which at a special meeting the eleven directors of the Astronomical Society of the Pacific would vote on the nominations.

11. For the entire list of Bruce medalists, 1898–1968, see *Publications of the Astronomical Society of the Pacific,* 80:250–251 (1968). Harlow Shapley received the medal in 1939.

12. *Harvard Alumni Bulletin,* November 20, 1931, p. 231.

CHAPTER VIII

1. The last great display of the Leonid meteors, November 13–14, 1866. This annual shower, which has been traced back to the year 902 and reaches a maximum about every 33 years, produced phenomenal rains of meteors in 1799, 1833, and 1866. Since then, the number observed at maximum has markedly diminished.

2. Señor de Romaña and his family became close personal friends of the Baileys; a few years later they gave the name Romaña to a second son, who died in infancy. For more than 75 years the family have maintained their ties with the Observatory. Fernando L. de Romaña assisted the Harvard expedition observing the solar eclipse of November 6, 1966, in Peru and that of March 7, 1970, in Mexico.

3. Irving Widmer Bailey (1884–1967) graduated from Harvard in 1907 and later became professor of plant anatomy in the Bussey Institution in Jamaica Plain, Massachusetts. In 1911 he married Helen Harwood, sister of Margaret Harwood who after serving as an assistant at the Observatory became the first director of the Maria Mitchell Observatory in Nantucket, Massachusetts.

4. Solon I. Bailey, "History of the First Peruvian Expedition, 1889–1891," *Annals,* 34 (1895), chap. 1, p. 12.

5. *Ibid.,* p. 30.

6. *Ibid.,* p. 45.

7. Solon I. Bailey, "A Catalogue of 7922 Southern Stars Observed with the Meridian Photometer during the Years 1889–91," constituting Volume 34 of the *Annals* (1895).

8. William H. Pickering to James Monroe at MIT, May 27, 1889.

9. Edward C. Pickering to E. W. Hooper, February 16, 1891.

10. William H. Pickering, "Investigations in Astronomical Photography," *Annals*, 32, part I (1895), p. 109.

11. Asaph Hall to E. C. Pickering, September 6, 1877.

12. William H. Pickering, "Photographs of the Surface of Mars," *Sidereal Messenger*, 9:254–255 (1890).

13. William H. Pickering, "Mars," *Astronomy and Astro-Physics*, 11:668–675, 849–852 (1892).

14. Visiting the Arequipa Station on his way to and from Chile to view the eclipse of April 16, 1893, Schaeberle had tried the Boyden telescope and was thus able to make a direct comparison between it and the Lick instrument.

15. E. S. Holden, "The Lowell Observatory, in Arizona," *Publications of the Astronomical Society of the Pacific*, 6:160–169 (1894); see p. 166.

16. William H. Pickering, "The Planet Jupiter and its Satellites," *Astronomy and Astro-Physics*, 12:193–202 (1893); "The Rotation of Jupiter's Outer Satellites," 12:481–494 (1893); "Jupiter's Satellites," 12:390–397 (1893).

17. J. M. Schaeberle and W. W. Campbell, "The Forms of Jupiter's Satellites," *Publications of the Astronomical Society of the Pacific*, 3:355–358 (1891); E. E. Barnard, "On the Forms of the Discs of the Satellites of Jupiter as Seen with the 36-inch Equatorial of the Lick Observatory," *Astronomy and Astro-Physics*, 13:272–278 (1894).

18. A. E. Douglass, "Swift's Comet (a 1892)," *Astronomy and Astro-Physics*, 12:202–205 (1893).

19. William H. Pickering, "Astronomical Possibilities at Considerable Altitudes," *Astronomische Nachrichten*, 129:97–100 (1892); "The Mountain Station of the Harvard College Observatory," *Astronomy and Astro-Physics*, 11:353–357 (1892); "The Boyden Station of the Harvard College Observatory," *Astronomy and Astro-Physics*, 11:357–362 (1892).

20. Marshall Bailey graduated in 1894 from the College of Physicians and Surgeons, Baltimore (which shortly thereafter became part of the University of Maryland). In 1895 he was appointed Medical Advisor for Harvard University students, a post he held until his retirement in 1928. He died September 20, 1946.

21. E. C. Pickering to Anne Pickering, November 4, 1892: "I am much concerned to learn from your letter to Lizzie that you so much regret your return to the United States . . . If you wish to blame anyone why not blame me, where from your standpoint blame certainly belongs. Yet when you know all the facts I am sure, if you will look at the matter fairly, you will see that I had no alternative. You certainly must not blame the Baileys who are entirely innocent in this matter. You are probably not aware how near you came to an early recall and abandonment of the expedition. If Prof. Bailey did not wish to go, I doubt if we should continue it on its present basis."

22. Edward C. Pickering, "A New Star in Norma," *Astronomy and Astro-physics,* 13:40–41 (1894).

23. Edward C. Pickering, *A New Star in Carina,* Harvard College Observatory Circular No. 1, October 30, 1895; *A New Star in Centaurus,* HCO Circular No. 3, December 20, 1895; *A New Star in Sagittarius,* HCO Circular No. 42, March 14, 1899.

24. Solon I. Bailey, "ω Centauri," *Astronomy and Astro-Physics,* 12:689–692 (1893).

CHAPTER IX

1. See A. Lawrence Lowell, *Biography of Percival Lowell* (New York: Macmillan, 1935).

2. C. W. Eliot to E. C. Pickering, November 22, 1894: "But Mr. Percival Lowell is undoubtedly an intensely egoistic and unreasonable person, and in my opinion his frame of mind towards the Observatory is a hopeless one. Fortunately he is generally regarded in Boston among his contemporaries as a man without good judgment. So strong was this feeling a few years ago that it was really impossible for him to live in Boston with any comfort."

3. John A. Brashear (1840–1920), who began life as a workman, with the encouragement of S. P. Langley at the Allegheny Observatory in Pittsburgh became a telescope maker of great skill. In the opinion of some astronomers of the time, his instruments rivalled those of the Clarks.

4. Percival Lowell, "Mars," *Astronomy and Astro-Physics,* 13:538–553, 645–652, 740, 814–821 (1894).

5. A new paper by Schiaparelli, "The Planet Mars," had been published in *Natura ed Arte,* February 15, 1893. According to William Pickering's translation, Schiaparelli wrote that "the intervention of intelligent beings might explain the geometrical appearance of the gemination, but it is not at all necessary for such a purpose. The geometry of nature is manifested in many other facts, from which are excluded the idea of any artificial labor whatever. The perfect spheroids of the heavenly bodies and the ring of Saturn were not constructed in a turning lathe, and not with compasses has Iris described within the clouds her beautiful and regular arch. And what shall we say of the infinite variety of those exquisite and regular polyhedrons in which the world of crystals is so rich?" *Astronomy and Astro-Physics,* 13:714–723 (1894), p. 722.

6. Douglass (1867–1962) stayed with the Lowell Observatory until 1901. He then entered politics and became probate judge in Coconino County. He continued his astronomical work as professor of physics at the University of Arizona and in 1918 became professor of astronomy and first director of the Steward Observatory. One of his most remarkable accomplishments, however, was the identification of climatic cycles in the past, from studies of tree growth in the forests of yellow pine in northern Arizona. To determine whether the sun's activity had varied in previous centuries, he examined the relation between the pattern of tree rings and the amount of rainfall. He thus originated the science of dendrochronology, a method of determining the ages of trees and hence of many archaeological ruins.

7. The Lowell Observatory has continued to be one of the few to emphasize the study of objects in the solar system and, under the more conservative directors who succeeded Lowell, has produced fundamental information about the planets.

8. Percival Lowell, *Mars and Its Canals* (New York: Macmillan, 1906), p. 376.

9. Alfred C. Lane to Harlow Shapley, March 21, 1930.

10. Harlow Shapley to Alfred C. Lane, March 24, 1930.

11. R. A. Lyttleton, *Mysteries of the Solar System* (Oxford, The Clarendon Press, 1968), Chapter 7.

12. S. C. Chandler, "Second Catalogue of Variable Stars," *Astronomical Journal,* 13:89–110 (1893).

13. S. C. Chandler, "On the Observations of Variable Stars with the Meridian-Photometer of the Harvard College Observatory," *Astronomische Nachrichten,* 134:355–360 (1894).

14. Ogden Nicholas Rood (1831–1902), professor of physics at Columbia University.

15. G. Müller and P. Kempe, "Remarks on Professor E. C. Pickering's Article 'Comparison of Photometric Magnitudes of the Stars,'" *Astronomische Nachrichten,* 137:430–433 (1895).

16. Simon Newcomb, "Aspects of American Astronomy," *Astrophysical Journal,* 6:289–309 (1897), p. 301.

17. A few months after Bailey left Peru, in the spring of 1898, Clymer received a paper purporting to contain the names of the persons who committed the robbery. A lawyer advised him not to prosecute, since to do so might arouse ill feeling and result in future injury to the station.

18. *The Observatory,* 19:428 (1896).

19. Robert de Courcy Ward (1867–1931), who graduated from Harvard in 1889 and received his M.A. in 1893, was editor of the *Meteorological Journal* (1892–1896) and later became the first professor of climatology at Harvard. Some of his meteorological investigations were published in the Harvard *Annals.* They included "Thunderstorms in New England during the Years 1886 and 1887," *Annals,* 31, pt. 2, p. 261 (1893). Robert's son, Henry de Courcy Ward, who graduated from Harvard in 1920, served as a member of the visiting committee during the administrations of both Shapley and Menzel.

20. Harlow Shapley, *Through Rugged Ways to the Stars* (New York: Scribner, 1969), p. 90.

CHAPTER X

1. Mrs. Pickering died on August 29, 1906. There were no children.

2. A fireproof structure (the present Building C) completed in 1893 to house the collection of photographic plates. Contributions from friends of the Observatory supplied the necessary funds.

3. See N. Lockyer, *Smithsonian Institution Annual Report* (1900), p. 123.

4. Gould died in 1896. By his will he bequeathed the editorship of the journal to S. C. Chandler, as set forth in Article 7: "To my friend, Seth C. Chandler, now of Cambridge, I give whatever amount my ledger shall show, to the satisfaction

of my executors, to remain available from the special fund reserved for the maintenance of the Astronomical Journal,—also all published volumes and numbers of said Journal remaining in store, and likewise the additional sum of three thousand dollars; in order that he may use the same, according to his discretion, for the purpose of continuing and maintaining said Journal as a means of encouraging and promoting astronomical research in the United States." The Journal is now published for the American Astronomical Society by the American Institute of Physics.

5. See Chapter XII, note 29.

6. In regard to income, Harvard at this time occupied a place near the top among American observatories, and was surpassed only by the U.S. Naval Observatory, which received $85,000. A few years later the situation had changed greatly. In response to a request from S. A. Mitchell, director of the Leander McCormick Observatory of the University of Virginia, Pickering sent (May 3, 1916) estimates of the yearly budgets of the leading sister institutions: Mt. Wilson, some $220,000; the Naval Observatory, more than $100,000; Harvard, $50,000; Lick, $30,500; Yerkes, $30,000; Allegheny, $15,000; Princeton, $10,000.

7. See William Lassell, "Discovery of a New Satellite of Saturn," *Monthly Notices of the Royal Astronomical Society,* 8:195–197 (1848); G. P. Bond, "Discovery of a New Satellite of Saturn," *ibid.,* 9:1–2 (1848).

8. Several versions of this poem are known among astronomers but efforts to locate the original have so far failed. The verses quoted here (except for the bracketed lines, supplied by D. H. Menzel) are those printed by the New York *Times,* January 22, 1967.

9. The date on the letter does not include the year but it cannot have been later than 1901. Myers (1843–1901) was one of the founders of the Society for Physical Research. In 1894 Pope Leo XIII had taken the first step toward the eventual canonization of Joan of Arc.

10. De Lisle Stewart, *An Asteroid of Great Eccentricity,* Harvard College Observatory Circular No. 63, November 19, 1901.

11. Henrietta S. Leavitt, "1777 Variables in the Magellanic Clouds," *Annals,* 60, pt. 4, 87–108 (1908), p. 107.

12. Henrietta S. Leavitt, *Periods of 25 Variables in the Small Magellanic Cloud,* Harvard College Observatory Circular No. 173, March 3, 1912.

13. D. H. Menzel, "Distribution of Two Thousand New Nebulae, and Distance of the Large Magellanic Cloud," *Popular Astronomy,* 30:627–628 (1922).

14. "A Photographic Atlas of the Moon," *Annals,* 51 (1903), p. 23.

15. Bailey to Paul A. Jenkins, of the *Popular Science Monthly,* October 18, 1921.

16. D. B. McLaughlin, "Who First Explained the Nova Bands?" *Popular Astronomy,* 44:299–300 (1936).

17. *Astronomy and Astro-Physics,* 13:201–204 (1894), p. 203.

18. Among others who served as assistants during the remainder of the Pickering regime were William B. Clymer (1895–1900), De Lisle Stewart (1896–1902), Royal H. Frost (1896–1908), Edmund S. Manson 1902–1907), Benjamin F. Wyeth

(1902–1907), Frank E. Hinkley (1908–1911, 1915–1918, 1919–1921), Herbert E. Blackett (1908–1911), C. J. G. Vogel (1910–1911), Harold I. Peckham (1912–1915), Leon Campbell (1911–1915), and L. C. Blanchard (1916–1918).

19. W. E. Castle (1867–1962) to E. C. Pickering, November 3, 1908. Castle was a brilliant biologist and made basic studies in embryology and genetics.

20. An experiment involving a somewhat closer connection between astronomy and biology was proposed by J. A. Dunne, who served as an assistant at the Observatory from 1888 to 1909. In a letter to Pickering (March 29, 1907) he suggested that "it might be a good plan to attempt to procure the spectrum of the light of a firefly and possibly of a glow-worm, which experiment can be tried next June or July. The experiment, perhaps, may not strictly be an astronomical one, yet as the source of light from these little bodies is one lacking in heat, it would be interesting to learn if their spectra resembled at all that of the aurora." The records do not show that the experiment was ever carried out.

21. The Boyden station was finally transferred from Arequipa to Bloemfontein in 1927, during Shapley's administration.

22. For the meeting of the American Astronomical Society held at Evanston, Illinois, August 25–28, 1914. Pickering attended, as president.

23. Annie J. Cannon, *Science,* n.s., 41:380–382 (1915).

24. Muñiz continued his duties until 1925 and in the last years was assisted by his son, Juan E. Muñiz.

25. After Pickering's death, Bailey was appointed acting director and served from 1919 to 1921, when Shapley became the new director and Bailey returned to Peru to supervise the station. In the autumn of 1923, shortly before turning the station over to his successor, Professor John S. Paraskevopoulos, Bailey received the degree of Doctor of Science from the University of San Augustin and was appointed honorary professor of astronomy in the same University. In 1925, at the age of seventy, Bailey retired. The succeeding years he spent in writing his *History and Work of the Harvard College Observatory, 1839 to 1927* (New York: McGraw-Hill, 1931), a project he had begun more than twenty years earlier. He died on June 5, 1931, only a few weeks after the publication of his *History.*

26. See the *Annals,* 78, pt. 1 (1913), pt. 2 (1917), and pt. 3 (1919).

27. Shapley to Henry Norris Russell, April 2, 1927.

28. Some of the most important of Professor King's work is described in the *Annals,* 59 (1912) and 80, pts. 5 and 6 (1916).

29. The two catalogues appear in the *Annals,* 48, no. 3 (1903) and 55 (1907).

30. Some years later Professor Campbell, in collaboration with Luigi Jacchia, produced a comprehensive monograph, *The Story of Variable Stars* (Philadelphia: Blakiston, 1941).

CHAPTER XI

1. Women contributed to funds for the purchase of Winlock's meridian circle in 1868–1870 [*Annals,* 8:11 (1877)] and, as noted in the annual reports for 1878–79, 1882–83, and 1891–92, to every subscription campaign undertaken by Pickering. In 1877 the Observatory received the Charlotte Harris bequest; in 1898, $45,000

from the Misses Charlotte and Eliza Haven; and in 1912, upon the death of her husband, Mrs. A. Lawrence Rotch assumed support of the Blue Hill Observatory, which he had founded and with which the Harvard Observatory had long been associated in meteorological observations. The gifts of Mrs. Margaret Jewett and Mrs. Charles Hinchman for fellowships and other assistance are noted below. A fund in memory of Adelaide Ames, a valued assistant of Harlow Shapley's, was established in 1932, and the Elizabeth E. Bemis Fund in 1934. The bequest of Mary Fifield King for the unrestricted use of the Observatory was received in 1949.

2. Maria Mitchell (1818–1889) in 1865 became the first professor of astronomy at the newly established Vassar College and instituted the first course in astronomy for women. See *DAB,* 13:57–58; Helen Wright, *Sweeper in the Sky* (New York: Macmillan, 1949); Mary Whitney, "Maria Mitchell," *Sidereal Messenger,* 9:49–51 (1890); Phebe Mitchell Kendall, *Maria Mitchell: Life, Letters, and Journals* (Boston, 1896).

3. President Everett's determination to secure the medal for Miss Mitchell is revealed in a printed but unpublished pamphlet, "Correspondence Relative to the Award of the King of Denmark's Comet Medal to Miss Maria Mitchell, of Nantucket, for the Discovery of a Telescopic Comet on the 1st of October, 1847" (Cambridge, 1849).

4. "As you invited the contributions of the *unlearned* to the observations on the comet," Miss Quincy wrote George Bond in 1861, "I send you some sketches which I took from a northeast window in my room . . . I presume I am millions of miles out of the right path . . . but . . . you will excuse my eccentricities! which I have no doubt will divert you." Of another phenomenon she wrote, "Although it seems *rather* presumptuous to communicate *my* observations on the meteor of July 20th to *you,* yet as your sister said this morning you welcome any communication on the subject you shall have mine," and then proceeded to give the most detailed information about what she had seen. In response to his praise for the acuteness of her observations, she expressed herself as "highly gratified you can denominate my account of that wonderful aerolite *intelligible*—The attempt caused such meteoric agitation in my ideas, that I hardly did justice to the little information I do possess on the subject." Besides such amusing reports, her delightful letters to the Bonds, Winlock, and Pickering supply a number of details about the Observatory, its benefactors, and her relation with the Herschel family.

5. See Mary Byrd, "Anna Winlock," *Popular Astronomy,* 12:254–258 (1904). For an example of Rogers's typically generous acknowledgments of Miss Winlock's part in their joint work, see "A Catalogue of 130 Polar Stars," *Memoirs of the American Academy of Arts and Sciences,* 11:227–299 (1886).

6. A complete list of staff members from the beginning to 1931 appears in Bailey, *The History and Work of Harvard Observatory, 1839 to 1927* (New York: McGraw-Hill, 1931), pp. 274–276. Pickering's policy of introducing women to the staff and of encouraging them to study astronomy on a graduate level was pursued with great vigor by his successor, Harlow Shapley. During his administration

(1921–1953) he interested himself particularly in such a graduate program at Radcliffe, which he believed should become the leading college in the world for training women in astronomy. "The high standing of the Harvard Observatory," he wrote at the outset of his directorship, "depends largely on the activity and faithfulness of its women associates during the last thirty years. I hope that at some time in the future their work will be officially recognized." Like Pickering before him, he lost no opportunity to win that recognition for such notable associates as Miss Cannon, Miss Leavitt, Miss Maury, and others who later distinguished themselves at the Observatory. "There are more women scientists here," he once wrote, "than in any other department of the University." For Radcliffe's early role in the collegiate education of women see Arthur Gilman, *Harvard Graduate Magazine* (September 1907); Joseph Warner, *ibid.* (March 1894); *Radcliffe Magazine* (June 1905); Lucy Paton, *Elizabeth Cary Agassiz* (Boston: Houghton Mifflin, 1919.)

7. Miss Klumpke, a native of San Francisco, was awarded the degree of Doctor of Science at the University of Paris in 1893, having written her thesis on Saturn's rings. She had charge for a time of the great reflector in the Paris Observatory, then moved to the Bureau of Measurements, where she measured the positions of stars on photographic plates for the catalogues of all stars down to the 11th magnitude. See *Astronomy and Astro-physics,* 13:260–261 (1894); Poggendorf's *Biographisch-Literarisches Handwörterbuch* (Barth: Leipzig, 1904), p. 762.

8. Pickering's pamphlet and his deep interest in variable stars led to the organization in 1911 of the American Association of Variable Star Observers. See R. Newton Mayall, "The Story of the AAVSO," *Review of Popular Astronomy* (September–October, November–December 1961.) The present membership includes many women, of whom seven have been presidents of the organization. The director is Mrs. Margaret Mayall, for many years a close associate of the Observatory and of Miss Cannon in the Henry Draper Memorial work, the *Extension* of which, in the *Annals,* 112 (1949), she completed after Miss Cannon's death.

9. See *DAB,* 6:462–463; Annie J. Cannon, "Williamina Paton Fleming," *Astrophysical Journal,* 34:314–317 (1911); *Scientific American,* (June 3, 1911); *Science,* n.s., 33:987–988 (1911); *Who's Who in America, 1910–1911; Monthly Notices of the Royal Astronomical Society,* 72:261–264 (1912). Pickering's eloquent tribute to her, "In Memoriam," was issued by the Observatory on October 20, 1911. Mrs. Fleming's spectacular career inspired a number of articles about the women in the Observatory. See, for example, Helen Leah Reed, "Women's Work at the Harvard Observatory," *New England Magazine* (April 1892), pp. 166–176; Mabel Loomis Todd, "A Great Modern Observatory," *Century Magazine* (March 1896), pp. 290–300; "Women Astronomers at Harvard," *Harvard Alumni Bulletin* (January 13, 1927), pp. 420–422: Anne P. McKenney, "What Women Have Done for Astronomy in the United States," *Popular Astronomy,* 12:171–182 (1904).

10. Read at the Congress of Astronomy and Astrophysics, Chicago, August, 1893, published in *Astronomy and Astro-Physics,* 12:683–689 (1893).

11. See Dorrit Hoffleit, "Antonia C. Maury," *Sky and Telescope,* 10:106 (March 1952); *New York Times,* January 10, 1952; *Who Was Who in America, 1951–1960,* p. 564.

12. Although, in her supposedly farewell letter (December 21, 1894) to Pickering, Miss Maury wrote that she would be sorry "to go away feeling that my attitude was misunderstood by anyone whose kindness I so much appreciate and for whom I have so much respect," she evidently continued to feel hurt long afterward. In 1924, when she was ill and unable to return to the Observatory, her brother, Dr. John Draper (he had assumed the name of his grandfather), wrote Shapley, "I am very grateful to you for your interest in her. Your recognition and friendly support has been a great inspiration to her in the latter years of her work. It stands in striking contrast to the environment into which she was cast by your predecessor."

13. See notes 25 and 27.

14. See *DAB*, 10:83; Solon I. Bailey, "Henrietta Swan Leavitt," *Popular Astronomy*, 30:197–199 (April 1922); Bailey, *History*, 264–265, and references throughout; *New York Herald*, December 13, 1921.

15. In 1925 Professor Mittag-Leffler, a member of the Swedish Academy of Sciences and editor of *Acta Mathematica*, unaware of Miss Leavitt's death, wrote her that he wished to nominate her for the Nobel prize because of her discovery. Shapley replied, "Miss Leavitt's work on the variable stars in the Magellanic Clouds, which led to the discovery of the relation between period and apparent magnitude has afforded me a very powerful tool in measuring great stellar distances. To me personally it has also been of the highest service, for it was my privilege to interpret the observation of Miss Leavitt, place it on a basis of absolute brightness, and, extending it to the variables of the globular clusters, use it in my measures of the Milky Way."

16. J. C. Kapteyn (1851–1922), the noted Dutch astronomer, completed his *Cape Photographic Durchmusterung* in 1900, a catalogue of the positions and brightness of nearly half a million southern stars. In 1902, complimenting him on "one of the greatest astronomical achievements," Pickering proposed his undertaking a comparable Durchmusterung of northern stars and offered to supply as many photographs taken at either Arequipa or Cambridge as would be needed. He was unable to secure funds for the purpose, however, as some members of the board administering the Carnegie Institution were sternly opposed to Pickering's aid to foreign astronomers (see Chapter XII). In 1906 Kapteyn presented his "Plan of Selected Areas" for the study of the magnitudes, proper motions, parallaxes, spectral classes, and radial velocities of stars in 252 regions, and a number of observatories eventually (in 1912) joined the program. Pickering, Hale, Schwarzschild, and others were members of the committee to carry out the plan, and Pickering provided a number of plates. See Pickering's preface to *Annals*, 101 (1918).

17. See Leon Campbell, "Annie Jump Cannon," *Popular Astronomy*, 49:345–347 (1941); Bailey, *History* (index); *Harvard Alumni Bulletin* (November 5, 1925), p. 166; *ibid.*, "Women Astronomers at Harvard" (January 12, 1927), pp. 420–422: "Women Find Their Place in the Sky," *New York Times*, December 20, 1936; *Who Was Who in Science* (Chicago: Marquis, 1968); Edna Yost, *Women in Science* (New York and Philadelphia: Stokes, 1943), chap. 2; *New York Times*, April 14, 1941 (obituary), April 15, 1941 (editorial). The Delaware State Museum

has dedicated a special room to Miss Cannon, which depicts her career, and a biography based on her diaries and papers is in preparation by her former assistant, Mrs. Margaret Mayall.

18. See Annie Cannon, "Sarah Frances Whiting," *Popular Astronomy*, 35:539–545 (1927); for Durant, *DAB* 5:541–542; for the history of the College, Florence Converse, *The Story of Wellesley* (Boston: Little Brown, 1915); Alice Payne Hackett, *Wellesley, Part of the American Story* (New York: Dutton, 1949). Lady Huggins, impressed by Miss Whiting's achievements in promoting women's education in astronomy, bequeathed to Wellesley her library, some choice personal possessions, and a few astronomical instruments, including a spectroscope once belonging to Professor Miller, her husband's collaborator in the study of the chemistry of stars; see *Popular Astronomy*, 23:698–699 (1915). The bibliography on the higher education of women and opposition to it is vast, but among many useful volumes, see Thomas Woody, *The History of Women's Education in the United States* (2 vols., first published 1923, reissued New York: Octagon Books, 1966), esp. vol. 2 and bibliography.

19. Professor W. P. Atkinson to Rogers; William B. Rogers, *Life and Letters* (2 vols., Boston, 1896), II, 275–276.

20. Mr. Agassiz's benefactions to the Observatory, besides payment of salaries of the women who assisted Miss Cannon, included contributions toward instruments, expeditions, and facilities at the Observatory, special aid to visiting research assistants, use of the Agassiz cottage at Harvard, Massachusetts, and the establishment of the Agassiz Research Fellowship that brought to the Observatory many advanced students who became distinguished astronomers.

21. Mrs. Jewett, beginning in 1920, contributed substantially toward publication of the Henry Draper Catalogue and in 1928 established the Margaret Weyerhauser Jewett Research Fellowship in Astronomy "to be awarded to an advanced student of exceptional promise."

22. Professor H. H. Turner of Oxford, a warm friend of Miss Cannon's (see her obituary of him in *Popular Astronomy*, 39:59–66, 1931), wrote Shapley, April 27, 1925: "It is only within the last few years that a degree of any kind has been possible for a woman here . . . I am very glad that we can welcome her here. She will be the *first* woman honorary doctor." At this time the only other American to receive such a degree in science from Oxford was George Hale.

23. J. C. Kapteyn wrote Miss Cannon, June 10, 1921, "As far as I know you and Schwartszchild are the only persons upon whom the Doctorate in Mathematics and Astronomy 'honoris causa' of the Groningen University was ever bestowed. I hope that you will find in this tribute at least some small return for a work of such devotion, energy, and organization, a work which even in Astronomy has hardly a parallel, a work which is so urgently demanded for the further progress of science and which will earn for you the gratitude of all who try to penetrate somewhat further in mysteries of the Stellar Universe."

24. Rogers died in 1898, after which the completion of the work was directed by Arthur Searle. See *DAB*, 16:114–115.

25. The leading spirit in establishing the fellowship was Mrs. Charles Hinchman

of Philadelphia. The committee to administer it included Professor Mary Whitney (first chairman, succeeded by Miss Cannon in 1912), Miss Young of Mt. Holyoke, Pickering, Mrs. Hinchman, secretary, and two Vassar alumnae, Miss Elizabeth Coffin and Miss Florence Cushing.

26. "The Variable Stars in the Scutum Cloud," *Annalen van de Sterrewacht te Leiden,* 21:387–464 (1962).

27. Miss Cannon's account of the presentation of the fellowship to Pickering appears in the *15th Annual Report* of the Nantucket Maria Mitchell Association (February 1, 1916–January 31, 1917). Her last report was made in March 1941. After her death Dr. John D. Duncan of the Whitin Observatory, Wellesley, became chairman; he in turn was succeeded by Dirk Brower of Yale.

28. Miss Block married John Paraskevopoulos (at one time in charge of the Arequipa station and of other expeditions in Latin America) and accompanied him in 1927 to the new Boyden station at Mazelspoort, near Blomfontein, South Africa.

29. The recipients of the fellowship from 1917 to 1951 are listed in the *50th Annual Report* of the Association (February 1, 1931–January 3, 1952). Since then six additional awards have been made.

30. Cecilia Payne (Gaposchkin) received a Ph. D. from Radcliffe College in 1925 and holds advanced degrees from Cambridge University, as well as several honorary degrees. Appointed Phillips Astronomer at Harvard in 1938, in 1958, by a revision of the Phillips bequest, she was named Phillips Professor of Astronomy. She is the author of innumerable publications.

31. See Louise Barber Hoblit, "Mary E. Byrd," *Popular Astronomy,* 42:496–498 (1934). On her departure from Smith College, see W. W. Payne, "Miss Mary E. Byrd's Resignation," *Popular Astronomy,* 14:447–448 (1906).

32. See Caroline Furness, "Mary W. Whitney," *Popular Astronomy,* 30:597–608 (1922); 31:25–35 (1923); *DAB,* 20:163–164.

33. William Elwood Byerly (1849–1935) received the first Ph. D. awarded at Harvard. He succeeded James Mills Peirce as Perkins Professor of Mathematics in 1905 and retired in 1913. He was one of the founders of Radcliffe College, where a building is named for him. (*Harvard Gazette* minute, December 29, 1936.) Peirce (1834–1906), son of Benjamin Peirce and brother of Charles S. Peirce, became professor of mathematics in 1869 and Perkins Professor in 1885; see *DAB,* 14:405–406.

34. See Maude W. Makemson, "Caroline Ellen Furness," *Popular Astronomy,* 44:233–238 (1936).

35. The measurement and reduction of plates of the north polar region made at Helsingsfors as a part of the international astrophotographic program of charting the sky.

36. Proctor was born in England in 1837 and, after a somewhat unsuccessful financial though respectable astronomical career, came to this country in 1881, where he died in 1888. His lectures were widely attended, and his books aroused considerable interest and controversy. He was elected a Fellow of the Royal Astronomical Society in 1866.

37. Miss Proctor was born in Dublin, Ireland, in 1862. A close companion of her father, she early took a deep interest in astronomy and determined to follow in his footsteps as a popular writer on astronomy. Besides a long series of contributions to various periodicals and newspapers, she wrote several books, among them *Evening with the Stars* (New York: Harper, 1925) and *Romance of the Sun* (New York and London: Harper, 1927). See *National Cyclopedia of American Biography,* 10:282.

CHAPTER XII

1. *Astrophysical Journal,* 50:233–244 (1919).

2. The huge correspondence on the Naval Observatory reveals a long record of unsuccessful attempts by American astronomers to effect a reorganization that would put at the head of the establishment a recognized man of science instead of an officer of the Navy. Pickering at one time was chairman of a Board of Advisors, but their efforts to change the situation came to nothing. In 1902 all the members resigned.

3. Published in *Science,* n.s., 17:721–729 (1903).

4. This fund Pickering himself regularly contributed to, usually anonymously.

5. See *Carnegie Institution Year-Book,* No. 1 (Washington, 1903). Carnegie's own statement is on p. xiii.

6. See *DAB,* 3:266–269.

7. *Carnegie Institution Year-Book,* No. 1, pp. 87–160.

8. *Ibid.,* p. 132.

9. "An International Southern Telescope," read before the American Philosophical Society at the Franklin Bi-Centennial Meeting, April 1906; *Proceedings of the American Philosophical Society,* 45:44–53 (1906).

10. There can be little harm in revealing now that the benefactor was Henry H. Rogers of Standard Oil. See *DAB,* 16:95–96.

11. Keeler had been sent to Pickering by Charles Hall Rockwell, an able amateur astronomer, of whom it was said that if he had done "only this one good deed in his life [in having discovered such a genius], it would entitle him to a place on the roll of fame." *Popular Astronomy,* 8:262–265; 409–417; 476–481 (1900). See also *DAB,* 10:278–279.

12. Frank Washington Very (1852–1927) went to Allegheny as Langley's assistant on Pickering's recommendation in 1878. He remained there until 1895, when he followed Langley to the Smithsonian Institution. Besides useful observations of Mars, he designed and constructed a device for measuring intensities of the Fraunhofer lines in the solar spectrum.

13. For Langley's report on this expedition, see "Researches on Solar Heat and Its Absorption by the Earth's Atmosphere," *Professional Papers of the Signal Service,* no. 15 (U. S. War Department, 1884).

14. *Science,* n.s., 49:151–155 (1919).

15. Letter to Bailey, February 5, 1919.

16. In 1912 Russell wrote Pickering to request materials on which Shapley could work in Princeton. Pickering was happy to comply: "You and Shapley are

making such good use of [the material sent], that I hope you will let me know if you desire copies of other volumes." On March 9, 1912, Russell reported that Shapley was nearly through with the Harvard plates "and has got some very satisfactory elements." He continued to send other details of this research, and on August 7, 1912, he suggested that, if Pickering had time for a little talk with Shapley at the forthcoming meeting at Allegheny, it would be a favor. "He is the best student I ever had."

17. *Science,* n.s., 49:154–155 (1919).

18. See *Astrophysical Journal,* 98:241–243 (1943).

19. The threat to close down this observatory in 1915, because of the failure of the United States to continue its support, led to a vigorous campaign by the American Astronomical Society for a restoration of funds. Pickering, as president of the Society, wrote a number of letters on its behalf. The appropriation was consequently restored.

20. See S. E. Morison, ed., *The Development of Harvard University* (Cambridge: Harvard University Press, 1930), pp. 277–291.

21. Reported to Pickering by Eliot in 1908.

22. See *Science,* n.s., 39:680–682 (1914), for a report on the meeting in Washington and a complete list of members.

23. *Science,* n.s., 64:342–347 (1926); Edwin G. Conklin *et al.,* "James McKeen Cattell in Memoriam," *Science,* n.s., 99:151–165 (1944).

24. Morison, *Development of Harvard University,* pp. 290–291.

25. Pickering wrote to Cattell on May 20, 1918, "I think there are others who would like the positions I have held. I had rather resign when my friends want me to stay, than to stay when they want me to resign."

26. See A. H. Dupree, *Science and the Federal Government* (Cambridge: Harvard University Press, 1957), pp. 308–314.

27. Dupree's chapter, "The Impact of World War I," pp. 302–325, is especially useful for this period of preparation and the problems of scientific skills. On the Counsulting Board, see especially pp. 306–308.

28. Cattell had written to members of Congress on August 23, 1917, asking support of a measure then before the Senate and House to prohibit sending conscripts to fight in Europe against their will. The Trustees of Columbia University expelled both Cattell and Henry W. L. Dana on October 1, after which Charles A. Beard resigned.

29. Donald H. Menzel, "The Beginnings of the IAU," read at the meeting of the IAU at Liége, Belgium, on June 29, 1969.

30. *Popular Astronomy,* 27:7 (1919).

31. H. H. Turner, *Monthly Notices of the Royal Astronomical Society,* 80:361 (1920).

32. Pickering played the flute and owned a gramophone and a collection of more than 500 records. In a letter (October 26, 1914) to Major Henry L. Higginson, founder of the Boston Symphony Orchestra, Pickering spoke of the need for recording the best music and offered to assist in technical problems. He had modified his own machine, he said, so that it was an instrument with "an usually

good sound base," but the orchestral records available were generally poorly selected and not well performed—the only recording of the Moonlight Sonata was played by a brass band! "As you have done so much to bring the best music attainable, to Bostonians, I have hoped that you would do the same for the world by having records made by the Boston Symphony Orchestra of some of the great compositions of the past. Orchestral music depends so much on the conductor, that a study of a Beethoven Symphony as interpreted by different great leaders would be most instructive."

33. Pickering to R. T. A. Innes (of the Union Observatory at Johannesburg, South Africa), January 16, 1919.

34. H. H. Turner, *Monthly Notices of the Royal Astronomical Society,* 80:360 (1920).

INDEX

Abbe, Cleveland, 199, 218, 270
Abbott, Charles G., 378
Abney, William, 201, 205, 211
Absolute magnitudes, 239–241, 369
Adams, George, 34
Adams, John, 28
Adams, John Couch, 63, 64, 101, 278, 279, 331, 383; *see also* Neptune
Adams, John Quincy, 23; promotion of Observatory, 31–36, 42–43; at Cincinnati Observatory, 38; on Visiting Committee, 54–56; visit to Observatory, 69
Agassiz, Alexander, 183, 218
Agassiz, George, 408
Agassiz, Louis, 29, 92, 93, 106, 114, 120, 126, 130, 131; *see also* Lawrence Scientific School
Agassiz Station, 97
Airy, George, 37, 47, 64, 87, 89, 90, 125
Alcor (Zeta Ursae Majoris), 81
Alcyone, 60
Alexander, Stephen, 127
Almagest, 184
Almanacs, 1, 3, 5
Almucantar, 332
Alpha Aquilae, 221, 223
Alpha Aurigae, 221, 232
Alpha Lyrae, 76, 81, 84, 213, 223, 224, 226, 236
Alpha Scorpii, 221, 233
Altair (Alpha Aquilae), 221, 223
American Academy of Arts and Sciences: cooperation in solar eclipse expedition of 1780, 26, of 1869, 162; contributions toward 15-inch telescope, 49–51; Pickering elected to, 179;

American Academy of Arts and Sciences: (*continued*)
Pickering address to, 223; *Memoirs* of, 28, 80, 81, 88, 96, 99, 125–126, 275
American Almanac, 56, 82
American Association for the Advancement of Science, 99, 114, 138, 178, 181, 207, 418, 436, 437
American Association of Variable Star Observers, 381
American Astronomical Society, 193, 363
American Ephemeris and Nautical Almanac, 138, 162, 215, 250
American Journal of Science, 99, 178, 218
American Philosophical Society, 21, 109, 162, 436
Amory, William, 148, 183, 251
Andromeda Nebula, 68, 96, 118, 206, 228
Annals of the Harvard College Observatory, 43, 56, 88, 89, 150; *see also references under individual staff members*
Antares (Alpha Scorpii), 221, 233
Appalachian Mountain Club, 255
Appleton, William, 50
Arago, François, 29, 48, 49, 63, 64, 72, 73
Arcturus (Alpha Boötis), 221
Arequipa station, 296 ff., 345 ff., 375 ff., 378
Argelander, Friedrich W. A., 107, 125, 157, 183, 185, 188, 203
Arizona, proposed station in, 325 ff.
Asteroids, 48, 132, 343, 366
Astrand, J., 279

485

Index

Index

487

Index

Index

Index

Index

Mitchell, Maria, Association (Nantucket), fellowship, 411, 412
Mitchell, Samuel A., 434
Mitchell, William, 47, 53, 84, 90, 105, 109, 113, 115, 137, 138
Mizar; *see* Zeta Ursae Majoris
Monthly Notices, 178, 188, 244
Moon: early photography of, 71–75, 77–81; photographic map of, 371; photometry of, 80–81; surface features of, 73, 77–80, 371–374
Morison, Samuel Eliot, 1, 3, 5, 26, 57
Morley, Edward, 435
Morse, Samuel F. B., 72
Morton, Henry, 151, 164, 168
Mouchez, Amédée, 208, 429
Moulton, Forest R., 437
Mountain sites, advantages of, 84–87, 246–249, 253–254
Mt. Harvard station, 291–296
Mt. Wilson station, 256–266
Muller, G., 341
Muñiz, Juan, 323, 375, 378
Muzzy, Artemus B., 137

National Academy of Sciences, 126 ff., 140, 171, 206, 217, 218, 220, 228, 243, 247, 248, 250–251, 252, 377, 420, 430, 440
National Research Council, 430
Nature, 151, 229
Nebulae, 67–69, 95–96, 133, 217, 262, 392
Neptune, 62–64, 95, 101, 138, 204, 330, 331, 383
"New Astronomy," 145, 161, 332, 362
Newcomb, Simon, 138, 151, 164, 168, 176–177, 178, 182, 184, 215, 250, 279, 285, 305, 344, 363, 417, 422, 427
Newton, Isaac, 5, 9, 12, 143, 360
North polar sequence, 370, 402
Nova Aurigae, 333
Nova Normae, 320, 334.
Novae, theory of, 374–375
Nowell, Alexander, 4, 5

Observatories, *American*
Allegheny, 100, 159, 165, 177, 184, 215, 218, 363, 426, 428, 434
Chamberlin (Denver), 361
Cincinnati, 36, 38, 62, 65, 147, 163, 177, 184
Columbia University, 435

Observatories: (*continued*)
Dearborn, 124, 145, 147, 159, 160, 177, 182, 184, 415
Detroit, 177
Dudley, 70, 112, 119, 147, 159, 177, 196, 215, 223, 250, 252, 425, 435
Georgetown, 38
Goodsell (Carleton College), 324, 361, 414
Hawaii, 254
Hudson (Ohio), 37
Indiana University, 361
Ladd (Brown University), 361
Leander McCormick (Virginia), 207, 434
Lick, 87, 202, 247, 249, 256, 258, 263, 264, 278, 307, 308, 312, 332–333, 361, 411, 428
Litchfield (Hamilton College), 115–116, 183–184, 194, 208, 218
Lowell (Flagstaff), 328, 329, 330, 331, 361, 372
Maria Mitchell (Nantucket), 411, 412, 415
Mills (Santiago), 361
Mt. Wilson, 266, 267, 354, 361, 370, 380, 425
Naval, 38, 56, 119, 147, 164, 168, 176, 182, 184, 205, 207, 208, 215, 218, 223, 226, 250, 304, 326, 419
Palomar, 65, 361
Payne (Mt. Holyoke), 416
Philadelphia High School, 38, 50
Princeton, 215, 432
Shattuck (Amherst), 38
Shelby College (Kentucky), 138, 164
Smith College, 361
Smithsonian Astrophysical, 146, 198, 254, 361
Sproul (Swarthmore), 434
Tuscaloosa (Alabama), 38
University of Illinois, 361
University of Michigan, 147, 159, 184
University of Minnesota, 361
University of North Carolina, 37
University of Pennsylvania, 361
University of Wisconsin, 361
Vassar, 215, 414, 416, 435
Warner (Rochester), 184, 195, 215
West Point, 37
Whitin (Wellesley), 361, 394
Williams College (later Field Memorial), 37
Yale, 37, 208

491

Index

Index

Index

Science, 229, 437, 440, 441
Science Observer, 195, 196, 333
Scott, Thomas, 165
Seagrave, Frank, 189
Searle, Arthur, 149–150, 156, 158, 165, 169, 172, 176, 183, 187, 188, 190, 195, 323, 336, 337, 360, 379, 386, 409, 411, 415, 444
Searle, George, 149, 172, 199, 415
Sears, David, 50, 51, 54
Sears Fund, 119, 142, 148
Secchi, Father Pietro, 163, 232, 233, 234, 236
Seed, M. A., Dry Plate Co., 289
Sewell, Stephen, 23
Shackleton, Ernest Henry, 364
Shackleton, William, 364
Shaler, Nathaniel, 157, 165, 172
Shapley, Harlow, 62, 164–165, 193, 198, 268, 285, 330, 354, 369–370, 374, 380, 402, 408, 432, 443, 444
Short, James, 18
Sidereal Messenger (Mitchel's), 62, 101, 178
Sidereal Messenger (Payne's), 361
Silliman, Benjamin, Jr., 218
Simms, William, 37, 54, 148, 152, 153, 281; see also Troughton & Simms
Sirius, 213, 232, 397; companion star of, 123
Slipher, V. M., 331
Smith, William, 21
Smithson, James, 36
Smithsonian Institution, 126, 127, 130, 145, 194, 195, 196, 197. 218, 253, 334, 378, 385, 436
Solar constant, 428
Solar corona, 102, 163–164, 166–167, 169–170, 215–216
Solar photography, early, 71–74, 205
Solar spectrum, 143–145
Somerville, Mary, 383, 394
Southern California, University of, 256, 258, 261, 263–264
Sparks, Jared, President, 11, 93, 97
Spectroheliograph, 267, 429–430
Spectroscopic binaries, 243–244, 397–399
Spectrum analysis, 142–145, 149, 163, 204
Spence, E. F., 256, 258, 260, 267
Spring governor, 81, 88
Stanford, Leland, 263
Star catalogues, ancient, 184
Stars: composition of, 233–234; spectral classification of, 232–239

Stebbins, Joel, 427
Steil, Peter, 262, 265
Stellar evolution, theories of, 233–234, 241–242
Stellar magnitudes, 185–187, 199, 203, 369
Stellar spectra, catalogues of, 232, 234–238, 399, 406–409
Stewart, Balfour, 171
Stewart, De Lisle, 324, 355, 356, 366, 375
Stone, Ormond, 184, 207
Storin, Nellie C., 253
Storrow, Charles, 91
Strain, A. G., 260, 262, 265, 266
Strömgren, Elis, 442
Struve, Friedrich, 51–52, 60, 87, 105
Struve, Ludwig, 279
Struve, Otto, 148, 208, 223, 429
Sturgis, William, 124
Sturgis Fund, 142, 171, 172
Sun, distance to, 16, 20, 89, 114, 343–345
Swift, Lewis, 184, 195, 196, 215, 216, 279

Taft, Robert, 73
Tacchini, Pietro, 429
Talbot, H. Fox, 72, 73
Tariff on astronomical instruments, 64–65, 154–156
Tatnall, R. T., 363
Telegraph code, astronomical, 194, 195, 197, 332
Themis, 365, 367
Thompson, Benjamin, 11
Thompson, Elizabeth, 277
Thompson Fund, 277, 420, 432, 434, 439
Thoreau, Henry, 90
Time service, 52–53, 64, 87, 159–161, 193–194, 326
Tombaugh, Clyde, 330
Trans-Neptunian planet, 330–331, 374
Troughton & Simms, 30, 37, 38, 52, 55, 142, 145, 146, 153
Trouvelot, Léopold, 171, 172
Trowbridge, John, 435, 436
Turner, H. H., 273 ff., 276, 279, 342, 343, 349, 352, 364, 425, 427, 439, 443
Tuttle, Charles, 76, 103
Tuttle, Horace, 117, 120
29 Canis Majoris, 245

U. S. Signal Corps, 254

494

Index